Contents

Preface

Every book is initially just a concept; it takes months of research and hard work to give it the final shape in which the readers receive it. In its early stages, this book also went through rigorous reviewing. The notable contributions made by experts from across the globe were first molded into patterned chapters and then arranged in a sensibly sequential manner to bring out the best results.

This multi-contributor descriptive book provides in-depth knowledge regarding the field of earthquake engineering. With increasing seismic activities and major earthquakes, the need for intense research and better understanding of the nature of seismisms and their effects on structures is rapidly increasing. In the wake of frequent disasters, earthquake engineering research has been expanding as significant data has become available from a large array of seismic instruments, large scale experiments and numerical simulations. This book opens with results from some of the contemporary seismic research works including 3D wave propagation in different soil media, seismic loss assessment, probabilistic hazard analysis, geotechnical problems including soil-structure interaction. It also focuses on the seismic behavior of structures including historical and monumental structures, bridge embankments and different types of bridges and bearings.

It has been my immense pleasure to be a part of this project and to contribute my years of learning in such a meaningful form. I would like to take this opportunity to thank all the people who have been associated with the completion of this book at any step.

Editor

Seismic Risk, Hazard, Wave Simulation and Geotechnical Aspects

Seismic Risk of Structures and the Economic Issues of Earthquakes

Afshin Kalantari

Additional information is available at the end of the chapter

1. Introduction

As one of the most devastating natural events, earthquakes impose economic challenges on communities and governments. The number of human and economic assets at risk is growing as megacities and urban areas develop all over the world. This increasing risk has been plotted in the damage and loss reports after the great earthquakes.

The 1975 Tangshan (China) earthquake killed about 200,000 people. The 1994 Northridge, (USA) earthquake left 57 dead and about 8,700 injured. The country experienced around $42 billion in losses due to it. The 1995 earthquake in Kobe (Japan) caused about 6,000 fatalities and over $120 Billion in economic loss. The August 1996 Izmit (Turkey) earthquake killed 20,000 people and caused $12 billion in economic loss. The 1999 Chi-chi (Taiwan) earthquake caused an estimated $8 billion in loss. The 2006 Gujarat (India) earthquake saw around 18,000 fatalities and 330,000 demolished buildings [1]. The Sichuan (China) earthquake, on May 12th 2008 left 88,000 people dead or missing and nearly 400,000 injured. That earthquake damaged or destroyed millions of homes, leaving five million homeless. It also caused extensive damage to basic infrastructure, including schools, hospitals, roads and water systems. The event cost around $29 billion in direct loss alone [2]. The devastating earthquake of March 2011 with its resulting tsunami along the east coast of Japan is known to be the world's most costly earthquake. The World Bank estimated the cost at $235 billion while government estimates reported the number at $305 billion. The event left 8,700 dead and more than 13,000 missing [3].

As has been shown, earthquake events have not only inflicted human and physical damage, they have also been able to cause considerable economic conflict in vulnerable cities and regions. The importance of the economic issues and the consequences of earthquakes attracted the attention of engineers and provided new research and working opportunities for engineers, who up until then had been concerned only with risk reduction options through engineering strategies [4].

Seismic loss estimation is an expertise provided by earthquake engineering and the manner in which it can be employed in the processes of assessing seismic loss and managing the financial and economical risk associated with earthquakes through more beneficial retrofit methods will be discussed. The methodology provides a useful tool for comparing different engineering alternatives from a seismic-risk-point of view based on a Performance Based Earthquake Engineering (PBEE) framework [5]. Next, an outline of the regional economic models employed for the assessment of earthquakes' impact on economies will be briefly introduced.

1.1. The economic consequences of earthquakes

The economic consequences of earthquakes may occur both before and after the seismic event itself [6]. However, the focus of this chapter will be on those which occur after earthquakes. The consequences and effects of earthquakes may be classified in terms of their primary or direct effects and their secondary or indirect effects. The indirect effects are sometimes referred to by economists as higher-order effects. The primary (direct) effects of an earthquake appear immediately after it as social and physical damage. The secondary (indirect) effects take into account the system-wide impact of flow losses through inter-industry relationships and economic sectors. For example, where damage occurs to a bridge then its inability to serve to passing vehicles is considered a primary or direct loss, while if the flow of the row material to a manufacturing plant in another area is interrupted due to the inability of passing traffic to cross the bridge, the loss due to the business's interruption in this plant is called secondary or indirect loss. A higher-order effect is another term as an alternative to indirect or secondary effects which has been proposed by economists [7]. These potential effects of earthquakes may be categorized as: "social or human", "physical" and "economic" effects. This is summarized in Table 1 [8].

The term 'total impact' accordingly refers to the summation of direct (first-order effects) and indirect losses (higher-order effects). Various economic frameworks have been introduced to assess the higher-order effects of an earthquake.

With a three-sector hypothesis of an economy, it may be demonstrated in terms of a breakdown as three sectors: the primary sector as raw materials, the secondary sector as manufacturing and the tertiary sector as services. The interaction of these sectors after suffering seismic loss and the relative effects on each other requires study through proper economic models.

2. The estimation of seismic loss of structures in the PBEE framework

The PBEE process can be expressed in terms of a four-step analysis, including [9-10]:

- Hazard analysis, which results in Intensity Measures (IMs) for the facility under study,

- Structural analysis, which gives the Engineering Demand Parameters (EDPs) required for damage analysis,
- Damage analysis, which compares the EDPs with the Damage Measure in order to decide for the failure of the facility, and;
- Loss Analysis, which evaluates the occurrence of Decision Variables (DVs) due to failures.

	Social or human effects	Physical effects	Economic effects
Primary effects (Direct or first-order)	Fatalities Injuries Loss of income or employment opportunities Homelessness	Ground deformation and loss of ground quality Collapse and structural damage to buildings and infrastructure Non-structural damage to buildings and infrastructure (e.g., component damage)	Disruption of business due to damage to industrial plants and equipment Loss of productive work force, through fatalities, injuries and relief efforts Disruption of communications networks Cost of response and relief
Secondary effects (indirect or higher-order)	Disease or permanent disability Psychological impact of injury, Bereavement, shock Loss of social cohesion due to disruption of community Political unrest when government response is perceived as inadequate	Reduction of the seismic capacity of damaged structure which are not repaired Progressive deterioration of damaged buildings and infrastructure which are not repaired	Losses borne by the insurance industry, weakening the insurance market and increasing the premiums Losses of markets and trade opportunities,

Table 1. Effects from Earthquakes [8]

Considering the results of each step as a conditional event following the previous step and all of the parameters as independent random parameters, the process can be expressed in terms of a triple integral, as shown below, which is an application of the total probability theorem [11]:

$$v(DV) = \iiint \ G[DV|DM]|dG[DM|EDP]|dG[EDP|IM]|d\lambda[IM] \tag{1}$$

The performance of a structural system or lifeline is described by comparing demand and capacity parameters. In earthquake engineering, the excitation, demand and capacity parameters are random variables. Therefore, probabilistic techniques are required in order to estimate the response of the system and provide information about the availability or failure of the facility after loading. The concept is included in the reliability design approach, which is usually employed for this purpose.

2.1. Probabilistic seismic demand analysis through a reliability-based design approach

The reliability of a structural system or lifeline may be referred to as the ability of the system or its components to perform their required functions under stated conditions for a specified period of time. Because of uncertainties in loading and capacity, the subject usually includes probabilistic methods and is often made through indices such as a safety index or the probability of the failure of the structure or lifeline.

2.1.1. Reliability index and failure

To evaluate the seismic performance of the structures, performance functions are defined. Let us assume that $z=g(x_1, x_2, ...,x_n)$ is taken as a performance function. As such, failure or damage occurs when $z<0$. The probability of failure, p_f, is expressed as follows:

$$P_f = P[z<0] \tag{2}$$

Simply assume that $z=EDP-C$ where EDP stands for Engineering Demand Parameter and C is the seismic capacity of the structure.

Damage or failure in a structural system or lifeline occurs when the Engineering Demand Parameter exceeds the capacity provided. For example, in a bridge structural damage may refer to the unseating of the deck, the development of a plastic hinge at the bottom of piers or damage due to the pounding of the decks to the abutments, etc.

Given that EDP and C are random parameters having the expected or mean values of μ_{EDP} and μ_C and standard deviation of σ_{EDP} and σ_C, the "safety index" or "reliability index", β, is defined as:

$$\beta = \frac{\mu_c - \mu_{EDP}}{\sqrt{\sigma_c^2 + \sigma_{EDP}^2}} \tag{3}$$

It has been observed that the random variables such as "EDP" or "C" follow normal or log-normal distribution. Accordingly, the performance function, z, also will follow the same distribution. Accordingly, probability of failure (or damage occurrence) may be expressed as a function of safety index, as follows:

$$P_f = \phi\,(-\,\beta) = 1 - \phi\,(\beta) \tag{4}$$

where $\phi(\ )$ is a log-normal distribution function.

2.1.2. Engineering demand parameters

The Engineering Demand Parameters describe the response of the structural framing and the non-structural components and contents resulting from earthquake shaking. The parameters are calculated by structural response simulations using the IMs and corresponding earthquake motions. The ground motions should capture the important characteristics of earthquake ground motion which affect the response of the structural framing and non-structural components and building contents. During the loss and risk estimation studies, the EDP with a greater correlation with damage and loss variables must be employed.

The EDPs were categorized in the ATC 58 task report as either direct or processed [9]. Direct EDPs are those calculated directly by analysis or simulation and contribute to the risk assessment through the calculation of P[EDP | IM]; examples of direct EDPs include interstory drift and beam plastic rotation. Processed EDPs - for example, a damage index - are derived from the values of direct EDPs and data on component or system capacities. Processed EDPs could be considered as either EDPs or as Damage Measures (DMs) and, as such, could contribute to risk assessment through P[DM | EDP]. Direct EDPs are usually introduced in codes and design regulations. For example, the 2000 NEHRP Recommended Provisions for Seismic Regulations for Buildings and Other Structures introduces the EDPs presented in Table 2 for the seismic design of reinforced concrete moment frames [12-13]:

Reinforced concrete moment frames
Axial force, bending moment and shear force in columns
Bending moment and shear force in beams
Shear force in beam-column joints
Shear force and bending moments in slabs
Bearing and lateral pressures beneath foundations
Interstory drift (and interstory drift angle)

Table 2. EDPs required for the seismic design of reinforced concrete moment frames by [12-13]

Processed EDPs are efficient parameters which could serve as a damage index during loss and risk estimation for structural systems and facilities. A Damage Index (DI), as a single-valued damage characteristic, can be considered to be a processed EDP [10]. Traditionally, DIs have been used to express performance in terms of a value between 0 (no damage) and 1 (collapse or an ultimate state). An extension of this approach is the damage spectrum, which takes on values between 0 (no damage) and 1 (collapse) as a function of a period. A detailed summary of the available DIs is available in [14].

Park and Angin [15] developed one of the most widely-known damage indices. The index is a linear combination of structural displacement and hysteretic energy, as shown in the equation:

$$DI = \frac{u_{max}}{u_c} + \beta \frac{E_h}{F_y u_c}$$ (5)

where u_{max} and u_c are maximum and capacity displacement of the structure, respectively, E_h is the hysteresis energy, F_y is the yielding force and β is a constant.

See Powell and Allahabadi, Fajfar, Mehanny and Deierlein, as well as Bozorgnia and Bertero for more information about other DIs in [16-19].

2.2. Seismic fragility

The seismic fragility of a structure refers to the probability that the Engineering Demand Parameter (EDP) will exceed seismic capacity (C) upon the condition of the occurrence of a specific Intensity Measure (IM). In other words, seismic fragility is probability of failure, P_f, on the condition of the occurrence of a specific intensity measure, as shown below:

$$\text{Fragility=P [EDP>C|IM]}$$ (6)

In a fragility curve, the horizontal axis introduces the IM and the vertical axis corresponds to the probability of failure, P_f. This curve demonstrates how the variation of intensity measure affects the probability of failure of the structure.

Statistical approach, analytical and numerical simulations, and the use of expert opinion provide methods for developing fragility curves.

2.2.1. Statistical approach

With a statistical approach, a sufficient amount of real damage-intensity data after earthquakes is employed to generate the seismic fragility data. As an example, Figure 1 demonstrates the empirical fragility curves for a concrete moment resisting frame, according to the data collected after Northridge earthquake [20].

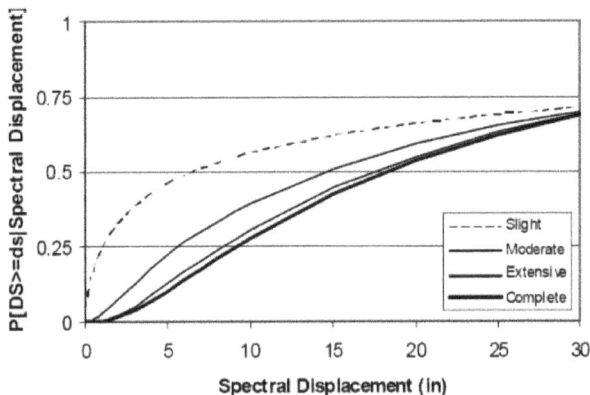

Figure 1. Empirical fragility curves for a concrete moment resisting frame building class according to the data collected after the Northridge Earthquake, [20].

2.2.2. Analytical approach

With an analytical approach, a numerical model of the structure is usually analysed by nonlinear dynamic analysis methods in order to calculate the *EDPs* and compare the results with the capacity to decide about the failure of the structure. The works in [21-24] are examples of analytical fragility curves for highway bridge structures by Hwang et al. 2001, Choi et al. 2004, Padgett et al., 2008, and Padgett et al 2008 .

Figure 2 demonstrates the steps for computing seismic fragility in analytical approach.

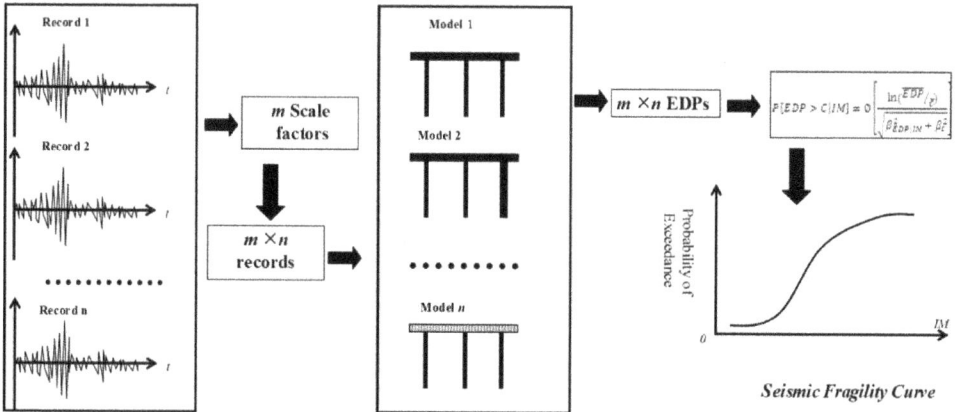

Figure 2. Procedure for generating analytical fragility curves

To overcome the uncertainties in input excitation or the developed model, usually adequate number of records and several numerical models are required so that the dispersion of the calculated data will be limited and acceptable. This is usually elaborating and increases the cost of the generation of fragility data in this approach. Probabilistic demand models are usually one of the outputs of nonlinear dynamic analysis. Probabilistic demand models establish a relationship between the intensity measure and the engineering demand parameter. Bazorro and Cornell proposed the model given below [25]:

$$\overline{EDP} = a(IM)^b \tag{7}$$

where $\overline{EDP}$ is the average value of *EDP* and a and b are constants. The model has the capability to be presented as linear in a logarithmic space such that:

$$\ln(\overline{EDP}) = \ln(a) + b\ln(IM) \tag{8}$$

Assuming a log-normal distribution for fragility values, they are then estimated using the following equation:

$$P[EDP > C|IM] = \emptyset \left[\frac{\ln(\overline{EDP}/\hat{c})}{\sqrt{\beta^2_{EDP|IM} + \beta^2_c}} \right] \tag{9}$$

The parameter β introduces the dispersion in the resulting data from any calculations. An example of analytical fragility curves for highway bridges is shown in Figure 3.

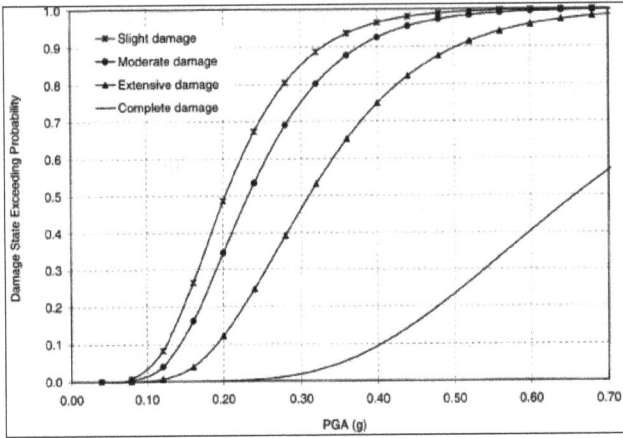

Figure 3. Fragility curve for the 602-11 bridge for 4 damage states [21]

2.2.3. Expert opinion approach

Given a lack of sufficient statistical or analytical data, expert opinion provides a valuable source for estimating the probability of the failure of typical or specific buildings for a range of seismic intensity values. The number of experts, their proficiency and the quality of questionnaires, including the questions, their adequacy and coverage, can affect the uncertainty of the approach and its results.

2.3. Seismic risk

The expected risk of a project, assuming that the intensity measure as the seismic hazard parameter is deterministic, is calculated by equation 10, below:

$$R = P \times L \tag{10}$$

where P is the probability of the occurrence of damage and L indicates the corresponding loss. The equation shows that any factor which alters either the probability or the value of the resulted loss affects the related risk. Diverse damage modes and associated loss values, L_i (i=1 to a number of probable damage modes), with a different probability of occurrence, P_i, may be envisaged for a structure. The probable risk of the system, R, can be estimated as a summation of the loss of each damage mode:

$$R = \sum P_i \times L_i \tag{11}$$

Loss functions are usually defined as the replacement cost - corresponding to each damage state - versus seismic intensity. The loss associated with each damage mode, presented schematically in Figure 4, is usually collected through questionnaires, statistical data from post-earthquake observations or else calculated through numerical simulations. ATC 13 provides an example of the collection of earthquakes' structural and human damage and loss data for California [26].

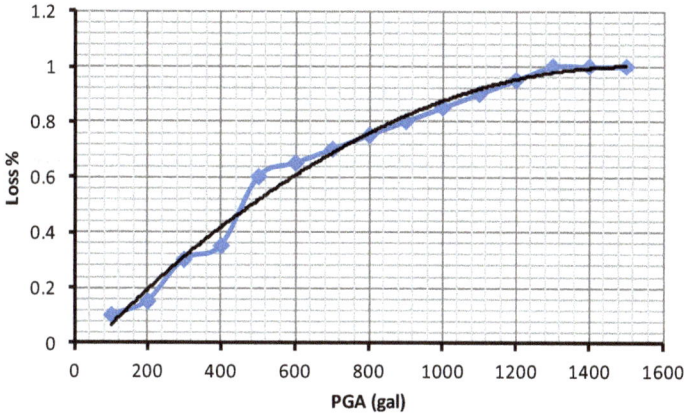

Figure 4. Seismic loss data

A summary of calculations required for estimating the risk of a project under a specific seismic intensity level may be illustrated by an "event tree" diagram.

3.3.1. Event tree diagram

An Event tree diagram is a useful tool for estimation of the probability of occurrence of damage and corresponding loss in a specific project due to a certain seismic event. The procedure requires information about seismic intensity, probable damage modes, seismic fragility values and the vulnerability and loss function of the facility under study.

As an example, suppose that partial seismic damage, structural collapse, partial fire and extended fire are considered to be the loss-generating consequents of an earthquake for a building. Figures 5 and 6 are the event tree diagrams, which demonstrate the procedure followed to calculate the corresponding risk for the seismic intensity of two levels of PGA=300gal and 500gal. To select the probability of the occurrence of each damage mode, (i.e., the probability of the exceedance of damage states) the fragility curves can be utilized. Each node is allocated to a damage mode. The probability of the incidence or non-incidence of each damage mode is mentioned respectively on the vertical or horizontal branch immediately after each node. The probability of the coincidence of the events at the same root is calculated by multiplying the probability of incidence of the events on the same root. The final total risk, R, is then calculated as the summation of all R_is.

Figure 7.a demonstrates the distribution of risk values for different damage modes. In addition, it can be seen how increasing seismic intensity increased the risk of the project. Figure 7.b shows the distribution of the probability of the occurrence of different loss values and how an increase of seismic intensity from 300gal to 500gal affects it in this structure. As mentioned, the calculations in an event tree diagram are performed for a special level of hazard. The curves present valuable probabilistic data about the points on the seismic loss curve. A seismic loss curve may be developed according to the information from event trees for a range of probable seismic intensities of the site. Figure 8 shows a schematic curve for the seismic loss of a project. The curve is generated by integrating the seismic risk values for each damage mode. It provides helpful data for understanding the contents and elements of the probable loss for each level of earthquake hazard.

Collapse		Partial Seismic Damage		Fire		Extended Fire			P_i	L_i	$R_i=P_i \times L_i$
0.95		0.8		0.9				ND	0.684	0%	0%
0.05		0.2		0.1							
						0.9		PF	0.0684	15%	1.03%
						0.1		CF	0.0076	65%	0.49%
				0.9				PD	0.171	25%	4.28%
				0.1							
						0.9		PD+PF	0.0171	40%	0.68%
						0.1		PD+CF	0.0019	75%	0.14%
								CO	0.05	100%	5.0%
											$\sum R_i=11.62\%$

ND: No Damage, F: Partial Fire, CF: Complete Fire, PD: Partial Damage, CO: Collapse

Figure 5. Event Tree, PGA=300gal

Collapse		Partial Seismic Damage		Fire		Extended Fire			P_i	L_i	$R_i=P_i \times L_i$
0.8		0.6		0.8				ND	0.3840	0%	0
0.2		0.4		0.2							
						0.8		PF	0.0768	15%	1.15%
						0.2		CF	0.0192	65%	1.25%
				0.8				PD	0.2560	25%	6.40%
				0.2							
						0.8		PD+PF	0.0512	40%	2.05%
						0.2		PD+CF	0.0128	75%	0.96%
								CO	0.2000	100%	20
											$\sum Ri=31.81\%$

Figure 6. Event Tree, PGA=500gal

The information provided by an event tree simply increases the awareness of engineers and stakeholders about the importance and influence of each damage mode on the seismic risk of the project and demonstrates the distribution of probable loss among them.

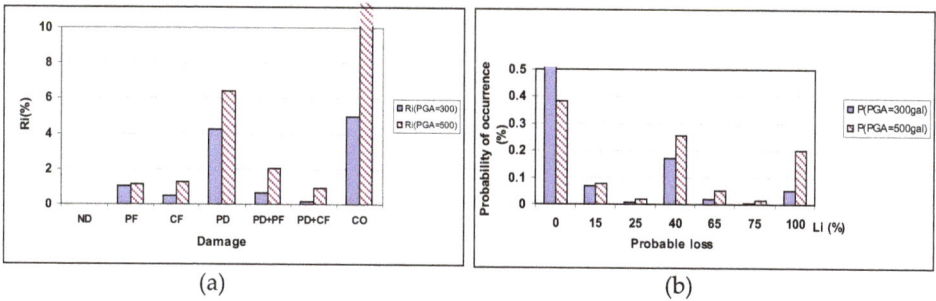

(a)   (b)

Figure 7. a) Distribution of seismic risk values vs. damage, b) Probability of occurrence vs. probable loss

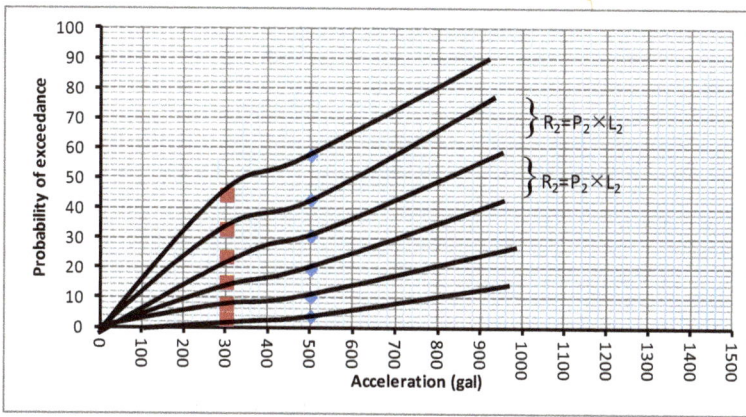

Figure 8. Seismic loss curve

The total probable loss calculated by event trees provides valuable information for estimating the annual probable loss of facilities, as shown in the next part.

3. The employment of seismic hazard analysis for the assessment of seismic risk

If the uncertainties in the seismic hazard assessment of a specific site could be avoided, a deterministic approach could provide an easy and rational method for this purpose. However, the nature of a seismic event is such that it usually involves various uncertainty sources, such as the location of the source, the faulting mechanism and the magnitude of the event, etc. The probabilistic seismic hazard analysis offers a useful tool for the assessment of annual norms of seismic loss and risk. [27]

3.1. Probabilistic seismic hazard analysis

In an active area source, k, with a similar seismicity all across it, the seismicity data gives the maximum magnitude of m_{uk} and a minimum of m_{lk} and the frequency of the occurrence of

v_k. Similar assumptions can be extended for a line source from which the Probability Density Function (PDF) of magnitude for a site, $f_{Mk}(m_k)$, can be constructed, as is schematically demonstrated by Figure 9.a [27].

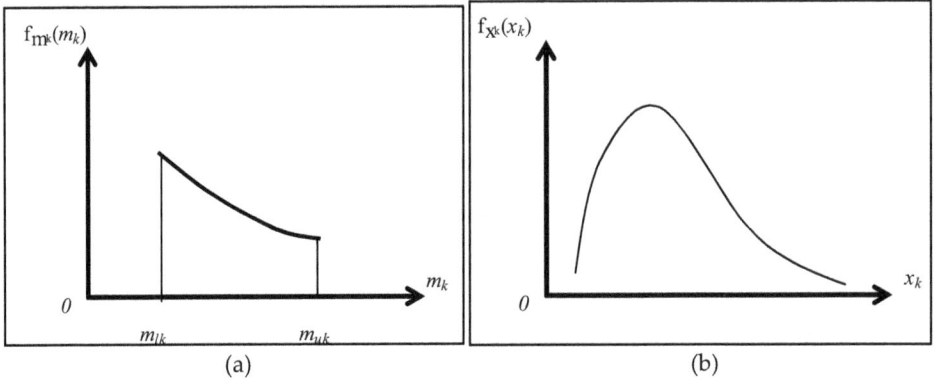

Figure 9. Variability of seismic intensity as a function of magnitude and distance

if in the active zone under study, an area or line source can be assumed as a point, the probability density function of the focal distance of the site, x, f_{Xk} (x_k) can be developed, as schematically demonstrated in Figure 9.b.

3.1.1. Ground motion prediction models

Ground motion prediction models - or attenuation functions - include the gradual degradation of seismic energy passing through a medium of ground up to site. The ground motion prediction models, schematically shown in Figure 10, have been provided according to the statistical data, characteristics of the ground, seismic intensity and distance, etc.

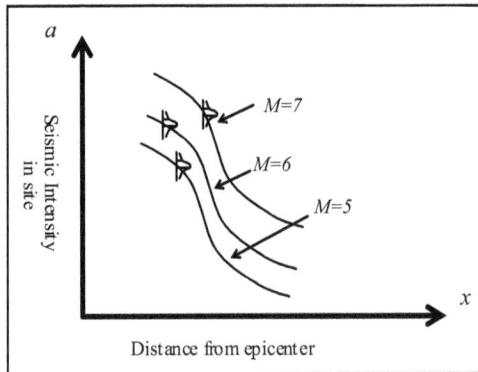

Figure 10. a) Schematic ground motion prediction models for a site

The ground motion prediction models are usually empirical relations, which do not match the real data exactly. The dispersion between the real data and the empirical attenuation relations may be modelled by a probability density function f(a|m, x) which shows the distribution density function of intensity *a* if a seismic event with a magnitude of *m* occurs at a distance x from the site. Figure 11 shows how f(a|m, x) changes when an intensity measure *a* varies.

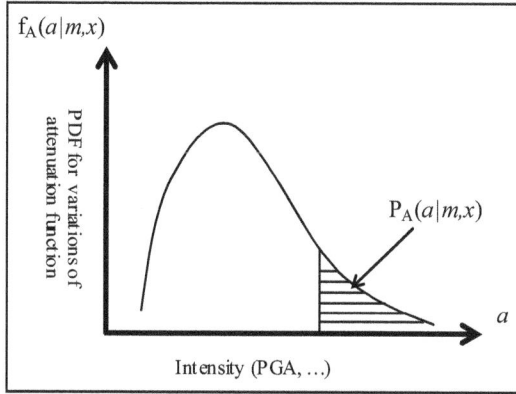

Figure 11. Probability of exceedance from a specific intensity using a probability density function

According to the above-mentioned collected data, the annual rate of earthquakes with an intensity (acceleration) larger than *a*, *v(a)* can be calculated from the following equation:

$$v_{(a)} = \sum_k v_k \int_{x_k} \int_{m_{lk}}^{m_{uk}} P_A\left(a \middle| m_k, x_k\right) f_{M_k}\left(m_k\right) f_{X_k}\left(x_k\right) dm_k dx_k \tag{12}$$

Where, $P_A(a|m_k, x_u)$ stands for the probability of occurrence of an earthquake with an intensity larger than *a* at a site with an attenuation relation of $f_A(a|m,x)$.

Poison process is usually employed to model the rate of the occurrence of earthquakes within specific duration. For an earthquake with an annual probability of occurrence of v(a), the probability of the occurrence of *n* earthquakes of intensity greater than *a* within *t* years is given by:

$$P(n, t, a) = \frac{\left(v(a)t\right)^n \exp\left(-v(a)t\right)}{n!} \tag{13}$$

Meanwhile, the annual probability of exceedance from the intensity *a*, P(a) can be expressed as:

$$P(a) = 1 - P(0, 1, a) = 1 - \exp\left(-v(a)\right) \tag{14}$$

The time interval of earthquakes with an intensity exceeding *a* is called the return period and is shown as T_a. The parameter can be calculated first knowing that the probability of *T* is longer than t:

$$P(T \geq t) = P(0, t, a) = \exp\left(-v(a)t\right) \tag{15}$$

then the probability distribution function of T_a becomes:

$$F_T(t) = 1 - P(T_a \geq t) = 1 - \exp\left(-v(a)t\right) \tag{16}$$

Accordingly, the probability density function of T, f_T, is derived by taking a derivation of the above F_T function:

$$f_T(t) = v(a)\exp\left(-v(a)t\right) \tag{17}$$

The return period is known as the mean value of T and can be calculated as:

$$T_a = E(t) = \int t f_{T_a}(t)dt = 1 / v(a) \tag{18}$$

The probability density function, $f_A(a)$, the accumulative probability, $F_A(a)$, and the annual probability of exceedance function, $P(a)$, for intensity a (for example PGA), are related to each other, as shown below:

$$F_A(a) = \int_{-\infty}^{a} f_A(a)da \tag{19}$$

$$P(a) = \int_{a}^{\infty} f_A(a)da \tag{20}$$

$$P(a) = 1 - F_A(a) \tag{21}$$

A hazard curve, as shown below, refers to a curve which relates the annual probability of exceedance of an intensity a, $P(a)$, to the intensity value a. Two seismic hazard curves were employed in Figure 12 to schematically demonstrate two sites with relatively low and high seismic hazard.

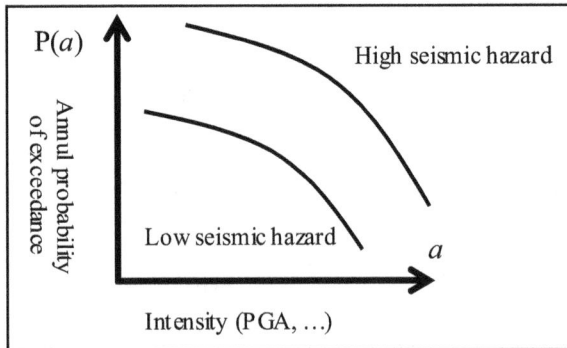

Figure 12. Seismic hazard curve. A demonstration of relatively low and high seismic hazard by means of seismic hazard curves

A probabilistic hazard analysis for a site has resulted in the following plots of a probability density function and accumulative distribution.

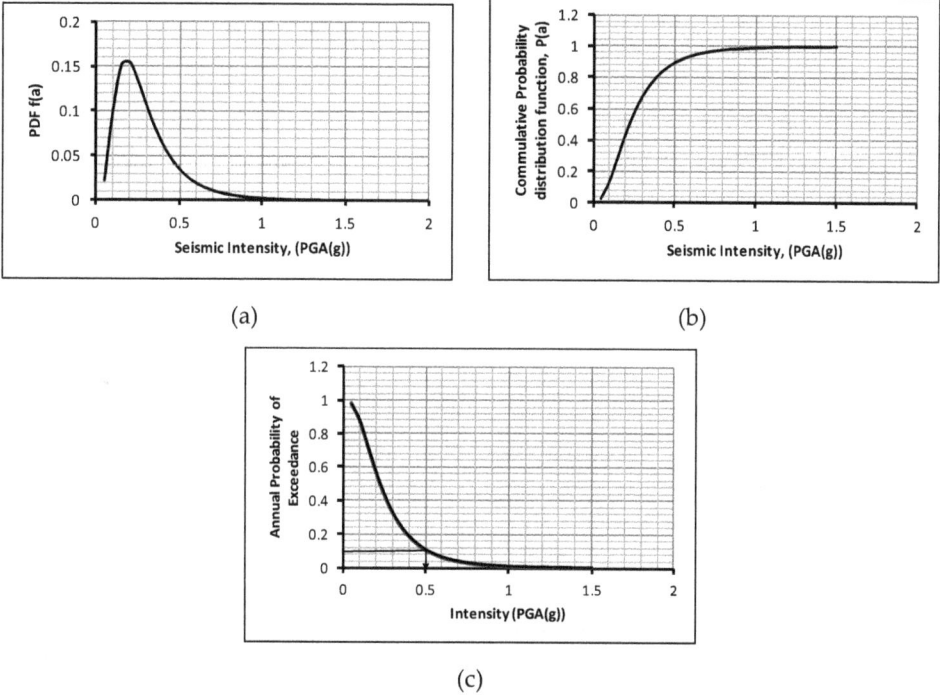

(a)   (b)

(c)

Figure 13. Seismic hazard data, a) PDF of intensity, b) cumulative probability of occurrence $\int_{-\infty}^{a} f_A(a)\, da$, and c) annual probability of exceedance, where the seismic hazard curve = $\int_{a}^{+\infty} f_A(a)\, da$

3.2. Annual seismic loss and risk

By applying the data available from seismic hazard and loss curves, an annual seismic risk density and seismic risk curve can be estimated.

A seismic loss curve is a useful tool for comparing the seismic capacity of different facilities. Seismic hazard and loss curves with basic information about the site and facility play a key role in the evaluation of seismic risk assessment and management procedures. The "annual seismic risk density" and "seismic risk" curves constitute two important measures which can be derived from the above data. The steps to obtain annual seismic risk density curves are shown in Figure 14. The probability density function for seismic intensity (e.g., PGA) is found using a seismic hazard curve using equations 18-20. Accordingly, the annual seismic risk density is derived by multiplying this result with the corresponding loss values, as shown in Figure 14.d below [27].

(a)   (b)

(c)   (d)

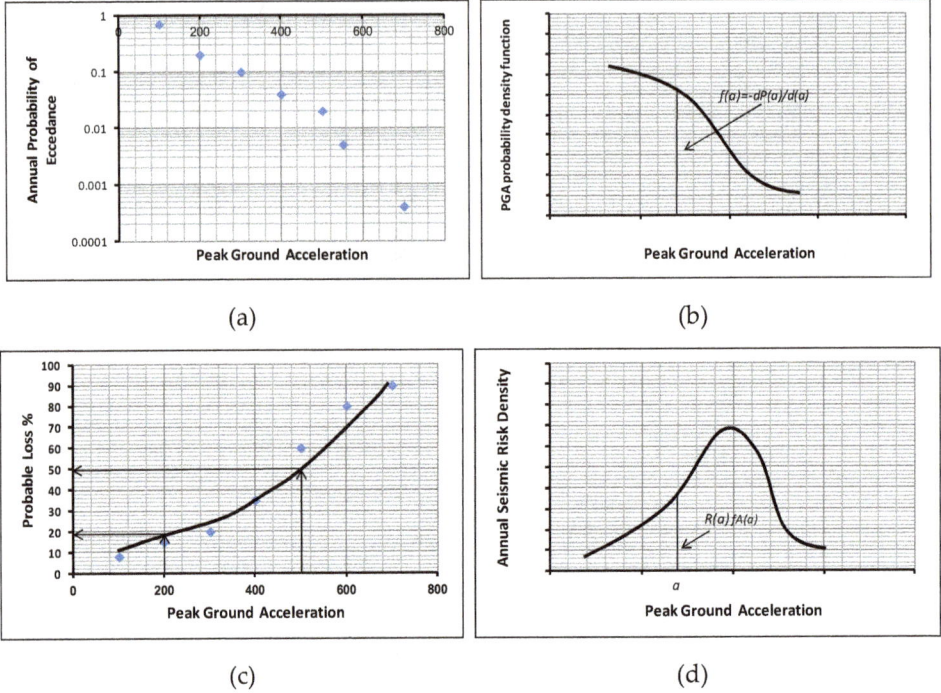

Figure 14. Generating the annual seismic risk density from seismic hazard and loss curves, a) seismic hazard curve, b) probability density function, c) seismic loss curve and d) annual seismic risk density.

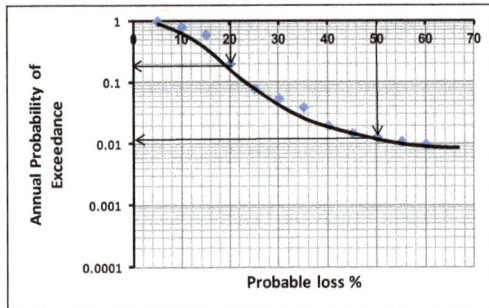

Figure 15. Seismic risk curve

The seismic risk curve, as shown in Figure 15, is calculated using seismic hazard probability and loss values corresponding to similar intensities.

The seismic risk and annual risk density contain helpful information for risk management efforts. As an example, insurance premiums are calculated using this data for various seismic loss limits which can be decided by the client and insurance company.

4. Regional economic models

Perhaps the most widely used modelling framework is the Input-Output model. The method has been extensively discussed in the literature (for example, in [28-30]). The method is a linear model, which includes purchase and sales between sectors of an economy based on technical relations of production. The method specially focuses on the production interdependencies among the elements and, therefore, is applicable for efficiently exploring how damage in a party or sector may affect the output of the others. HAZUS has employed the model in its indirect loss estimation module [31].

Computable General Equilibrium (CGE) offers a multi-market simulation model based on the simultaneous optimization of behaviour of individual consumers and firms in response to price signals, subject to economic account balances and resource constraints. The nonlinear approach retains many of the advantages of the linear I-O methods and overcomes most of its disadvantages [32].

As the third alternative, econometric models are statistically estimated as simultaneous equation representations of the aggregate workings of an economy. A huge data collection is required for the model and the computation process is usually costly [33].

As another approach, Social Accounting Matrices (SAMs) have been utilized to examine the higher-order effects across different socio-economic agents, activities and factors. Cole, in [34-36], studied the subject using one of the variants of SAM. The SAM approach, like I-O models, has rigid coefficients and tends to provide upper bounds for estimates. On the other hand, the framework can derive the distributional impacts of a disaster in order to evaluate equity considerations for public policies against disasters. A summary of the advantages and disadvantages of the models mentioned has been presented in Table 3 [37].

The economic consequences of earthquakes due to the intensity of the event and the characteristics of the affected structures may be influential on a large-scale economy. As an example, the loss flowing from the March 2011 earthquake and tsunami in east Japan could amount to as much as $235 billion and the effects of the disaster will be felt in economies across East Asia [3]. To study how the damage to an economic sector of society may ripple into other sectors, regional economic models are employed. Several spatial economic models have been applied to study the impacts of disasters. Okuyama and Chang, in [30], summarized the experiences about the applications of the three main models - namely Input-Output, Social Accounting and Compatible General Equilibrium - to handle the impact of disaster on socio-economic systems, and comprehensively portrayed both their merits and drawbacks. However, they are based on a number of assumptions that are questionable in, for example, seismic catastrophes.

Studies have been recommended to address issues such as double-counting, the response of households and the evaluation of financial situations. According to the National Research Council, 'the core of the problem with the statistically based regional models is that the historical relationship, embodied in these models, is likely to be disrupted in a natural disaster. In short, regional economic models have been developed over time primarily to forecast future economic conditions or to estimate the effects of a permanent change (e.g., the opening or closing of a manufacturing plant). The random nature and abruptness of a natural disaster do not fit the event pattern upon which regional economic models are based [38].

	Strengths	Weaknesses
IO	- simple structure - detailed inter-industry linkages - wide range of analytical techniques available - easily modified and integrated with other models	- linear structure - rigid coefficients - no supply capacity constraint - no response to price change - overestimation of impact
SAM	- more detailed interdependency among activities, factors and institutions - wide range of analytical techniques available - used widely for development studies	- linear structure - rigid coefficients - no supply capacity constraint - no response to price change - data requirement - overestimation of impact
CGE	- non-linear structure - able to respond to price change - able to cooperate with substitution - able to handle supply capacity constraints	- too flexible to handle changes - data requirement and calibration - optimization behaviour under disaster -underestimation of impact
Econometrics	- statistically rigorous - stochastic estimate - able to forecast over time	- data requirement (time series and cross section) - total impact rather than direct and higher-order - order of impacts distinguished

Table 3. The advantages and disadvantages of the regional economic models for a seismic impact assessment [37]

Yamano et al., in [39], examined the economic impacts of natural disasters using the originally estimated finer geographical scale production datasets and the redefined interregional input–output table. For more effective estimates of the direct losses of the disasters, the precise geographical information of industrial distribution was required because most of the economic data was published according to political boundaries, which may be too aggregated to provide practical information for disaster preventions and retrofit policies. The direct losses were captured by the output data at the district level (500square meters) by sector and population density. The map of economic hotspots was obtained after estimating the economic importance of each district. They showed that the advantages of finer geographical scale datasets and the total economic losses are not proportional to the distributions of the population and industrial activities. In other words, the disaster prevention and retrofit policies have to consider the higher-order effects in order to reduce the total economic loss [39].

It has been shown that in having both virtues and limitations, these alternate I-O, CGE or econometric frameworks may be chosen according to various considerations, such as data collection/compilation, the expected output, research objectives and costs. Major impediments to analysing a disaster's impact may involve issues related to data collection and estimation methodologies, the complex nature of a disaster's impact, an inadequate national capacity to undertake impact assessments and the high frequency of natural disasters.

5. Conclusions

In this chapter, a summary of the methodology for performance-based earthquake engineering and its application in seismic loss estimation was reviewed. Describing the primary and secondary effects of earthquakes, it was mentioned that the loss estimation process for the direct loss estimation of structures consists of four steps, including hazard analysis, structural dynamics analysis, damage analysis and seismic loss analysis. EDPs, as the products of structural dynamic analysis, were explained and the methodologies' seismic fragility curves were briefly introduced. Employing a probabilistic hazard analysis, the method for deriving the annual probability of seismic risk exceedance and seismic risk curves was presented. Considering the importance of both secondary effects and interactions between different sectors of an economy due to seismic loss, those regional economic models with common application in the evaluation of economic conditions after natural disasters (e.g., earthquakes) were mentioned.

Author details

Afshin Kalantari
International Institute of Earthquake Engineering and Seismology, Iran

6. References

[1] The Third International Workshop on Earthquake and Megacities, Reducing Vulnerability, Increasing Sustainability of the World's Megacities. Shanghai, China, 2002.

[2] UNICEF, Sichuan Earthquake One Year Report. May 2009.

[3] The recent earthquake and Tsunami in Japan: implications for East Asia, World Bank East Asia and pacific Economic Update 2011, Vol. 1.

[4] Financial Management of Earthquake Risk, Earthquake Engineering Research Institute (EERI), May 2000.

[5] Bertero R. D. and Bertero V. V., Performance-based seismic engineering: the need for a reliable conceptual comprehensive approach, Earthquake Engineering and Structural Dynamics, 2002; 31, pp. 627–652.

[6] Dowrick D. J., Earthquake Risk Reduction. John Wiley & Sons Ltd, 2005.

[7] Okuyama Y., Impact Estimation Methodology: Case Studies in: Global Facility for Disaster Reduction and Recovery, www.gfdrr.org.

[8] The Institute of Civil Engineers Megacities Overseas Development Administration, Reducing Vulnerability to Natural Disasters. Thomas Telford Publications, 1995.

[9] ATC-58 Structural and Structural Performance Products Team, ATC-58 Project Task Report, Phase 3, Tasks 2.2 and 2.3, Engineering Demand Parameters for Structural Framing Systems, Applied Technology Council, 2004.

[10] ATC-58 Nonstructural Performance Products Team, ATC-58 Project Task Report, Phase 3, Task 2.3, Engineering Demand Parameters for Nonstructural Components, Applied Technology Council, 2004.

[11] Moehle J. P., A Framework for Performance-Based Earthquake Engineering. Proceedings, Tenth U.S.-Japan Workshop on Improvement of Building Seismic Design and Construction Practices, ATC-15-9 Report, Applied Technology Council, Redwood City, California, 2003.

[12] FEMA, 2000a, NEHRP Recommended Provisions for Seismic Regulations for Buildings and Other Structures, Part 1, Provisions, prepared by the Building Seismic Safety Council, published by the Federal Emergency Management Agency, Report No. FEMA 368, Washington, DC.

[13] FEMA, 2000b, NEHRP Recommended Provisions for Seismic Regulations for Buildings and Other Structures, Part 2, Commentary, prepared by the Building Seismic Safety Council, published by the Federal Emergency Management Agency, Report No. FEMA 369, Washington, DC.

[14] Williams M. S. and Sexsmith R. G., Seismic damage indices for concrete structures: a state-of the-art review. Earthquake Spectra, 1995, Volume 11, No. 2, pp. 319-349.

[15] Park Y. J. and Ang A. H.-S, Mechanistic seismic damage model for reinforced concrete. Journal of Structural Engineering, 1985, Vol. 111, No. 4, pp. 722-739.

[16] Powell G.H. and Allahabadi R., Seismic damage prediction by deterministic methods: concepts and procedures. Earthquake Engineering and Structural Dynamics, 1988, Vol. 16, pp. 719-734.

[17] Fajfar P., Equivalent ductility factors, taking into account low-cycle fatigue. Earthquake Engineering and Structural Dynamics, 1999, Vol. 21, pp. 837-848.

[18] Mehanny S. and Deierlein G. G., Assessing seismic performance of composite (RCS) and steel moment framed buildings, Proceedings, 12th World Conference on Earthquake Engineering, Auckland, New Zealand, 2000.

[19] Bozorgnia Y. and Bertero V. V., Damage spectra: characteristics and applications to seismic risk reduction. Journal of Structural Engineering, 2003, Vol. 129, No. 10, pp. 1330-1340.

[20] Sarabandi1 P., Pachakis D., King S. and Kiremidjian A., Empirical Fragility Functions from Recent, Earthquakes, 13WCEE, 13th World Conference on Earthquake Engineering, Vancouver, B.C., Canada, August 1-6, 2004, Paper No. 1211.

[21] Hwang H., Liu J. B. and Chiu Y., Seismic Fragility Analysis of Highway Bridges. Technical Report, MAEC RR-4 Project, Center for Earthquake Research and Information, The University of Memphis, 2001.

[22] Choi E., DesRoches R. and Nielson B., Seismic fragility of typical bridges in moderate seismic zones. Engineering Structures, 2004, 26, pp. 187–199.

[23] Padgett J. E and DesRoches R., Methodology for the development of analytical fragility curves for retrofitted bridges. Earthquake Engineering and Structural Dynamics, 2008, 37, pp. 1157–1174.

[24] Padgett J. E., Nielson B. G. and DesRoches R., Selection of optimal intensity measures in probabilistic seismic demand models of highway bridge portfolios, Earthquake Engineering and Structural Dynamics, 2008, 37, pp. 711–725.

[25] Bazzurro P. and Cornell C. A., Seismic hazard analysis for non-linear structures. I &2, ASCE Journal of Structural Engineering 1994; 120, 11, pp. 3320–3365.

[26] Earthquake Damage Evaluation Data for California, ATC 13, Applied Technology Council, Founded by Federal Emergency Management Agency, 1985.

[27] Seismic Risk Management of Civil Engineering Structures, Hoshiya M., Nakamura T. San Kai Do, 2002 (in Japanese).

[28] Cochrane H. C., Predicting the economic impacts of earthquakes. in: H.C. Cochrane, J. E. Haas and R. W. Kates (Eds) Social Science Perspectives on the Coming San Francisco Earthquakes—Economic Impact, Prediction, and Reconstruction, Natural Hazard Working Paper No.25, University of Colorado, Institute of Behavioural Sciences, 1974.

[29] Rose A. J., Benavides S. E., Chang P., Szczesniak P. and Lim D., The Regional Economic Impact of an Earthquake: Direct Effects of Electricity Lifeline Disruptions. Journal of Regional Science, 1997, 37, 3, pp. 437-458.

[30] Okuyama Y. and Chang S., (Editors), Modelling Spatial and Economic Impacts of Disasters, Springer, 2004.

[31] Federal Emergency Management Agency, Earthquake Loss Estimation Methodology (HAZUS). Washington, DC; National Institute of Building Science, 2001.

[32] Rose A. and Liao S. Y., Modelling regional economic resilience to disasters: a computable general equilibrium analysis of water service disruptions, Journal of Regional Science, 2005, 45, pp. 75-112.

[33] Rose A. and Guha G. S., Computable general equilibrium modelling of electric utility lifeline losses from earthquakes, in: Y. Okuyama and S. E. Chang, (Eds) Modelling Spatial and Economic Impacts of Disasters, 2004, pp. 119-141, Springer.

[34] Cole S., Lifeline and livelihood: a social accounting matrix approach to calamity preparedness, Journal of Contingencies and Crisis Management, 1995, 3, pp. 228-40.

[35] Cole S., Decision support for calamity preparedness: socioeconomic and interregional impacts, in M. Shinozuka, A. Rose and R. T. Eguchi, (Eds) Engineering and Socioeconomic Impacts of Earthquakes, 1998, pp. 125-153, Multidisciplinary Center for Earthquake Engineering Research.

[36] Cole S., Geohazards in social systems: an insurance matrix approach, in: Y. Okuyama and S.E. Chang (Eds) Modelling Spatial and Economic Impacts of Disasters, 2004, pp. 103-118, Springer.

[37] Okuyama Y., Critical review of methodologies on disaster impact estimation, Background Paper to the joint World Bank - UN Assessment on the Economics of Disaster Risk Reduction, 2009.

[38] National Research Council (Committee on assessing the cost of natural disasters, The impact of disasters, A framework for loss estimation. Washington, DC; National Academy Press, 1999.

[39] Yamano N. Kajitani Y. and Shumuta Y., Modelling the Regional Economic Loss of Natural Disasters: The Search for Economic Hotspots. Economic Systems Research, 2007, Volume 19, Issue 2, Special Issue: Economic Modelling for Disaster Impact Analysis, pp. 163-181.

Chapter 2

Assessment of Seismic Hazard of Territory

V. B. Zaalishvili

Additional information is available at the end of the chapter

1. Introduction

The new complex method of seismic hazard assessment that resulted in creation of the probabilistic maps of seismic microzonation is presented in this chapter. To study seismicity and analyze seismic hazard of the territory the following databases are formed: macroseismic, seismologic databases and the database of possible seismic source zones (or potential seismic sources - PSS) as well. Using modern methods (over-regional method of IPE RAS - Russia) and computer programs (SEISRisk-3 – USA) in GIS technologies there were designed some probabilistic maps of seismic hazard for the Republic North Ossetia-Alania in intensity units (MSK-64) at a scale of 1:200 000 with exceedance probability being of 1%, 2%, 5%, 10% for a period of 50 years, which corresponds to recurrence period of 5000, 2500, 1000, 500 years. Moreover, first the probabilistic maps of seismic hazard were made in acceleration units for the territory of Russia. The map of 5% probability is likely to be used for the large scale building, i.e. the major type of constructions, whereas the map of 2% probability should be used for high responsibility construction only. The approach based on physical mechanisms of the source is supposed to design the synthesized accelerograms generated using real seismic records interpretation.

For each of the zoning subject the probabilistic map of the seismic microzonation with location of different calculated intensity (7, 8, 9, 9*) zones is developed (the zones, composed by clay soils of fluid consistency, which can be characterized by liquefaction at quite strong influences, are marked by the index 9*). The maps in acceleration units show the similar results.

The complex approach based on the latest achievements in engineering seismology, can significantly increase the adequacy or foundation for assessments and reduce the inaccuracy in earthquake engineering and construction.

Realization of investigations on mapping of seismic hazard such as detailed seismic zoning (DSZ) based on the most advanced field research methods and analysis of every subject of the Northern Caucasus separately on a scale of 1:200 000 gives the possibility to merge a bit unavailable, at first glance, schemes into geologically and geophysically quite reasonable map of DSZ for the Northern Caucasus with equal scale system of the source zones.

2. Assessment of seismic hazard. General and detailed seismic zoning

The seismic hazard of some territory represents a possible potential or a level of expected hazard, caused by geological structure features, tectonic movements, geophysical fields, macroseismical catalog, engineer-geological and hydrogeological structure etc. The adequate assessment of seismic hazard, at the same time is one of the important problems of engineer seismology. Unlike short-range and middle-range earthquake forecast, the involved assessment of seismic hazard, presented as seismic zoning maps, in fact is a long-ranged forecast of the earthquake strength and place.

One can mark out three types of analysis three consecutive stages of seismic zoning:

1. general seismic zoning – GSZ or SZ, is realized in 1:5 000 000 or 1:2 500 000 scale
2. detailed seismic zoning DSZ, was originally carried out for the most studied regions of perspective construction in 1:1 000 000, 1:500 000 scale or very rarely in 1: 200 000 scale.
3. seismic microzonation – SMZ, in 1:25 000 scale or greater, contained in engineer investigation system.

The results of seismic zoning have to be the appropriate map creation GSZ, DSZ and SMZ. DSZ differs from GSZ in investigation scale. At the same time, in DSZ process may and must be studied all potential sources of possible earthquakes, which may be not taken in account, e.g. they have relatively small seismic potential during GSZ analyzing. It has to be mentioned, that in the real conditions the consequences of seismic hazard generation with that types of sources may have, if not great, but noticeably negative effect. At the same time both types of zoning are very similar, nothing to say about minuteness.

The third stage or stage of seismic hazard assessment in SMZ type has absolutely other physical meaning, in spite of similar name with GSZ and DSZ. The SMZ using allows to take into account the seismic properties of site soils, including physicomechanical and dynamical properties of soil.

The SMZ map traditionally is a normative part of Building Codes, and regularly is revised. At the same time during the map design only huge geology-geophysical zones are taken into account, which the seismicity determined.

The assessment of seismic hazard of the site is carried out using necessity and probabilistic methods. The probabilistic analysis of seismic hazard assessment includes alternative models of seismic sources, the earthquake returne periods, the seismic signal attenuation and distance dependence, and much vagueness, caused by careless information of some parameters, and by random character of seismic events. In the necessity analysis of seismic hazard assessment the vagueness is not considered, only the extreme seismic effect is estimated on the real site, using near earthquake sources with fixed magnitudes. There are many domestic and foreign algorithms and programs for this purpose.

Practically all the previous maps of seismic zoning, from the first map (1937) in the former USSR till the last but one map (1978) were necessity. They not take into account the main characteristic of seismic regime of seism active territory, although in the middle of 40th S.V. Medvedev (Medvedev, 1947) proposed to bring in seismic hazard zones internal differentiation including the strong earthquake return periods and assumed constructions durability. Then U.V. Riznichenko created algorithms and programs for seismic "shakeability" estimation (Riznichenko, 1966). But all these progressive development of domestic seismologists, like their other ideas were not brought in use. (Seismic zoning of USSR territory, 1980). At the same time these ideas were brought in use abroad, after analogous paper of Cornel K.A. (Cornell, 1968). And then western countries begun to create seismic zoning map in exceeding (or nonexceeding) probability of seismic hazard in given times intervals.

The vagueness conditions, are always presented in nature, so the necessity method in the seismic zoning is incompetent. The seismic zoning process must use only probabilistic methods. The risk is always presented, but it must be estimated and reduced to minimum. These ideas are presented in the new more progressive maps of Russia general seismic zoning - GSZ -97. For the first time in Russia was proposed to use the probability map kit GSZ -97 for different constructions (Ulomov, 1995). General map GSZ -97 is presented on fig. 1.

Wide spread usage of GSZ is caused by insufficient development of DSZ and distinct labor-intensiveness of its realization for researchers. Prof. Ulomov and his colleges use modern methods instead of ancient and out of date approach. In the same time the GSZ materials using sometime is impossible due impossibility to use more detailed information of regional and local materials including tectonical materials. The map generalization is enough for state overall planning, but is not enough for reliable estimation of real objects seismic conditions.

The process of Detailed seismic zonation is very complicated and expensive complex of geology tectonical, geophysical and seismical investigation for quantitative estimation of seismic effect in any site of perspective region (Aptikaev, 1986).

That type of investigation consists of all methods used in DSZ, but estimated quantitatively the source (background) seismic effects only on concerned site GSZ (more precisely for mean soil conditions or 2nd seismic category soils on site).

So, it is necessary to develop DSZ approach. The modern DSZ has clear and argumented content. There is huge Strong Motion Data Base with many records of soil velocity and acceleration, including South Caucasus Countries. Now, there are many modern computer programs, reliable digital velocity and acceleration registrators, now we may obtain many records of earthquakes. So, it is possible to realize DSZ purpose using reliable data. And, in spite of updating initial seismicity (UIS) for DSZ we have tye possibility to estimate site seismic hazard.

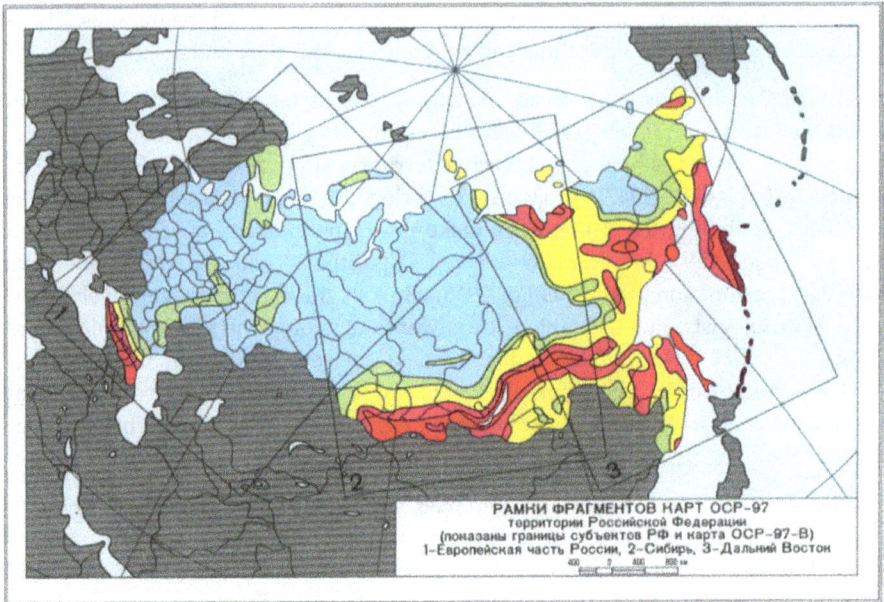

Figure 1. Map of General seismic zonation GSZ-97 of Russia

It must be told, that UIS-DSZ methodic always formed parallel with GSZ methodic, but the scale differs, and some additional methods.

There are some methods that may be used in GSZ and DSZ for seismic generic structures (SGS) identifications, it is identification of zones of danger earthquake appearance (Nesmeianov, 2004).

2.1. Seismogeological method

Using the first epicentral zone investigation in the end of XIX and beginning of XX centuries Abich G. and Lagorio A.E. find out the dependence between earthquake and tectonic structures. Mushketov I.V. writing about Vern earthquake 1887 year, told that Turkestan earthquake is connected with discontinuous disturbance (Mushketov , 1889). He wrote earthquakes "culminate on the boundary of the most huge and new disturbance" (Mushketov, 1891). Besides, he wrote that some groups of earthquakes are connected with lines, transversal to common stretch of rugosity, e.g. connected them with transversal structures in modern terminology. K.I. Bogdanovich analyzing Kebi earthquake (1911) consequences in the North Tien Shan, introduced new term – seismotectonic element, and for the first time proved the seismic shock migration inside seism active zone.

So, seismogeological method was able to connect strong earthquakes with tectonic structures. Those bonds later were named as geological seismicity criterion and were used in other methods.

2.2. Seismotectonical methods

Seismotectonical method was introduced in the end of 40th years of XX century by Gubin I.E. when investigate Garm region on the Pamirs - Tien Shan border. He connected earthquakes with discontinuous zones some tenth km wide. He wrote, that "seismogenity degree" is stable all over the zone, "seismicity degree" may be ascribed to other similar zones, "if this structures, using geological data, are connected by mutual evolution process with equal intensity". This method (Gubin seismotectonics law) says, that in a given geological medium in the active structures of the same type and size, maximum earthquakes, originate from the rock displacement along the active rupture, have equal magnitudes and sources. Seismotectonical method accents on geological seismicity criterion – the velocity of young rupture displacement.

2.3. Seismostructural method

Seismostructural method developed in the mid-50's, by V. Belousov, A. Goriachev, I. Kirillova, B. Petrushevsky, I.A. Rezanov, A.A. Sorsky, but most fully reflected in works of B. Petrushevsky. Earthquakes associated with large structural complexes-blocks allocated by using the historical-structural analysis and discontinuous joints.

Large-scale analysis blocks allowed to associate with them (and the underlying faults) varied range of depths of earthquakes (most profound on the articulation of the Pacific with Eurasian and the American continents). Picture of the strong earthquakes focuses with different three-dimensional structures of the Earth's crust was further developed in the works of G.P. Gorshkov (Gorshkov, 1984). However, this promising direction needs to be fleshed out.

2.4. Tektonophysical method

Tektonophysical method developed in the second half of the 50-ies by M.V. Gzovsky. The method connects the earthquake with the maximum tangential stresses area, which is in conjunction with the maximum gradients of average speeds of tectonic movements and breaks. The energy of the earthquake was put by M.V. Gzovsky in dependence on a number of factors. But a precise calculation is impossible because the mechanical properties of the Earth's crust and its viscosity in Maxima tangential stresses can be evaluated only in qualitative terms.

2.5. Method of allocating quasihomogeneous zones

In the late 50-ies started to be developed method of allocating quasihomogeneous zones of earthquakes for one or all geological and geophysical criteria, some of which have tectonical nature. However, these criteria have not been effective in a number of regions.

Since the choice of number and encoding parameters and their combinations are endless, equally infinite may be variants of map Mmax. In connection with this were analysed practically all existing geologic-structural, seismic and geophysical maps for the territorial zoning using seismotectonic capacity (combined geological criteria reflecting the characteristics of the medium properties and the intensity of tectonic process), described in conventional units on a reference site. Based on mathematical patterns is forecasting of magnitude Mmax with reference site to the rest territory. Let's note the approach developed by Reisner and Ioganson, where reference sites were used in all of the zones of the planet. The analysis involved areas with variety of tectonic properties, where the seismicity criteria are mixed. Naturally, the common criteria were often not the fault criteria (thickness of the Earth crust, the heat flux density, height, isostatic gravity anomalies, the depth of the consolidated Foundation, etc.) The method later became known as the "extraregional method".

2.6. Method of seismoactive nodes

Structural refinement of earthquakes has allowed to V.M.Reiman at the turn of the 50 's and 60's to make an idea using the Central Asian material of disjunctive nodes in which the strong earthquakes concentrate or sejsmogenetic nodes. Later became actively used the term the seismically active sites. The best method was developed by E.Y. Rantsman (Rantsman , 1979), which extended the scheme to many orogenetic regions of the world. E.Y. Rantsman links with sites the earthquakes epicenters, stressing that "the earthquakes focuses can reach hundreds of miles away and go far beyond the morfostructural nodes". To classify the structures of seismicity was proposed the complex system of formalized criteria (distance from the edges of the site, type of terrain, maximum height, and area of friable deposits) and mathematical apparatus. Study of seismogenerating structures made it possible to include cross rises in the number of structures that make up the nodes (Nesmeyanov, Barkhatov, 1978).

2.7. Paleoseismological method

Paleoseismological method (V.P. Solonenko, V.S. Khromovskikh, A. Nikonov, etc.) allows using paleoseismodislocations to trace possible seismic sources zones (PSS zones) and estimate their magnitude and seismic intensity. To evaluate these parameters the seismotectonic dislocation is used. Currently, there are many formulas (public and regional), describing the statistical associations between seismodislocations (length, amplitude displacement) and seismological parameters (magnitude, the depth of the epicenter, intensity of seismic vibrations) of earthquake. To determine the occurrence frequency of earthquakes it is necessary to have a reliable assessment of the age of paleo-seismic dislocations. All possible approaches are used: geological-geomorphological, archaeological, historical data and radiocarbon dating of sediments, broken by seismodislocation and later. Dendrohronological method is used, which takes into account the changes in the growth of trees associated with earthquakes, as well as lihenometrical method of dating seismogenical samples and dedicated to its some species of lichens.

2.8. Detailed seismic zoning

As an example, the assessment of seismic hazard, let's consider some estimations on the DSZ level in the territory of North Ossetia (Zaalishvili & Rogozhin, 2011).

On the basis of an analysis of methods for identification of PSS zones was elected out of regional seismotektonic method to objectively identify the sejsmogenic sources. Despite some shortcomings, the entire method is characterized by the quantitative indicators and has strong decision-making apparatus. The method has been used by prof. E.A. Rogozhin in North Ossetia when solving various scientific tasks. In addition, using this method some similar tasks were solved not only for Russian but also overseas territories (Israel, Italy etc.) (Rogozhin, 1997, 2007; Rogozhin et al., 2001; Rogozhin et al., 2008). At the same time, this does not preclude obtaining reliable results and other known methods.

The methodology used in most probabilistic seismic hazard analysis was first defined by Cornell and as usually accepted it consists of four steps (Reiter 1990, Kramer 1996): 1. Definition of earthquake source zones (SSZ), 2. Definition of recurrence characteristics for each source, 3. Estimation of earthquake effect and 4. Determination of hazard at the site. The probabilistic hazard maps for the territory of under study was compiled and we shell describe in brief this works according to the above noted steps.

2.8.1. Definition of earthquake sources

As a rule, today probabilistic assessment of seismic hazard is used all over the world for the identification of seismic loads for the engineering projects. The probabilistic approach is a more systematized method for the assessment of quantity, sizes and location of future earthquakes (Bazzuro & Cornell, 1999; Cornell, 1968; McGuire, 1995) than any other methods. Formal procedures for the probabilistic assessment include the determinations of spatio-temporal ambiguities for the expected (future) earthquakes. The computer program EQRISK of McGuire became the main stage in the method development (McGuire, 1976). The program became widespread and is very popular up to present day. In this connection the probabilistic assessment of seismic hazard is often called Cornell McGuire's method. The program includes integration on ambiguities distribution.

The Caucasian region is characterized by high intensity of dynamic geological processes (McClusky et al., 2000) and hazards, connected with them, of both natural and technogenic character. The most clearly expressed among these hazards is seismicity, which is accompanied with wide range of secondary processes. Earth surface ruptures, activation of known earlier inactive faults, landslip phenomenon, collapses, avalanches, creep and subsidence of the earth surface, activation of surface structures, soil liquefaction and other hazardous phenomena can be noted among them.

The investigations on determination and parameterization of the seismic source zones in recent decades has been realized by V.P.Solonenko, V.S.Khromovskikh, E.A.Rogozhin, V.I.Ulomov, V.G.Trifonov, I.P.Gamkrelidze and others (Gamkrelidze et al., 1998; Paleoseismology of Great Caucasus, 1979; Nechaev, 1998; Rogozhin et al., 2008; Trifonov, 1999; Ulomov et al., 1999).

On basis of the results of the active faults study located southward of the Great Caucasian ridge, parameters of seismic source zones were chosen according to data of I.P.Gamkrelidze work (Gamkrelidze et al., 1998) and to the north of the ridge they were chosen on data of

E.A.Rogozhin and others (Rogozhin, 1997). According to the results of the executed expert evaluation of seismic potential (Mmax) the maps of seismic sources zoning of the territory of North Ossetia (zones of possible seismic sources - PSS zones) were made up.

A new original method of more accurate ascertainment of the boundaries of seismogenic source (fault) active part and assessment of the potential of seismic source hazard (at works of detailed seismic zoning – DSZ) has been worked out in recent years (Rogozhin et al., 2008).

Let's consider the process of territory seismic hazard assessment for explanation of procedure usage by the example of the Central Caucasus (the territory of the Republic of North Ossetia-Alania).

PSS zones are referred to the active fault systems, singled out on a basis of interpretation of the materials of remote sensing and geological data. Decoding of multispectral three-channel space images of Landsat–4/5 (resolution 30 m) and Landsat–7 (resolution 15 m) was realized. Decoding of space satellite photos was executed in colored multispectral variant as well as in black-and-white variant. Different variants of the image synthesis were used for the analysis of polyzonal scanner pictures. Besides, identification of the lineaments was also executed separately on channels. Combined deductive – inductive approach was used for lineaments identification: integrated structures were decoded on the base of strongly generalized images with the following zooming in for detailing and vice versa local peculiarities of tectonic and exogenous structures with the following zooming out and generalization. The method of stepwise generalization was used with quantization on the scale levels 1:25000; 1:50000; 1:100000; 1:200000; 1:300000; 1:400000; 1:500000. In the scale range 1:25000 - 1:1 500000 space photomap on basis of snapshots Landsat-7 is used and in the range 1:500000-1:2 millions – space photomap, created on basis of Landsat–4/5 snapshots.

Extensive lineaments systems were identified with known faults, which were qualified on modern stage as active. The name of PSS zones was formulated on basis of faults and large settlements names. Morpho-kinematics of active faults is the base for qualification of seismic displacements kinematics in PSS zones. Hypocenters depth of expected earthquakes was calculated from the depth of fault plans, the depth on geophysical anomalies data and from the magnitude of expected events.

Maximum magnitude of expected earthquakes (seismic potential, Mmax) was assessed on the results of usage of the over-regional seismotectonic method of seismic hazard assessment, offered by G.I.Reisner. Usage of this method, foundation of which is described in the number of publications (Reisner & Ioganson, 1997; Rogozhin et al., 2001), showed that the Northern Caucasus is the region of very high seismic hazard.

In 2007 it was determined on data of field investigations that for the urbanized territories of North Ossetia the most hazardous are Vladikavkaz, Mozdok, Sunzha and Tersk PSS zones (table 1), (Fig.2) (Arakelyan et al., 2008; Rogozhin et al., 2008).Parameterization of seismic sources was made after creation of these maps, i.e. maximum possible magnitude Mmax for each seismic source was assessed. This is the most difficult problem in the process of parameterization of PSS zones. Mmax was determined on the data of a number of authors (Chelidze, 2003; Rogozhin, 2007).

The second essential parameter, which characterizes expected earthquakes, is sources depth range, where the majority of seismic events with corresponding magnitude generate. According to the numerous investigations, Caucasus is the region with upper crust part location of seismic sources – their depth doesn't exceed 20–25 km (deeper seismicity is observed in Tersk-Sunzha zone in the area of Grozniy city and in Caspian Sea). As sources distribution on depth for this region wasn't executed, average value of depth (equal to 10 km) was taken for calculations (see table 1).

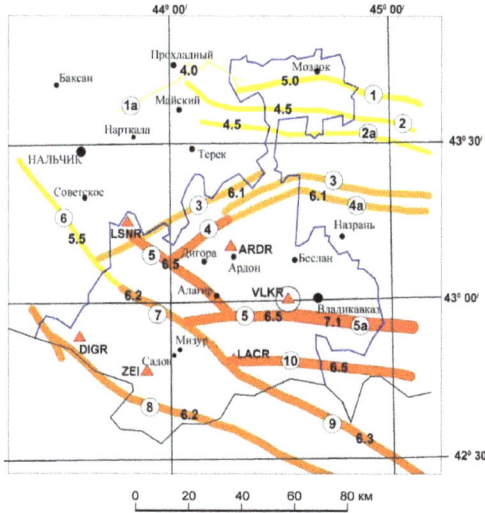

Figure 2. Map of PSS zones of the territory of the Republic North Ossetia-Alania (Rogozhin, 2007). Red triangles – basic seismic stations in the region. Blue and black lines are the state borders of North Ossetia.

№	PSS zone	Magnitude	H, km	Kinematics.
1	Mozdok eastern	5.0	10	reverse faulting
1a	Mozdok western	4.0	5	strike-slip
2	Tersk	4.5	5	reverse faulting
3	Sunzha northern	6.1	15	reverse faulting
4	Sunzha southern (western branch)	6.5	15	strike-slip
4a	Sunzha southern (eastern branch)	6.1	15	reverse faulting
5	Vladikavkaz (western branch)	6.5	15	reverse faulting
5a	Vladikavkaz (eastern branch)	7.1	20	reverse faulting
6	Nalchik	5.5	10	strike-slip
7	Mizur	6.2	15	strike-slip
8	Main ridge	6.2	15	reverse faulting
9	Side ridge	6.3	15	reverse faulting
10	Karmadon	6.5	15	reverse faulting

Table 1. PSS zones for North Ossetia characteristics (numbers in the rings on Fig.1).

2.8.2. Definition of reoccurrence characteristics

For the assessment of ratio parameters between reiterations during the process of execution of a number of investigations on the international projects the earthquake catalogue was checked and specified. The seismicity in each source zone was analyzed on basis of catalogue usage: New Catalogue... 1982, Corrected Catalogue of Caucasus, Institute of Geophysics Ac. Sci. Georgia (in data base of IG), the Special Catalogue of Earthquakes for GSHAP test area Caucasus (SCETAC), compiled in the frame of the Global Seismic Hazard Assessment Program (GSHAP), for the period 2000 BC - 1993, N.V. Kondorskaya (editor), (Ms>3.5) Earthquake catalogues of Northern Eurasia (for 1992-2000), Catalogue of NSSP Armenia, Special Catalogue for the Racha earthquake 1991 epicentral area (Inst. Geophysics, Georgia) and also the Catalogue of NORTH OSSETIA 2004–2006.

Corrected Catalogue of Caucasus contains data for more than 61000 of earthquakes, including 300 historical events (Byus, 1955a, 1955b, 1955c; New Catalogue of strong Earthquakes in the USSR..., 1982), which happened during 2000 years. This catalogue was checked and corrected. Some hypocentral parameters of earthquakes were recalculated.

Threshold of magnitude for the whole catalogue and a and b values of the frequency-magnitude law were determined for large tectonic zones, as their calculation for certain PSS zones was impossible because of data absence. Value of b of the frequency-magnitude law is determined by formula of Gutenberg-Richter:

$$\lg(N/T) = a - bM \qquad (1)$$

where a and b are parameters, the inclination and level of recurrence graph at M=0.

For each PSS zone (both linear and square) frequency of earthquake origination was studied on basis of observed seismicity. For study of Gutenberg-Richter ratio earthquakes were referred to the separate faults or PSS zones taking into account accuracy in epicenter determination. Because of the shortage of data about accuracy of location determination average model was accepted. This model supposes that mistakes have normal distribution with standard deviation equal to 3–4 km. Distances from each event to the all PSS zones were measured and only zones, which were on closer distances from the event than three standard deviations, were taken into account. Based on distances value, weighting coefficient was assigned to each zone, from the curve of density distribution of the standard deviation possibility.

2.8.3. Estimation of earthquake effect

Earthquake effect was estimated using two different parameters: macroseismic intensity and peak ground acceleration (PGA). Macroseismic intensity (MSK scale) was traditionally used for seismic zonation in former USSR. Macroseismic and instrumental data on 43 significant earthquakes occurred in Caucasus were revised to obtain the necessary information (Javakhishvili et al. 1998). Data on 37 earthquakes was selected and in some cases were compiled new isoseismal maps in the 1:500 000 scale. In a process of computations was observed a fact that the value of the attenuation coefficient in vicinity (within the limits of the first three isoseismals) of the source of the Ms>6 earthquake is very high (v~4.5-5.0), in comparison with small and moderate events (v~3.4). This fact has been tested on the other Caucasian strong earthquakes (Ms>6) and in general has been confirmed. In spite of the lack of data in the first approximation the equation of correlation in this case obtains the following form for small earthquakes:

$$I = 1.5M_s - 3.4lg(\Delta^2 + h^2)^{1/2} + 3.0 \tag{2}$$

and

$$I = 1.5M_s - 4.7lg(\Delta^2 + h^2)^{1/2} + 4.0 \tag{3}$$

for large events.

The attenuation model according to the (2) formula is given on fig.3.

It should be noted, that for hazard estimation we have used the second relationship. Besides that we have restricted maximal value in epicentral area for M=7, (6.5) earthquakes with intensity 9, M=6 (5.5) earthquakes with intensity 8, etc. this was done to avoid very high intensities in epicentral area. The epicentral areas were estimated using relationships for earthquake source sizes given in (Ulomov 1999).

On the other hand strong motion instrumental data in Caucasus and adjacent regions allows us to use PGA and spectral acceleration attenuation law for seismic hazard analysis. Since the installation of the first digital strong-motion station in the Caucasus area 451 acceleration time histories from 269 earthquakes were recorded (Smit et al. 2000). Based on the acceleration time histories recorded between June 1990 and September 1998 with the permanent and temporary digital strong-motion network in the Caucasus and adjacent area, 84 corrected horizontal acceleration time histories and response spectra from 26 earthquakes with magnitudes between 4.0 and 7.1 were selected and compiled into a new dataset. All time histories were recorded at sites where the local geology is classified as "alluvium". Therefore the attenuation relations derived in this study are only valid for the prediction of the ground motion at "alluvium" sites.

The calculation of the correlation coefficients and the residual root mean square was performed with the well known Joyner and Boore two step regression model. This method allows a de-coupling of the determination of the magnitude dependence from the determination of the distance dependence of the attenuation of ground motion. Using the larger horizontal component for spectra of the selected acceleration time histories, the values of coefficients were obtained for the coefficients at different frequencies. Because it is easy to obtain peak acceleration from corrected acceleration time histories, empirical attenuation models with peak ground acceleration as dependent parameter have always played an important role in different seismic hazard and earthquake engineering studies. The resulting equation for larger horizontal values of peak horizontal acceleration is:

$$\text{Log PHA} = 0.72 + 0.44 \, M - \log R - 0.00231 + 0.28 \, p, \tag{4}$$

$R = (D2 + 4.52)^{1/2}$,

where PHA is the peak horizontal acceleration in [cm/sec^2], M is the surface-wave magnitude and D is the hypocentral-distance in [km]. p is 0 for 50-percentile values and 1 for 84-percentile.

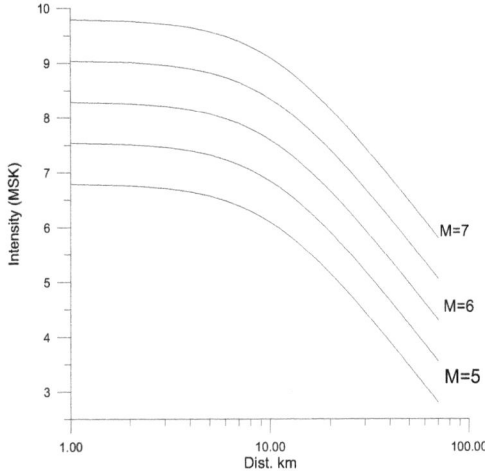

Figure 3. Attenuation model for intensity (MSK)

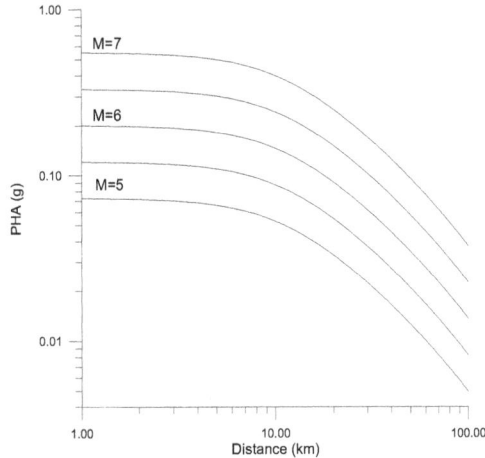

Figure 4. Attenuation model for acceleration

It is important to bear in mind that all equations given above represent a best fit of the selected dataset, and therefore represent mean values about which there is a considerable scatter. In the case of the attenuation model for the larger horizontal value of the peak horizontal acceleration the predicted mean plus one standard deviation is equal to 1.91

times the mean value. The scatter of the pha-models is the same as similar models for Europe and Western North-America (Smit et al., 2000). The attenuation is shown on fig. 4.

The comparison of the attenuation relationships for peak horizontal acceleration with similar relations for other areas shows a good agreement with the models from Western North-America. It is obvious, that the attenuation in Europe is lower compared to the Caucasus and adjacent area. The predicted peak values in the near-field are higher than the corresponding values obtained with other European models (Smit et al., 2000).

2.8.4. Determination of hazard

The probabilistic seismic hazard maps (the maps of detailed seismic zoning) have been constructed for the total area of North Ossetia in scale 1:200000 with exceedance probability for a period of 50 years (standard time of building or construction durability!) with 1%, 2%, 5%, 10% in GIS technologies, which corresponds to reoccurance of maximum probable earthquake for a period of 5000, 2500, 1000 and 500 years (Fig.5). The longer the period of time the higher the level of possible intensity. For a period of 500 years only a small part will be occupied by the zone of 7 intensity earthquake, for a period of 1000 years – 8 intensity and at 2500 years 9 intensity earthquake appearance, correspondingly.

Cornell approach, namely computer program SEISRisk- 3, developed in 1987 by Bender and Perkins (Bender & Perkins, 1987) was used for the calculations. The map of observed maximum intensity was compared with the maps of different periods of exposition and the most real map was chosen on a basis of the analysis of differences between the observed and calculated maps. According to these criteria the map of 5% probability with exceedance probability of 50 years can be recommended for seismic zoning of the territory of North Ossetia. Besides, for the first time probability maps of seismic hazard for Russian territory were made in acceleration units in scale 1:200 000 with exceedance probability for a period of 50 years - 1%, 2%, 5%, 10%.

According to the Musson (Musson, 1999) conception, it is necessary to use the data, which is maximum approximate to the real engineering-geological conditions, at assessments of territory seismic hazard. For the territory of North Ossetia the exposition equal to 1000 years is the most approximate to real conditions for mass building. It is necessary to consider greater exposition, for example, 2500 years etc. for unique buildings and constructions.

The maps of 5% probability are likely to be used for the large scale building, i.e. the major type of constructions, whereas the maps of 2% probability should be used for high responsibility construction only (Fig.5).

One can see great hazard in the south of North Ossetia on the map, where exists the increased level of seismic hazard (due to powerful Vladikavkaz fault, lying nearby).

As a matter of principle it is possible to make maps in scale 1:100 000 etc., but it actually makes no practical sense. Although accuracy of such maps must be higher, adequacy of the results can be considered as doubtful due to absence of reliable data on local peculiarities of past, i.e. historical earthquakes display. Laboriousness (irretrievable) at that increases multiply.

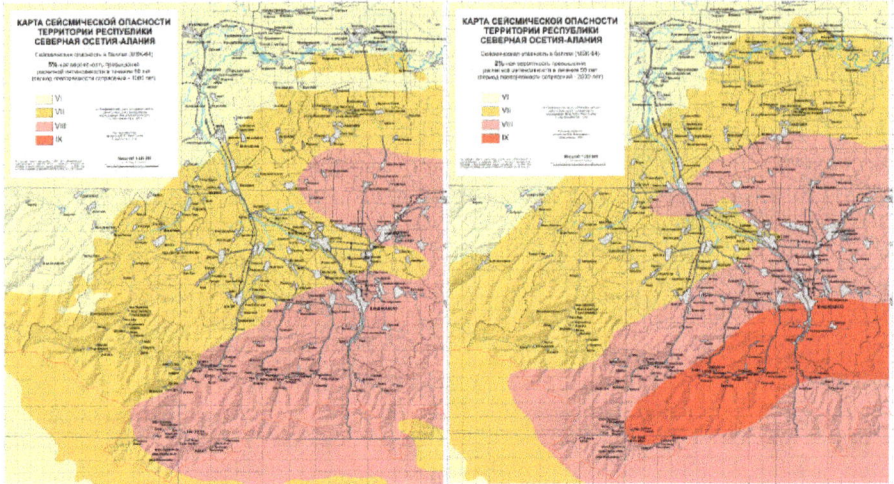

a)   b)

Figure 5. Probabilistic maps of seismic hazard (DSZ) in the intensities (MSK-64) with the exceedance probability 5% (a) и 2% (b) for North Ossetia territory and adjacent areas (Zaalishvili, 2006).

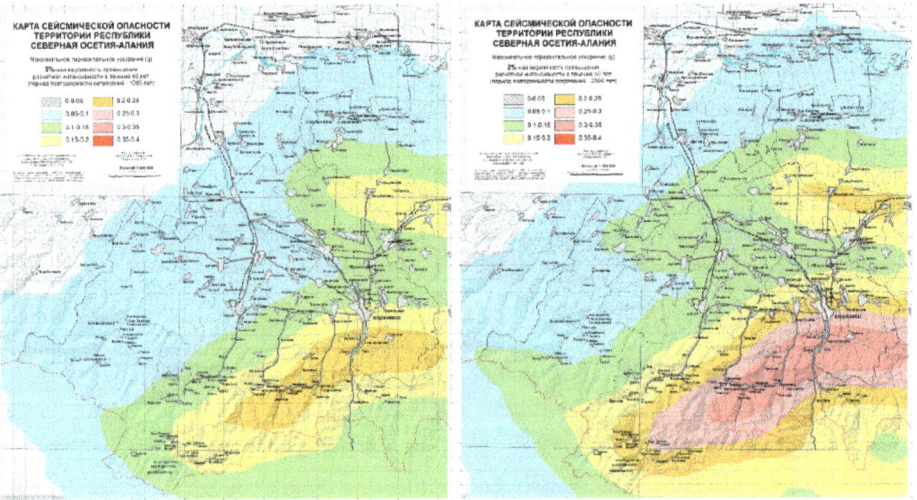

a)   b)

Figure 6. Probabilistic map of seismic hazard (DSZ) in accelerations (PGA)with exceedance probability 5% (a) and 2% (b) for North Ossetia territory (Zaalishvili, 2006).

The scientists from Vladikavkaz in collaboration with the colleagues from the Institute of Physics of the Earth of RAS not only offer to use large-scale maps but also decided to continue investigations and cover the whole Northern Caucasus in scale 1:200 000. So, maps of seismic hazard can be made up in scale 1:200 000 for the Republics of Chechnya, Ingushetia, Kabardino-Balkaria, Stavropol and Krasnodar areas and the other territories

(Fig. 7). Taking into account, that faults and other peculiarities of the territory exist out of any boundaries, including state boundaries, it is possible to make unusual but quite physically proved single general map of detailed seismic zoning of the territory of Northern Caucasus in scale 1:200000, moreover, one can make them for different exposition times and accordingly for different probabilities. So, created maps of detailed seismic zoning of North Ossetia conform to earthquake realization once in 500 years, 5% - in 1000 years and 2% - in 2500 years. The level of seismic hazard grows with the time increase etc.

It is possible to make detailed maps of seismic hazard for the whole Caucasus, including Azerbaijan, Armenia and Georgia, due to the features of spreading of hazardous seismic sources, which «neglect» states' boundaries. It is also possible to develop the maps jointly with Turkey and Iran and it's real to include such countries as Israel, Egypt, and Lebanon etc.

Maps of detailed seismic zoning can be called «long-term» prediction maps. It means that long-term prediction of hazardous phenomena is realized on their basis and, correspondingly the place of earthquake-proof building-stock is determined

Essentially, the long-term maps of expected intensities locations are that of described maps of detailed seismic zoning. Indeed, that evacuate people from the hazardous territory before expected earthquake is impossible, but it is real to prevent population burring under destroyed or, to be more precise, differently damaged buildings, which is formed on basis of such maps. The more educated society is the less seismic risk, i.e. economic and social losses. So, the priorities are clear.

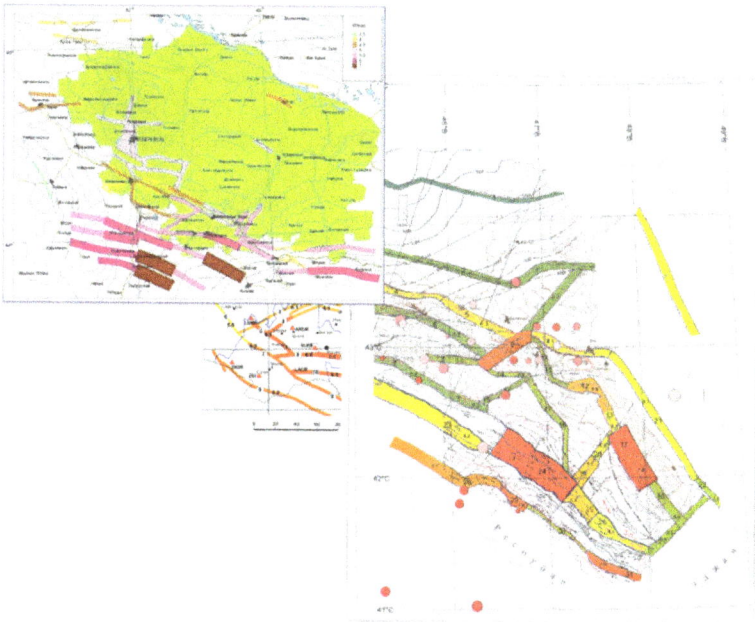

Figure 7. The mosaic of maps of hazardous potential seismic sources on the territories of the Northern Caucasus (model of the future joint map).

On basis of the given maps it is necessary to make up the maps of seismic microzonation (SMZ) of cities and large settlements of each certain subject of the Russian Federation with the usage of the most modern standard methods and tools, but in scale 1:10 000. The probabilistic maps of SMZ were first developed in the Center of Geophysical Investigations of Vladikavkaz Scientific Center RAS and RNO-A. Such maps of SMZ are direct and reliable base of earthquake-proof design and object construction.

Besides, it is necessary to note that at usage of the traditional units of macroseismic intensity the boundaries between different zones are characterized by sharp changes, which obviously do not correspond to the real situation of monotonous change of intensity for homogenous soil conditions of the investigated territory. No doubt, it will form evident inaccuracies at the assessment of the level of seismic hazard of this or that territory. The practical usage of artificial intensity subdivision, for example, in the form of 7.2 or 8.3 points is not validated enough from the theoretical point of view. So, firstly, it is not usually explained how these fractional assessments are obtained and, secondly, the following transition to the acceleration units (obviously, according to foreign data, as there are no acceleration records for forming reliable correlation in Russia), undoubtedly, forms considerable inaccuracy and it is hardly ever physically proved because of the formality of the parameter of «intensity» itself.

On the other hand, at seismic influence assessment at earthquake - proof design engineers use the acceleration values, (strictly speaking, conveniently) corresponding to specified intensities. Thus, it's assumed that design acceleration a = 0.1 g corresponds to the intensity 7 earthquake, 0.2g – to the intensity 8, 0.4g – to the intensity 9 etc. At the same time, network of digital stations dislocated on the Southern Caucasus installed in source zones of Spitak (Armenia, 1988), Racha (Georgia, 1991), Barisakho (Georgia, 1992), Baku (Azerbaijan, 2000), Gouban (Georgia, 1991), Tbilisi (Georgia, 2002) and other earthquakes collected seismic records for formation of database of accelerations for Caucasus. Namely it makes possible to design maps of the seismic hazard independently in units of PGA. Such maps for the territory of North Ossetia for exposition of 50 years with exceedance probability 1%, 2%, 5%, 10% in scale 1:200 000 were created (Fig. 6). It is obvious that at changing of smoothering step it is possible to obtain smooth variations of accelerations directly used as design impacts.

In contrast to the maps of general seismic zoning (GSZ) with a scale of M 1: 8000000 and, at the best, with the scale M 1:2500000 obtained maps of both types on a scale 1:200000 can be referred to the DSZ type maps.

Thus, these materials allow assessing seismic hazard on a detailed level, according to the known formulas to calculate the macroseismic field of seismic effects on a scale that may provide a reliable basis for SMZ.

3. Seismic microzonation of territory

Seismic microzonation (SMZ) actually is final stage of seismic hazard assessment. SMZ results are direct foundation for earthquake-proof construction. In the process of seismic microzonation sites with etalon ground conditions corresponding to specified seismic hazard level are specified. In Russia grounds with mean seismic properties for given territory are traditionally referred as etalon ground conditions. Usually these are soils with shear wave velocity of 250–700 m/s [SP 14.13330.2011]. In Georgia, for example, in dependence of specific engineering-geological situation etalon grounds in their seismic properties can be worst or mean for given territory. In USA firm rock grounds are referred as etalon. Seismic microzonation consists in intensity increments calculation caused by differences in ground conditions. Works on seismic microzonation are realized by instrumental and calculational methods.

3.1. Instrumental method of seismic microzonation

Instrumental method is the main SMZ method. Exactly it urges to solve a problem of forming earthquake intensity forecast. At the same time the calculation method, which allows to model any definite conditions of area and influence features, is often characterized by more reliability. It has great importance to soil thickness with high power. Combined usage of both methods significantly increases results validity.

3.1.1. Seismic microzonation on basis of strong earthquakes instrumental records

It is supposed at usage of strong earthquakes records for SMZ purposes, that at some strong seismic influence the observing soil behavior is adequate to the display of their potential seismic hazard at future strong earthquakes (Nikolaev, 1965). This fact was the reason of stimulation of a number of large international scientific-research projects on organization of long-term instrumental observations with the help of powerful measurement systems in the Earth's different regions with high seismic activity for the purpose of obtaining the strong movements of soils, which are the base of buildings and constructions (the groups SMART-1 and SMART-2 on the Taiwan island etc.).

At the same time, presence of unit record of a real strong seismic influence at its inestimable value for SMZ often can't give the adequate forecast of soil behavior at a next following strong earthquake. This problem can be solved by creation of a number of records of seismic influences, generated by hazardous for the zoned territory active fractures, i.e. by zones of possible earthquake source (PES).

3.1.2. Seismic microzanation with the help of weak earthquakes records

In the connection of the fact that strong earthquakes occur seldom, the intensity increments, as a rule, are assessed by records of weak earthquakes, when a linear dependence between the dynamic stress and the deformation takes place.

Soil conditions considerably change (fig. 8) the right shape of the original undistorted signal, incident from the crystal foundation. Complex shapes of isoseisms pointed out to the undoubtful link between the earthquake display intensity and soil conditions (Reiter, 1991).

Increase of the soil thickness depth (alluvium) considerably changes the character of earthquake records (Reiter, 1991) in the process of approaching the city (fig. 8).

Calculation of intensity increment with the help of weak earthquakes is realized by the formula (Medvedev, 1962; Recommendations on SMZ, 1974, 1985):

$$\Delta I = 3{,}3 \lg A_i / A_0, \tag{5}$$

where A_i, A_0 are the amplitudes of investigated and etalon soils vibrations.

The usage of tool in the form of registration of strong and weak earthquakes needs the organization of instrumental observations in a waiting mode.

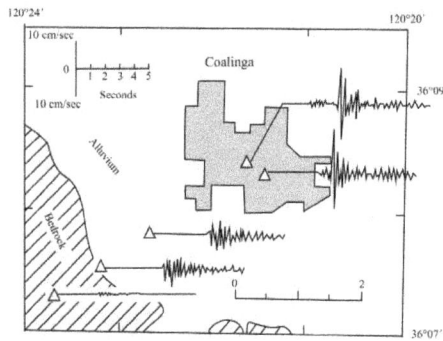

Figure 8. Scheme of California earthquake in Koaling sity

3.1.3. Seismic microzanation with the help of weak earthquakes records

In the connection of the fact that strong earthquakes occur seldom, the intensity increments, as a rule, are assessed by records of weak earthquakes, when a linear dependence between the dynamic stress and the deformation takes place.

Calculation of intensity increment with the help of weak earthquakes is realized by the formula (Medvedev, 1962, Recommendations on SMZ, 1974, 1985):

$$\Delta I = 3{,}3 \lg A_i / A_0, \tag{6}$$

where A_i, A_0 are the amplitudes of investigated and etalon soils vibrations.

The usage of tool in the form of registration of strong and weak earthquakes needs the organization of instrumental observations in a waiting mode.

3.1.4. Seismic microzonation using microseisms

The results of microseisms observations (Kanai, 1952) are used as subsidiary instrumental tool of SMZ. Predominant periods are determined at that in order to assess resonance properties of soils and amplitude level of microvibrations. Strictly speaking, the reference of microseism on their origin to the purely natural phenomena is not quite correct. Numerous artificial sources, influence degree of which can't be controlled, undoubtedly, take part in their forming along with the natural sources (fig. 8.6).

Intensity increment for strong earthquakes on microseism is calculated by the formula (Recommendations on SMZ, 1974, 1985):

$$\Delta I = 2 \lg A_i \, / \, A_0, \tag{7}$$

where A_i, A_0 are the maximum amplitudes of microvibrations for investigated and etalon soils.

Impossibility of the compliance of necessary standard conditions of microseism registration and large spread in values of maximum amplitudes limit the usage of microseism for calculation of soil intensity increment. The above mentioned causes the application of microseism tool only in complex with other instrumental tools.

Spectral features for different sites are estimated by means of H/V-rations (Nakamura, 1989).

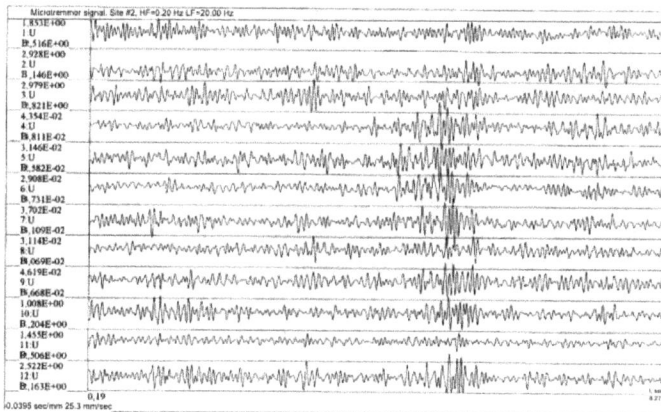

Figure 9. Microseisms records (10.07.1996, Voronezh Region, Russia)

3.1.5. Seismic microzonation using explosive impact

The intensity increment ΔI of the soils of the zoned territory is calculated by the formula (Medvedev, 1962; Recommendations on SMZ, 1974, 1985) at usage of weaker explosions:

$$\Delta I = 3{,}3 \lg A_i \, / \, A_0, \tag{8}$$

where A_i, A_0 are vibrational amplitudes of the investigated and etalon soils.

Execution of powerful explosions on the territory of cities, settlements or near the responsible buildings is connected with large and often insurmountable obstacles (technical and ecological problems, safety problems, labouriousness and economical expediency) and practically isn't used nowadays. This leads to the wide spreading of nonexplosive vibration sources.

3.1.6. Seismic microzonation using nonexplosive impulse impact

The features of SMZ methods development led to the situation when the tool of elastic wave excitation with the help of low-powered sources (for example, hammer impact with m = 8–10 kilograms) has become the most wide spread in the CIS countries, in order to determine S- and P-wave propagation velocities in soils of the typical areas of territory. Velocity values are used in order to calculate the intensity increment using the tool of seismic rigidities by S.V.Medvedev (Medvedev, 1962; Recommendations on SMZ, 1974, 1985):

$$\Delta I = 1,67 \lg (\rho_0 V_0 / \rho_i V_i), \tag{9}$$

where $\rho_0 V_0$ and $\rho_i V_i$ is the product of the soil consistency and P-wave (S-wave) velocity – seismic rigidities of the etalon and the investigated soil accordingly.

The intensity increment, caused by soil watering, is calculated by the formula

$$\Delta I = K e^{-0,04 h_{GL}^2} \tag{10}$$

where K = 1 for clay and sandy soils; K = 0,5 for large-fragmental soils (with sandy-argillaceous filler not less than 30%) and strongly weathered rocks; K = 0 for large-fragmental firm soils consisting of magmatic rocks (with sandy-argillaceous filler up to 30%) and weakly weathered rocks; h_{GL} is the groundwater level.

The simplicity and immediacy of practical application of S.V.Medvedevs' tool, which is called the tool of the "intensities", led to its widespread in CIS countries and countries of Eastern Europe, Italy, USA, India, and Chile in 1970-es. The tool of the "intensities" was advantageously different from other tools by the immediacy, simplicity in initial data obtaining and its processing and independence from seismic regime of the territory. It to a certain extent hampered the development and making up of new tools. Unfortunately, the calculation results of predicted values of intensity increment are often quite incorrect as data of macroseismic observations of destructive earthquake consequences shows (Shteinberg, 1964, 1965, 1967; Poceski, 1969; Stoykovic and Mihailov, 1973).

By means of the special investigations it was determined that the reliability of calculated intensity increments considerably increases at usage of modern powerful impulsive energy sources (fig. 9).

The lowering of final results quality is to a certain extent caused by the fact that in the tool of "intensities" the seismic effect dependence in soils on frequency or "frequency discrimination" of soils (Shteinberg, 1965) and also the origin of typical "nonlinear effects" at strong movements isn't taken into account. A.B.Maksimov tried to remedy this deficiency by developing the tool, where frequency peculiarities of soils were taken into account (Maksimov, 1969):

$$\Delta I = 0.8 \lg \rho_0 V_0 f_0^2 / \rho_i V_i f_i^2 \tag{11}$$

where f_0, f_i are predominant frequencies of etalon and investigated soils.

A.B.Maksimovs' tool didn't find wide distribution, as frequency differences of soil vibrations with sharply different strength properties (at usage of traditional for the seismic exploration of small depths low-powered sources) were insignificant and the calculation results on the formulas (9) and (11) were practically similar (Zaalishvili, 1986).

Intensity increment was determined by the following formula (Zaalishvili, 1986):

$$\Delta I = 0.8 \lg \rho_0 V_0 f_{wa0}^2 / \rho_i V_i f_{wai}^2 \qquad (12)$$

where f_{wa0}, f_{wai} are weight-average vibration frequencies of etalon and investigated soils.

Weight-average vibration frequency of soils was calculated at that on the formula [Zaalishvili, 1986]:

$$f_{CB} = \sum A_i f_i / \sum A_i \qquad (13)$$

where A_i and f_i are the amplitude and the corresponding frequency of vibration spectrum.

Figure 10. Surficial gasodinamical pulse source (SI-32)

3.1.7. Seismic microzonation using vibration impact

At usage of a vibration source (fig. 10) the calculation of intensity increment is realized with the help of the formula (Zaalishvili, 1986):

$$\Delta I = 2 \lg S_i / S_0, \qquad (14)$$

where S_i and S_0 are the squares of vibration spectra of investigated and etalon soils.

The developed tool was used at SMZ of the territories of cities Tbilisi, Kutaisi, Tkibuli, single areas of the Bolshoy Sochi city. The tools' feature consists in the fact that it allows to assess soil seismic hazard without any preliminary investigations: at realization of direct measurements of soil thickness response on standard (vibration or impulse) influence. Later the formula was successfully used at SMZ of the sites of Novovoronezh Nuclear power-plant (NPP) with the help of an impulsive source (Zaalishvili, 2009).

Figure 11. Vibration source (SV-10/100)

3.1.8. Seismic microzonation on basis of taking into account soil nonlinear properties

The comparison of the absorption and nonlinearity indices with the corresponding spectra of soil vibrations shows that at higher absorption the spectrum square prevails in LF field and at high nonlinearity it prevails in HF field of the spectrum. In other words, the presence of absorption is displayed in additional spreading of LF spectrum region, and the presence of nonlinearity – in spreading of HF range.

All the mentioned allowed to obtain the formula for calculation of intensity increment on basis of taking into account nonlinear – elastic soil behavior or elastic nonlinearity (at usage of vibration source) [Zaalishvili, 1996]:

$$\Delta I = 3 \lg A_{if_{wai}} / A_{0f_{wa0}}, \tag{15}$$

where $A_{if_{wai}}$, $A_{0f_{wa0}}$ is the product of spectrum amplitude on weight-average vibration frequency of investigated and etalon soils.

The formula (14) characterizes soil nonlinear–elastic behavior at the absence of absorption.

If the impulsive source is used at SMZ than the formula will have the form (Zaalishvili, 2009):

$$\Delta I = 2 \lg A_{if_{wai}} / A_{0f_{wa0}}. \tag{16}$$

3.1.9. Seismic microzonation on basis of taking into account soil inelastic properties

As soil liquefaction and uneven settlement of the constructions are observed at strong earthquakes (Niigata, 1966; Kobe, 1995), the most actual problem of SMZ is to assess possible soil nonelasticity adequately and physically proved at intensive seismic influences.

In order to assess directly nonelasticity of soil, the special scheme of the realization of experimental investigations (fig. 11, a) with gas-dynamic impulsive source GSK-6M (with two oscillators) was used. Selected location of the longitudinal profile allowed to influence alternately by two emitters from adjoining and somewhat far radiation zones. In the spectrum of soil vibrations, caused by near emitter, the HF component, which quickly attenuates with distance (fig. 11, b), predominates. In case of influence by distant emitter to the soil surface, the LF component predominates in the spectrum of vibrations (fig. 11, c). In other words, at nonlinear-elastic deformations the main energy is concentrated in the HF range of spectrum and at nonelastic – in the LF range. The signal spectrum has the symmetrical form in the far and practically linear-elastic zone.

Elastic linear and nonlinear vibrations are characterized for the given source by the constancy of the real spectrum square, which is the index of definite source energy value, absorbed by soil (which is deformed by the source). The analysis of strong and destructive earthquake records and also the analysis of specially carried out experimental influences showed that at nonelastic phenomena spectra square of corresponding soil vibrations is not the constant value. It can decrease and the more it decreases, the less the soil solidity and the greater the influence value (Zaalishvili, 2009).

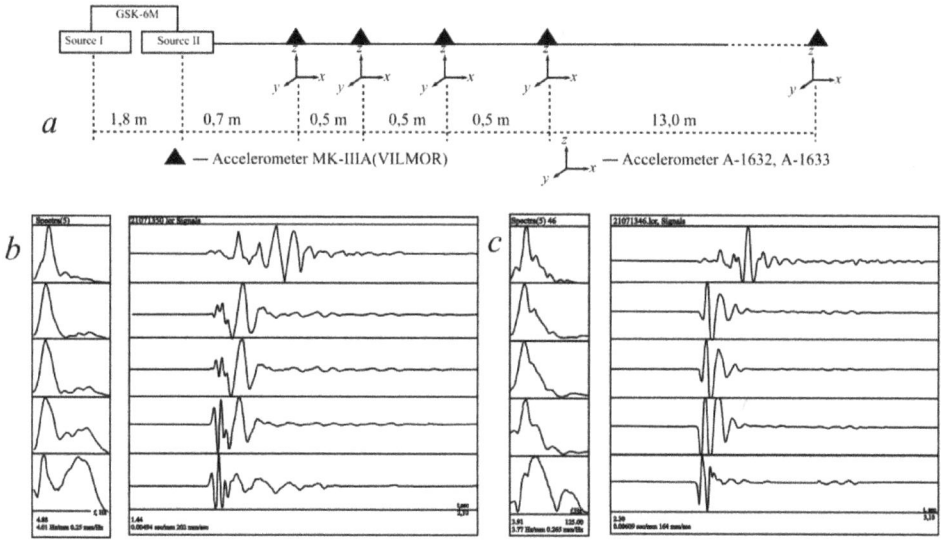

Figure 12. Investugation of site spectral features by means of GSK-6M seismic source: a) experiment scheme; b) record of second source impact; c) record of first source impact

At usage of vibratory energy source, the whole number of new formulas (Zaalishvili, 2009) in order to assess soil seismic hazard with taking into account the values of their nonelasticity were obtained:

$$\Delta I = 2{,}4 \lg \left[(S_{ri})_n (S_{r0})_d / (S_{ri})_d (S_{r0})_n \right], \tag{17}$$

where $(S_{ri})_{n,d}$ and $(S_{r0})_{n,d}$ are the squares of real spectra of investigated and etalon soils in near and distant zones of the source.

$$\Delta I = 3{,}3 \lg (A_i f_{awi})_n (A_0 f_{aw0})_d / (A_i f_{awi})_d (A_0 f_{aw0})_n, \tag{18}$$

where $(A_i\ f_{awi})_{n,d}$ and $(A_0\ f_{aw0})_{n,d}$ are the amplitudes and weight-average frequencies of investigated and etalon soils in near and distant zones of the source.

In case of powerful impulsive source usage the offered formulas will have a form:

$$\Delta I = 1,2\ [\lg\ (S_{ri})_n\ (S_{r0})_d\ /\ (S_{ri})_d\ (S_{r0})_n], \tag{19}$$

where $(S_{Pi})_{\delta A}$ and $(S_{P0})_{\delta A}$ are the squares of real spectra of investigated and etalon soils in near and distant zones of the source;

$$\Delta I = 2\ \lg\ [(A_i f_{awi})_n\ (A_0 f_{aw0})_d\ /\ (A_i f_{awi})_d\ (A_0 f_{aw0})_n], \tag{20}$$

where $(A_i\ f_{awi})_{n,d}$ and $(A_0\ f_{aw0})_{n,d}$ are the amplitudes and weight-average frequencies of investigated and etalon soils in near and distant zones of the source.

The formulas (17) and (18) are true only for loose dispersal soils. The formulas (17) and (18) were used at SMZ of the territory of Kutaisi city. Besides, with the help of the formulas (19) and (20) nonelastic deformation properties of soils in full-scale conditions on the site of Novovoronezh NPP-2 were defined more exactly (Zaalishvili, 2009). The formulas were obtained on basis of physical principle, which underlies the scheme, applied at the assessment of soil looseness measure (Zaalishvili, 1996, Nikolaev, 1987).

3.2. Calculational method of seismic microzonation

Calculational method of SMZ is used in order to analyse features of soil behavior with introduction of definite engineering–geological structure characteristics of investigated site as initial data: values of transverse wave velocities, index of extinction, modulus of elasticity, power of soil layers, their consistency etc. Calculational method includes thin-layer medium, multiple-reflected waves, finite-difference method, finite-elements analysis (FEA) and other techniques.

One can take nonlinear soil properties into account in the problems of earthquake engineering by means of instrumental and calculation methods. The instrumental method of SMZ is the main method. Nevertheless it is quite often necessary to solve such problems using calculational method, which allows to model practically any conditions, which are observed in the nature. At the same time the practice reqirements lead to the necessity of calculation of soil vibrations for the conditions of their nonlinear-elastic and nonelastic deformations. At the solution of such problem it is assumed that elastic half-space behaves as linear-elastic medium and the covering soil displays strong nonlinear properties at intensive seismic or dynamic influences (Bonnet & Heitz, 1994).

Instrumental stress-sstrain dependences can be used, for example one obtained for plastic clay soil shown in fig. 12. The conception of the so-called soil bimodularity, offered by A.V.Nikolaev (Nikolaev, 1987, Zaalishvili, 1996; 2000) is taken into account in the given dependence. Considerable differences in behavior of "weak" soils at compression and dilatation lie in the base of the phenomenon. Such soil is characterized at dilatation by quite small shear modulus.

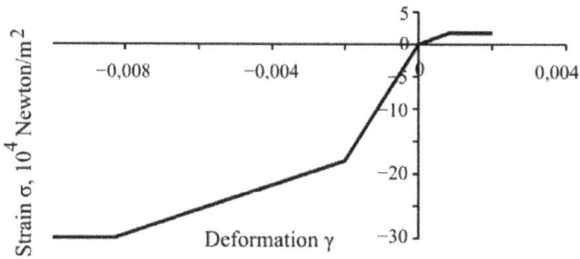

Figure 13. Instrumental stress-sstrain curve, showing property of soil bimodularity

The solution of the given nonlinear problem for soils in the analytic form is based, as a rule, on considerable assumptions due to the complication of adequate taking into account behavior features of such complex system as the soil (Bonnet & Heitz, 1994). Therefore the numerical solution of nonlinear problems on the modern stage of knowledge is the most proved if the data of field or laboratory investigations is taken into account in these or those connections. Thus, the correlations, which are determined by the experimental investigations, are the basis of the solution of calculation nonlinear problems. In other words calculation programs for the solution of calculation nonlinear problems essentially are analytical-empirical. The most adequate programs are exactly like these (SHAKE, NERA etc.).

3.2.1. Equivalent linear model. SHAKE and EERA programs

Equivalent linear model is one of the first models, which take nonlinear soil behavior into account. Equivalent linear approximation consists in modification of the model of Kelvin–Voight (for taking some types of nonlinearity into account) and, for example, is realized in the programs SHAKE (Schnabel et al., 1972) and EERA (Bardet et al., 2000).

Equivalent linear model is based on the hypothesis that shear modulus G and attenuation coefficient ξ are the functions of shearing strain γ (fig. 18.1). In the programs SHAKE and EERA (Equivalent-linear Earthquake site Response Analyses) the values of shear modulus G and attenuation coefficient ξ are determined (in the process of iteration) so that they correspond to the deformation levels in each layer.

3.2.2. IM model. NERA program

In 2001 realization principle, which was used in the program EERA, was applied in the programming of NERA (Nonlinear Site Response Analysis) (Bardet, Tobita, 2001), which allows to compute soil thickness nonlinear reaction on seismic influences. The program is based on the medium model, offered by Iwan (1967) and Mroz (1967), which is often called the IM model for short. As it is shown in the fig. 18.2, the model supposes the simulating of nonlinear curves strain-deformation, using a number of n mechanical elements, which have different stiffness k_j and sliding resistance R_j, where $R_1 < R_2 < ... < R_n$. Initially the residual stresses in all elements are equal to zero. At monotonically increasing load the element j deforms until the transverse strain τ reaches R_j. After that the element j keeps positive residual stress, which is equal to R_j.

The equation, describes dynamics of soil medium, is solved by the method of central differences.

3.2.3. Calculation of nonlinear absorptive ground medium vibrations using multiple reflected waves' tool of seismic microzonation

Let's suppose that we have the seismic wave, which falls on the soil thickness surface. Let's assume that soil thickness is nonlinear absorptive unbounded medium with the density ρ and S-wave propagation velocity vs. At small deformations the value of shear modulus G will be maximum for the given soils:

$$G = G_{max} = \rho v_S^2 \tag{21}$$

At the deformation increase the value G remains constant at first but at reaching some value (which is definite for each material or soil) the value G considerably changes, i.e. the soil begins to display its nonlinear properties. At the continued deformation increase the growth of stresses decelerates and then can remain unchanged until material destruction or hardening, i.e. until structural condition change.

As the main soil index, which characterizes its type and behavior at intensive loads, the value of plasticity PI was chosen. The parameters, which are necessary for calculations, are determined on basis of empirical ratios (Ishibashi, Zhang, 1993):

$$k(\gamma, PI) = 0.5 \left\{ 1 + \tanh \left[\ln \frac{0.000102 + n(PI)}{\gamma} \right]^{0.492} \right\} \tag{22}$$

where

$$n(P_I) = \begin{cases} 0.0 & \text{for} \quad PI = 0, \\ 3.37 \cdot 10^{-6} PI^{1.404} & \text{for} \quad 0 < PI \le 15, \\ 7.0 \cdot 10^{-7} PI^{1.976} & \text{for} \quad 15 < PI \le 70, \\ 2.7 \cdot 10^{-5} PI^{1.115} & \text{for} \quad PI > 70 ; \end{cases}$$

$$d = 0.272 \left\{ 1 - \tanh \left[\ln \left(\frac{0.000556}{\gamma} \right)^{0.4} \right] \right\} e^{-0.0145 PI^{1.3}} .$$

Then the change of shear modulus is determined on basis of the ratio

$$\frac{G}{G_{max}} = k(\gamma, PI)(\sigma)^d, \tag{23}$$

where G is the current shear modulus, σ is normal stress.

Seismic energy absorption is calculated by the formula

$$\xi = 0.333 \frac{1 + \exp\left(-0.0145 PI^{1,3}\right)}{2}\left[0.586\left(\frac{G}{G_{max}}\right)^2 - 1.547\frac{G}{G_{max}} + 1\right] \tag{24}$$

On basis of the given ratios and introduced by us ratios for determination of necessary indices (normal stress, deformation etc), nonlinear version of the program ZOND was worked out. From the database of strong motions AGESAS, which was formed by us (Zaalishvili et al., 2000), the accelerogram, which was recorded on rocks in Japan, with the characteristics (magnitude, epicentral distance, spectral features etc.) similar to the territory of Tbilisi city, was chosen as the accelerogram, given into the bedrock.

The analysis of the results of linear and nonlinear calculations models of definite areas of Tbilisi city territory confirms the adequacy of calculations to the physical phenomena, which were obtained in soils at intensive loads (fig. 13) (Zaalishvili, 2009). With the increase of seismic influence intensity the nonlinearity display increases. Absorption grows simultaneously. Hence the resulting motion at quite high influence levels can be lower than the initial level. It corresponds to the fact, which is known on the results of analysis of strong earthquake consequences, which happened in recent yares (for example, Northridge earthquake, 1994).

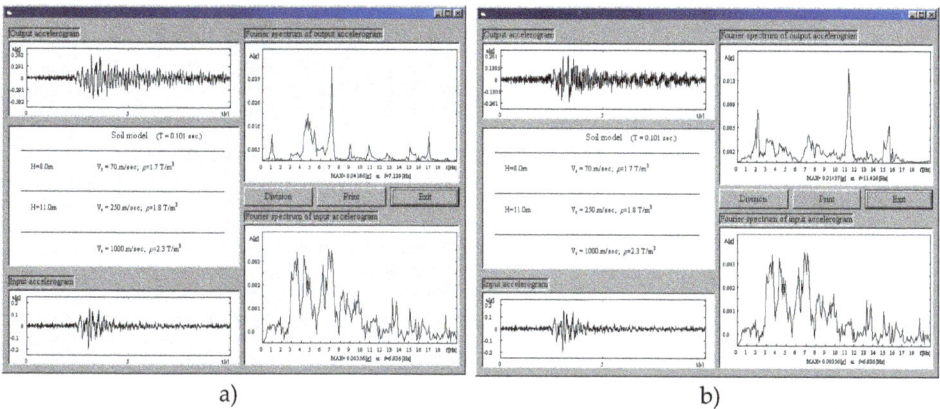

a)   b)

Figure 14. Results of calculations using multiple reflected waves' tool in linear (a) and nonlinear (b) cases.

3.2.4. Calculation of nonlinear soil response using FEM tool of seismic microzonation

The problem of the determination of soil massif response on dynamic influence with taking soil nonlinear properties into account can be solved by usage of finite element method (FEM) in the following way (Zaalishvili, 2009).

Soil medium is represented in the form of two-dimensional massif, which is approximate by triangular finite elements. The net, which consists of triangular elements, allows to describe quite accurately any relief form and form of the layer structure of soil massif with its physics-mechanical parameters. Within finite elemet the soil is homogeneous with inherent to it characteristics, which vary in time depending on influence intensity. Earthquake accelerogram of horizontal or vertical direction, which is applied, as a rule, to the foundation of soil massif, is used as the influence. Soil is in the conditions of plane deformation and is considered as an orthotropic medium. Axes of the orthotropy coincide with the directions of main strains.

The problem of nonlinear dynamics of soil massif is solved by means of the consecutive determination of mode of deflection of the system on the previous step. The system is linear-elastic on each step.

3.3. Instrumental-calculational method of seismic microzonation

In recent years a new «instrumental-calculational» method of SMZ (per se simultaneously having the features of both instrumental and calculational method) which includes tool of «instrumental-calculation analogies» has been developed in Russia in recent years (Zaalishvili, 2006). Its usage is based on direct usage of modern databases of strong motions.

As a basis at realization of tool instrumental database of strong movements, registered in definite soil conditions, is used. As a result of given database with the help of numerical calculations it is possible more or less safety to forecast behavior of these or those soils (or their combination) for strong (weak) earthquakes with typical characteristics for the investigated territory (magnitude, epicentral distance, focus depth etc.).

3.4. Relief influence on the earthquake intensity in SMZ problems

Morphological and morphometric features of relief meso- and macroforms influences on seismic intensity increment.

On basis of the analysis of numerous macroseismic observations the consequences of strong earthquakes, which took place on the territory of the former USSR, S.V.Puchkov and D.V.Garagozov offered the empirical formula for the intensity increment calculation (ΔI) depending on relief feature (Puchkov, Garagozov, 1973):

$$\Delta I = 3{,}3 \lg \left(W_{gr} / W_{et} \right) + 3{,}3 \lg \left(W_{top} / W_{fnd} \right) \tag{25}$$

where W_{gr}, W_{et} are the accelerations of vibratory motion on soil and etalon; W_{top}, W_{fnd} are the accelerations on the top of mountain construction and its foundation.

It was determined as a result of the instrumental and theoretical investigations that for the microrelief the increment of seismic intensity increases from the foundation of mountain-shaped feature to its top and can reach approximately 1.8 degree. For the locality mesorelilef the tendency of the increase of seismic vibration intensity from foundation to the top remains. The increment of seismic intensity for the relief mesoforms is about 0.3 degree. It was shown that weak hilly relief, with the inclinations less than 10°, does not influence on the seismic vibrations intensity.

The investigations of S.V.Puchkov and D.V.Garagozov (Puchkov, Garagozov, 1973) showed that at vibrations of mountain range, composed by volcanic tuf, the amplitude of seismic vibrations in S-waves increases on the height 15 m in 1.46 times in comparison with the foundation. For the massif, composed by loamy sand and loams on the same height marks the vibrational amplitude increased in 1.8 times for p-waves and in 3.2 times for S-waves.

Slope steepness considerably influences on the increment of seismic intensity. The increase of slope steepness, composed by incoherent gravel-pebble and sabulous-loamy grounds is conductive to the sharp worsening of engineering-geological and seismic conditions of the territory. So, for example, it is determined that slope steepness more than 19°–15° (for dry sandy-argillaceous and gravel-pebble differences) produces the intensity increment up to 1 degree and at variation of slope steepness from 10° to 40° the amplitudes of seismic vibrations increase approximately in 2.5 times.

It is known that the increase of slope steepness from 40° to 80° produces the increment of seismic intensity equal to 1.5 degree (Zaalishvili, Gogmachadze, 1989).

The correlation analysis of the dependence of seismic intensity increment on true altitude, slope steepness and relief roughness showed that the main factors, which change the value of seismic intensity, are the first two indices [Puchkov, Garagozov, 1973]. It conforms well to the investigation results of V.B.Zaalishvili, who introduced the new parameter of the relief coefficient (Zaalishvili, Gogmachadze, 1989) (fig. 14).

Later the data analysis allowed to offer us (I.Gabeeva & V.Zaalishvili) the empirical formula for the possible amplification calculation K and intensity increment ΔI, which are caused by the relief (Zaalishvili, 2006):

$$K = -0.1 + 0.68 \lg R \qquad (26)$$

where R= $\alpha \times H$ is the relief coefficient; α is the relief slope angle, degree; H is height, m.

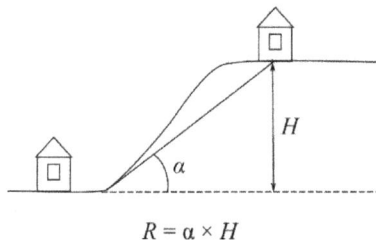

$$R = \alpha \times H$$

Figure 15. Relief coefficient R

The analysis of the experimental data shows that intensity increment can vary at that independently of the type of rocks, from 0 to 1.5 degree.

Finally, let's try to assess the amplification of vibrational amplitude, which is caused by relief, with the help of the calculation method of FEM (Zaalishvili, 2006).

The algorithm for the calculation of seismic reaction of soil thickness for the two-dimensional model was developed for this purpose (fig. 15) (Zaalishvili, 2009). The results of the executed earliear investigations were used for the program testing (Puchkov, Garagozov, 1973). Mountain structure had the form of frustum of a cone with the height 30 m and slope angle of the generatrix 30º. The element maximum size was equal to 5 m, S-wave propagation velocity was 300 m/s, the density 1800 kg/m³. The seismic influence was applied to the foundation of soil thickness in the form of instrumental accelerogram, modeling the vertically propagating SH wave.

It was determined that the vibrational amplitude considerably chances with the relief. The given dependence at that is various for the displacements, velocities and accelerations. The largest value of the amplification is observed for displacements and the maximum ratio of vibrational amplitudes, for example, in the point C to the point A, is 2.1 and for the point D – 3.2. It well satisfies the results of experimental observations where the ratio in the point C for the S-wave is equal to 2.3 and in the spectral region the maximum values are 1.8 (at T = 0.4 s) and 3.2 (at T = 0.7 s) for P- and S-waves accordingly. Spectral analysis also shows the resonance increase of vibrational amplitudes in the top part of the slope on the frequency 1.6 Hz (i.e. T=0.6 s).

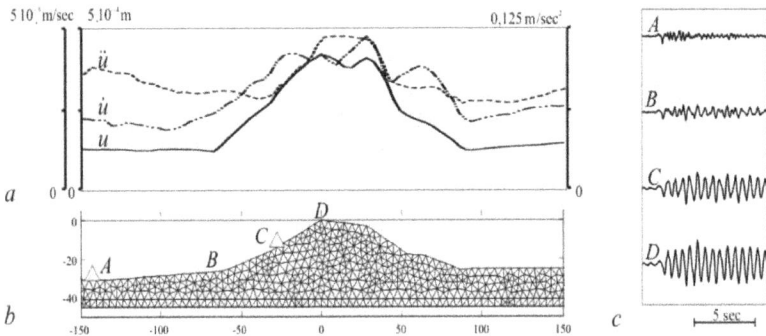

Figure 16. Final elements analysis (FEA) application example: a) Variation of amplitudes of displacement, velocity and acceleration along surface; b) calculational model; c) seismograms, calculated in points A, B, C, D.

Considerably fewer investigations are dedicated to the influence of the underground relief on the intensity. On the data of B.A.Trifonov (1979) the underground and buried topography of the rocks influences on seismic vibrations intensity, if the surface slope exceeds 0.3. At the vee couch of the rocks, which are covered by sedimentary thickness, the ratio between wave length and the sizes of vee stripping influences on seismic intensity change. Seismic intensity increment in the given case is formed by the wave interference and can be 1.5–2.0 degree (Bugaev & Kharlov, 1977; Bondarik et.al., 2007).

Thus, at the execution of SMZ works in the mountain regions or under the conditions of billowy relief, it is necessary to pay special attention to the influence of surface or underground relief on the intensity forming. It is necessary to continue the investigations in order to obtain statistically proved ratio for the calculation of intensity increment, caused by relief.

3.5. Seismic microzonation of Vladikavkaz city

If we consider 5% DSZ map as basis for seismic microzonation so seismic intensity of 8 corresponds to etalon grounds for whole territory.

Then, maps of seismic microzonation of cities must be created. According to the above mentioned maps of detailed zoning the maps of seismic microzonation with probability 1%, 2%, 5% or 10 %, correspondingly, were made up.

Though, that definitions of the word «zoning» are similar, actually they are quite different in essence. Unlike the maps of detailed seismic zoning, which give seismic potential (Mmax) and source features, the maps of seismic microzonation give assessments of soil condition influence (sands, rocks, pebbles, clays etc., their combination; watering; relief (as underground as surface); spectral distribution of incoming wave; predominant vibration frequencies on city square etc.) on forming of future earthquake intensity. As a rule, the scale of such maps is 1:10 000, in order to have the opportunity of taking them into account at building. Maps can be more detailed (1:5000 etc.) but this makes no sense as the type and physical condition of soils in space on the territory site can change fast. The most important thing is to assess intensity of possible earthquakes on areas with typical soil conditions for city territory.

Maps of seismic microzonation can be made up for the certain territories (cities and settlements, as a rule). It is impossible to make them up in entire format because of the necessity of geological conditions knowledge on larger territories, which are mostly not built up. We often don't have such data even for the modern cities! It's practically impossible because the resources will be lost for nothing! And absurdity! In the other words there is no the microzonation map even for the territory of North Ossetia let alone the whole Northern Caucasus.

Maps of seismic microzonation do not only show the place of earthquake-proof building up, but they also show on what intensity this or that building must be calculated and designed: on 6, 7, 8 or 9 points. And sometimes even on 10 points (for very soft grounds!). And this suggests investments of different financing for the realization of antiseismic measures (thicker armature, more connections etc.). Seismic risk can considerably be reduced at building-up zones with 7, 8 and 9 point of the calculated intensity by adequate site development on the territory of city, for example, as social losses will be minimal, though buildings will be damaged in this or that extent.

In the next stage we should carry out SMZ. It should be noted that as a basis the maps of different probability of exceedance will be used and as the initial intensity, the value of which corresponds directly to the intensity of the sites, composed by average soils or characterized by average soil conditions and, therefore, the maps will be referred to the 7, 8 or 9 points (and similarly for acceleration). The zones, composed by clay soils of fluid consistency, which can be characterized by liquefaction at quite strong influences, are marked by the index 9*. Intensity calculation here supposes the usage of special approaches in the form of direct taking soil nonlinearity into account (Zaalishvili, 2000). The usage of relevant methods and techniques of SMZ will allow to obtain the correspondent maps of SMZ.

Thus for maps with probability of exceedance 1%, 2%, 5% and 10% one can obtain corresponding maps of SMZ with probability of exceedance 1%, 2%, 5% and 10%, i.e. probabilistic maps of SMZ (Fig. 16).

For each of the zoning subject the probabilistic map of the seismic microzonation with location of different calculated intensity (7, 8, 9, 9*) zones is developed (the zones, composed by clay soils of fluid consistency, which can be characterized by liquefaction at quite strong influences, are marked by the index 9*). The maps in accelerations units show the similar results.

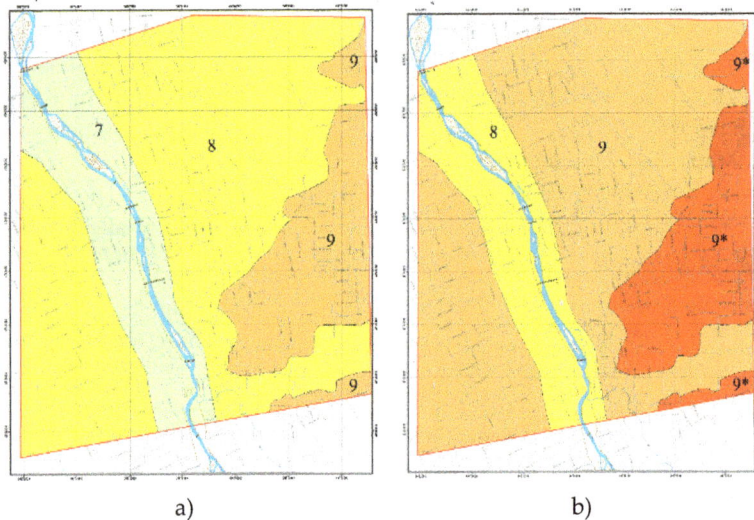

a)   b)

Figure 17. The maps of seismic intensity microzonation for probabilities of 5% (a) and 2% (b) for the central part of Vladikavkaz city territory (Zaalishvili et al., 2010).

Such maps of SMZ except of mentioned developments are also based on materials of local network of seismic observations "Vladikavkaz". Network was organized for the first time on the urbanized territory of the Northern Caucasus in July 2004. Stations are located on the sites with different typical for the city soils (clays of medium-hard and liquid consistence, gravels with filling material of less than 30% and more than 30%, and their assembly).

It must be noted that usage of the maps with high time exposition i.e. maximal magnitude (maximal intensity) for given territory (for return period of 50 years and exceedance probability 2% or 1%) physical nonlinearity of soils necessarily must be taken into account with the help of developed tools (Zaalishvili, 2009).

Unlike small-scale M 1:8 000 000 seismic hazard map of the territory of Russia (GSZ) maps of DSZ in scale 1:200 000 allow taking into account features of specific seismic sources (faults) directly. But the main thing is that such scale zoning is suitable for quite large territories. So it's seen that alignment of faults of different constituent entities of the Russian Federation of Northern Caucasus make a good sense (fig.7).

4. Specified seismic fault and design seismic motion

Analysis and consequent account of initial accelerograms transformation will become the basis for site effect analysis at strong seismic loadings (fig. 17) (Zaalishvili et al., 2010).

Methods of such modeling are based on accordance of spectral properties of modeled and real earthquake. In a whole modeling accuracy depending on the purposes of total motion usage and what characteristics defining structural system behavior must be reproduced.

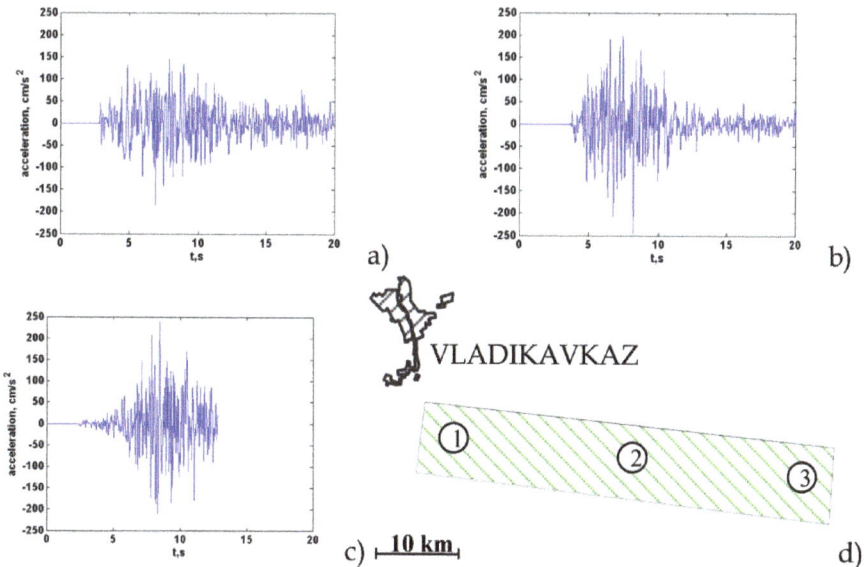

Figure 18. Synthetical accelerograms for different source locations: a – western part of fault; Sb – middle part of fault; c – eastern part of fault; d – scheme of sources of scenarios earthquakes

Earthquake source that is a region of rupture can be considered as point source only for much larger distances than fault size. At close distances effects of finite fault size become more significant. Those phenomena are mainly connected with finite rupture velocity, which causes energy radiation of different fault parts in different times and seismic waves are interference and causes directivity effects (Beresnev & Atkinson, 1997, 1998).

Let's compare amplitude spectra of obtained design accelerograms with spectrum of real earthquake from considered fault. Data analysis (fig. 18 and fig. 19) shows that spectra of calculated and real earthquakes in a whole are similar in their main parameters.

It must be noted that spectrum of vertical component of real earthquake is closer to design spectra. The last fact is quite obvious and is explained by proximity to earthquake source. Indeed, close earthquakes in general are characterized by predomination of vertical component. Record of TEA station (located in theater) was selected due to its location on dense gravel and has a minimal distortions caused by soil conditions.

Analysis of spectrum of weak earthquake shows that peaks are observed on 1.3 and 5.6 Hz (Fig. 18). In spectra of synthesize accelerograms mentioned amplitudes are also observed. At the same time medium response on strong earthquake, undoubtedly, differ from weak earthquake response (Fig. 19) (Zaalishvili, 2000).

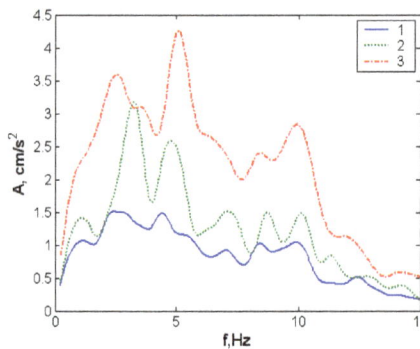

Figure 19. Spectra of design accelerograms at different source locations of earthquake M=7,1: 1 – western part of fault; 2 – middle part of fault; 3 – eastern part of fault

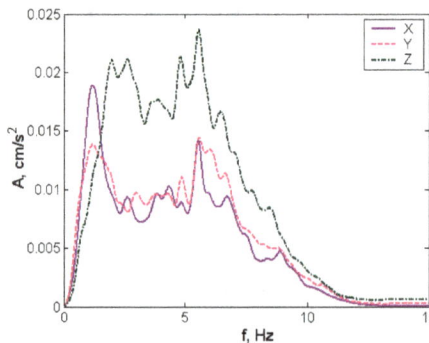

Figure 20. Spectra of accelerograms of weak earthquake with epicenter in the zone of Vladikavkaz fault. (25.08.2005 10:25 GMT, H = 8 km M= 2.5).

Usage of maps of detailed seismic zoning in units of accelerations at seismic microzonation level is possible only for calculation method giving results in units of accelerations. Today traditional instrumental method of seismic microzonation does not allow obtaining intensity increments in accelerations due to traditional orientation on macroseismic intensity indexes. The exclusion is the case of investigation of strong earthquakes accelerations when instrumental records are obtained (in presence of accelerometer) (Zaalishvili, 2000). At the same time investigations are conducted and the problem supposed to be solved.

On the other hand in recent years a new instrumental-calculation method was developed (Zaalishvili, 2006). New method is based on selection from database (including about 5000 earthquake records) soil conditions which are the most appropriate to real soil conditions of the investigated site. Then the selection of seismic records with certain parameters or their intervals follows (magnitude, epicentral distance, and source depth). Then maximal amplitudes are recalculated for given epicentral distances. Absorption coefficient can be calculated by attenuation model for given region.

Thus, a new complex method of seismic hazard assessment providing probability maps of seismic microzonation, which are the basis of earthquake-proof construction, is introduced. Undoubtedly such approach significantly increases physical validity of final results.

Considered procedures on the level of possible seismic sources zones exploration, maps of detailed seismic zoning and seismic microzonation may differ from described above. So paleoseismological investigations like «trenching» (Rogozhin, 2007), which allow determining more reasonable the recurrence and other features of seismic events realization are also possible when it is necessary.

Today, we have conditions for detailed seismic zoning maps development like the above mentioned but for all the territory of the Northern Caucasus on basis of the modern achievements of engineering seismology. It will give us a possibility to develop probabilistic maps of seismic microzonation with the help of powerful nonexplosive sources, methods taking into account physical soils nonlinearity (Zaalishvili, 2009).

Thus algorithm of seismic hazard assessment of the territory taking into account multiple factors forming seismic intensity was considered. Forms of typical seismic loadings for firm soils are given, which will be changed from site to site in dependence of differences in ground conditions (engineering-geological, geomorphological and gidrogeological conditions)

Author details

V. B. Zaalishvili

Center of Geophysical Investigations of RAS, Russian Federation

5. References

Aptikaev, F.F. et al. (1986) Methodological recommendations on detailed seismic zoning. *Questions of engineering seismology.* Issue 27. Moscow, 1986. 184-212. (in Russian)

Arakelyan, A. R., Zaalishvili, V. B., Makiev, V. D., Melkov, D. A. (2008) To the question of seismic zonation of the territory of the Republic of North Ossetia-Alania / Procs. of Ist International conference "Dangerous natural and man-caused processes on the mountaneous and foothill territories of Northern Caucasus", Vladikavkaz September 20-22, 2007. Vladilavkaz: VSC RAS and RNO-A, 2008, pp. 263-278 (in Russian).

Bardet J.P., Tobita T., NERA, A computer program for Nonlinear Earthquake site Response Analyses of layered soil deposits. Univ. of Southern California, Los Angeles, 2001. 44 p.

Bardet, J.P., Ichii, K., Lin, C.H., 2000. EERA, A Computer Program for Equivalent Linear Earthquake Site Response Analysis of Layered Soils Deposits. University of Southern California, Los Angeles

Bazzurro P. and Cornell C. A. (1999). Disaggregation of Seismic Hazard, Bull. Seism. Soc. Am. 89, 2, pp. 501-520

Bender, B. and Perkins, D. M. (1987). SEISRISK III: A Computer Program for Seismic Hazard Estimation. US Geological Survey Bulletin 1772, 48p.

Beresnev, I. A., Atkinson, G. M. (1997). Modeling finite fault radiation from ωn spectrum. Bull. Seism. Soc. Am., 87, 67–84.

Beresnev, I. A., Atkinson, G. M. (1998). FINSIM – a FORTRAN program for simulating stochastic acceleration time histories from finite faults. Seismological Research letters. Vol. 69. No. 1.

Bondarik, G.K., Pendin, V.V., Yarg, L.A. (2007) Engineering geodynamics. Moscow: "Universitet". 440 p. (in Russian)

Bonnet G., Heitz J.F. Non-linear seismic response of a soft layer // Proc. of the 10th European Conf. on Earthquake Eng.Vienna. 1994. Vol. 1. Pp.361–364.

Bugaev, E.G., Kharlov, E.M. (1977) Features of canion sides vibrations. *Seismic microzonation.* Moscow: "Nauka". pp. 91-98. (in Russian)

Byus, E.I. (1955a) Seismic conditions of Transcaucasus. Part I. Tbilisi: Academy of Sciences of USSR, 1948 (in Russian).

Byus, E.I. (1955b) Seismic conditions of Transcaucasus. Part II. Tbilisi: Academy of Sciences of USSR, 1952 (in Russian).

Byus, E.I. (1955c) Seismic conditions of Transcaucasus. Part III. Tbilisi: Academy of Sciences of USSR, 1955 (in Russian).

Chelidze T., Z. Javakhishvili (2003). Natural and technological hazards of territory of Georgia: implications to disaster management. Journal of Georgian Geophysical Society. Issue (A) Solid Earth, v. 8, pp. 3-18.

Cornell C. A. (1968) Engineering risk in seismic analysis. Bull. Seism. Soc. Am. 54 1968, pp. 583-1606

Cornell C. A. Engineering risk in seismic analysis. Bull. Seism. Soc. Am. 54 1968, 583-1606

Gamkrelidze, I., T. Giorgobiani, S. Kuloshvili, G. Lobjanidze, G. Shengelaia (1998). Active Deep Faults Map and the Catalogue for the Territory of Georgia // Bulletin of the Georgian Academy of Sciences, 157, No.1, pp. 80-85.

Gorshkov, G.P. (1984) Regional seismotectonics of the territory of south of USSR. Moscow: "Nauka", 1984. 272 p. (in Russian)

Ishibashi, I. and Zhang, X. (1993). "Unified dynamic shear moduli and damping ratios of sand and clay," Soils and Foundations, Vol. 33, No. 1, pp. 182-191.

Javakhishvili Z., Varazanashvili O., Butikashvili N. (1998). Interpretation of the Macroseismic field of Georgia. Journal of Georgian Geophysical Society. Issue (A) Solid Earth, v. 3, pp. 85-88.

Kanai K. Relation between the nature of surface layer and the amplitudes of earthguake motions // Bul. Earthquake Res. Inst. No 30. Tokyo Univ. 1952. Pp. 31–37.

Maksimov, A.B. (1969) Methodology of microzonation on the basis of detailed investigation of seismic properties of soils. Kandidate of phys.-math. sciences dissertation abstract. Moscow, 1969(in Russian)

McClusky S., S. Balassanian, C. Barku et al. (2000) Global Position System constraints on plate kinematics and dynamics of the Mediterranean and Caucasus // J. Geophys. Res. 2000, v. 105, No. B3, pp. 55695-5719.

McGuire R. (1976) FORTRAN computer program for seismic risk analysis, US Geological Survey, open file report, pp. 76-67.

McGuire R. (1995) Probabilistic Seismic hazard analysis and design earthquakes: closing the loop. vol. 83, No. 5, pp.1275-1284

Medvedev, S.V. (1947) On the question of taking into account seismic activity of region at construction. *Procs. of seismological institute of AS USSR.* No 119, 1947(in Russian)

Medvedev, S.V. (1962) Engineering seismology. Moscow: Gosstroyizdat, 1962. 284 p. (in Russian)

Mushketov, I.V. (1889) Venensk earthquake of May 28 (June 9) 1887. Procs of geological comm. 1889. Vol. 10. No 1. (in Russian)

Mushketov, I.V. Physical geology. St. Petersburg, 1891. Part. 1. 709 p. (in Russian)

Musson R. (1999) Probapilistic seismic hazard maps for the North Balkan region. 1999. Annali di Geofisica. vol. 42, No. 6, pp. 1109-1124.

Nakamura Y, A Method for Dynamic Characteristics Estimation of Subsurface using Microtremor on the Ground Surface. QR of RTRI, Volume 30, No. 1, 1989

Nechaev, Yu.V., Reisner, G.I., Rogozhin, E.A., et al. (1998) Geological-geophysical and seismological criteria of potencial seismicity of Western Caspian // Exploration and protection of subsurface resources. 1998, No. 2, pp. 13-16 (in Russian).

Nesmeyanov, S.A. (2004) Engineering geotectonics. Moscow: Nauka, 2004. 780 p. (in Russian)

Nesmeyanov, S.A., Barkhatov, I.I. (1978) Newest seismogenic structures of Western Gissaro-Alay. Moscow: Nauka, 1978. 120 p. (in Russian)

New Catalogue of strong Earthquakes in the USSR from Ancient times through 1977-1982, NOAA, USA, pp. 15-21

Nikolaev, A.V. (1965) Seismic properties of grounds. Moscow: Nauka, 1965. 184 p. (in Russian)

Nikolaev, A.V. (1987) Problems of nonlinear seismics. Moscow: Nauka, 1987. p. 5-20. (in Russian)

P. Smit, V. Arzmanian, Z. Javakhishvili, S. Arefiev, D. Mayer-Rosa, S. Balassanian, T. Chelidze (2000). The Digital Accelerograph Network in the Caucasus. In:

"Earthquake Hazard and Seismic Risk Reduction". Kluwer Academic Publishers, pp. 109-118.

Paleoseismology of Great Caucasus (1979). Moscow: Nauka, 1979, 188 p. (in Russian)

Poceski A. The Ground effects of the Scopje July 26, 1963 Earthquake, BSSA. 1969. Vol. 59. No 1. Pp.1–22.

Puchkov, S.V. Garagozov, D. (1973) Investigation of hilly relief of region on intensity of seismic vibrations during earthquakes. Problems of engineering seismology. Issue 15. Moscow: Nauka, 1973. pp. 90-93. (in Russian)

Rantsman, E.Ya. (1979) Places of eartquakes and morphostructure of mountainous countries. Moscow: Nauka, 1979. 171 p. (in Russian)

Recommendations on seismic microzonation (SMR-73). Influence of grounds on intensity of seismic vibrations. (1974) Moscow: Stroyizdat, 1974. 65 p. (in Russian)

Recommendations on seismic microzonation at engineering survey for construction (1985). Moscow: Gosstroy USSR, 1985. 72 p. (in Russian)

Reisner, G. I., Ioganson, L. I. Complex typification of earth crust as basis for fundamental and applied tasks solution. Article 1 and 2. Bull. MOIP, 1997. Geology dept., vol. 72. issue 3. pp. 5-13 (in Russian).

Reiter L. Earthquake hazard analysis. New York: Columbia Univ. Press, 1991. 245 p.

Riznichenko, Yu.V. (1966) Calculation of points of Earth surface shaking from earthquake in surrounding area. *Bull. of AS of USSR. Physics of the Earth.* 1966. 5. pp. 16-32. (in Russian)

Rogozhin, E.A. (1997) Geodynamics and seismotectonics. in *Problems of evolution of* tectonosphere. Moscow, 1997. pp. 84-92. (in Russian)

Rogozhin, E.A., Reisner, G.I., Ioganson, L.I. (2001) Assessment of seismic potencial of Big Caucasus and Apennines by independent methods // Modern mathematical and geological models in applied geophysics tasks: selected scientific works. Moscow: UIPE RAS, 2001, pp. 279-300 (in Russian).

Rogozhin, E.A. (2007) PSS zones and their characteristics for the territory of the Republic of North Ossenia-Alania. Procs. Of VI international conference "Innovative technologies for sustainable development of mountainous territories" May 28-30 2007. Vladikavkaz: "Terek", 2007. P. 283. (in Russian)

Rogozhin E. A., A. N. Ovsyuchenko, A. V. Marakhanov, S. S. Novikov, B. V. Dzeranov, D. A. Melkov (2008) // Research report "Investigations of marks of possible occurrence of seismic activity in the zone of Vladikavkaz fault". Vladikavkaz, 2008, vol. 1, book 8, 33 p., (in Russian).

Schnabel, P. B., Lysmer, J., and Seed, H. B. (1972) " SHAKE: A Computer Program for Earthquake Response Analysis of Horizontally Layered Sites", Report No. UCB/EERC-72/12, Earthquake Engineering Research Center, University of California, Berkeley, December, 102p.

Seismic zoning of USSR terrytory. Methodological basics and regional description of the map of 1978. Moscow: Nauka, 1980. 308 p. (in Russian)

Shteinberg, V.V. (1964) Analysis of grounds vibrations from close earthquakes /*Procs. of IPE RAS* No 33 (200). Moscow, 1964. pp. 11-24. (in Russian)

Shteinberg, V.V. (1965) Influence of layer on amplitude-frequency spectrum of vibrations on the surface. Seismic microzonation. / Questions of engineering seismology. Moscow: Nauka, 1965. pp. 34-35. (in Russian)

Shteinberg, V.V. (1967) Investigation of spectra of close earthquakes for prognosis of seismic impact. – Vibrations of earth dams / Questions of engineering seismology Moscow: Nauka, 1967. pp. 123-150. (in Russian)

SP 14.13330.2011. Construction works in seismic regions. Actualized version of SNiP II-7-81*. Minregion of Russia. – M. : «TsPP Ltd», 2011. – 167 p.

Stoykovic, Mihailov V. Some results of the investigations in the seismic microzoning of Banja Luka // Proc. 5th World Conf. on Earthquake Eng. Vol. 1. Rome, 1973. Pp. 1703–1708.

Trifonov, V. G. (1999) Neotectonics of Eurasia. Moscow: Nauchniy Mir, 1999, 252 p. (in Russian)

Ulomov, V.I. (1995) About main thesis and technical recommendations on creation of new map of seismic zoning of the territory of Russian Federation. Seismicity and seismic zoning of Northern Eurasia. Moscow: UIPE RAS, 1995. Issue 2/3. pp. 6-26. (in Russian)

Ulomov, V. I., Shumilina, L. S., Trifonov, V. G. et al. (1999) Seismic Hazard of Northern Eurasia // Annali di Geofisica, vol. 42, No. 6, pp. 1023-1038.

Zaalishvili, V.B. (1986) Seismic microzonation on the data of artificial vibrations of ground thickness. Candidate of phys.-math. sciences dissertation abstract. Tbilisi, 1986a. (in Russian)

Zaalishvili, V.B., Gogmachadze, S.A. (1989) Influence of relief on wave field of pulse and vibrational sources. Investigation of fields of pulse and vibrational sources for the means of seismic microzonation: Report of ISMIS AS GSSR. Tbilisi, 1989. pp. 25-40. (in Russian)

Zaalishvili, V.B. (1996) Seismic microzonation on the basis of nonlinear properties of grounds by means of artificial sources. Doctor of phys.-math. sciences dissertation abstract. Moscow: MSU, 1996. (in Russian)

Zaalishvili V., Otinashvili M., Dzhavrishvili Z. (2000) Seismic hazard assessment for big cities in Georgia using the modern concept of seismic microzonation with consideration soil nonlinearity. INTAS/Georgia/97-0870. Periodic report. 2000. 170p.

Zaalishvili, V. B. (2000) Physical bases of seismic microzonation. Moscow: UIPE RAS, 2000. 367 p. (in Russian).

Zaalishvili, V.B. (2006) Basics of seismic microzonation. VSC RAS&RNO-A. Vladikavkaz, 2006. 242 p. (in Russian)

Zaalishvili, V. B. (2009) Seismic microzonation of urban territories, settlements and large building sites. Moscow: Nauka, 2009, 350 p. (in Russian).

Zaalishvili, V. B., Melkov, D. A., Burdzieva, O. G. (2010) Determination of seismic impact on the basis of specific engineering-seismological situation of region // "Earthquake engineering. Buildings safety", 2010 No.1. pp. 35-39 (in Russian).

Zaalishvili V.B., Rogojin E.A. (2011) Assessment of Seismic Hazard of Territory on Basis of
    Modern Methods of Detailed Zoning and Seismic Microzonationю The Open
    Construction and Building Technology Journal, 2011, Volume 5, pp. 30-40.

Chapter 3

A Cognitive Look at Geotechnical Earthquake Engineering: Understanding the Multidimensionality of the Phenomena

Silvia Garcia

Additional information is available at the end of the chapter

Not even windstorm, earth-tremor, or rush of water is a catastrophe.
A catastrophe is known by its works; that is, to say, by the occurrence of disaster.
So long as the ship rides out the storm, so long as the city resists the earth-shocks,
so long as the levees hold, there is no disaster.
It is the collapse of the cultural protections that constitutes the proper disaster.
(Carr, 1932)

1. Introduction

Essentially, disasters are human-made. For a catastrophic event, whether precipitated by natural phenomena or human activities, assumes the state of a disaster when the community or society affected fails to cope. Earthquake hazards themselves do not necessarily lead to disasters, however intense, inevitable or unpredictable, translate to disasters only to the extent that the population is unprepared to respond, unable to deal with, and, consequently, severely affected. Seismic disasters could, in fact, be reduced if not prevented. With today's advancements in science and technology, including early warning and forecasting of the natural phenomena, together with innovative approaches and strategies for enhancing local capacities, the impact of earthquake hazards somehow could be predicted and mitigated, its detrimental effects on populations reduced, and the communities adequately protected.

After each major earthquake, it has been concluded that the experienced ground motions were not expected and soil behavior and soil-structure interaction were not properly predicted. Failures, associated to inadequate design/construction and to lack of phenomena comprehension, obligate further code reinforcement and research. This scenario will be repeated after each earthquake. To overcome this issue, *Earthquake Engineering* should change its views on the present methodologies and techniques toward more scientific, doable, affordable, robust and adaptable solutions.

A competent modeling of engineering systems, when they are affected by seismic activity, poses many difficult challenges. Any representation designed for reasoning about models of such systems has to be flexible enough to handle various degrees of complexity and uncertainty, and at the same time be sufficiently powerful to deal with situations in which the input signal may or may not be controllable. Mathematically-based models are developed using scientific theories and concepts that just apply to particular conditions. Thus, the core of the model comes from assumptions that for complex systems usually lead to simplifications (perhaps oversimplifications) of the problem phenomena. It is fair to argue that the representativeness of a particular theoretical model largely depends on the degree of comprehension the developer has on the behavior of the actual engineering problem. Predicting natural-phenomena characteristics like those of earthquakes, and thereupon their potential effects at particular sites, certainly belong to a class of problems we do not fully understand. Accordingly, analytical modeling often becomes the bottleneck in the development of more accurate procedures. As a consequence, a strong demand for advanced modeling an identification schemes arises.

Cognitive Computing CC technologies have provided us with a unique opportunity to establish coherent seismic analysis environments in which uncertainty and partial data-knowledge are systematically handled. By seamlessly combining learning, adaptation, evolution, and fuzziness, CC complements current engineering approaches allowing us develop a more comprehensive and unified framework to the effective management of earthquake phenomena. Each CC algorithm has well-defined labels and could usually be identified with specific scientific communities. Lately, as we improved our understanding of these algorithms' strengths and weaknesses, we began to leverage their best features and developed hybrid algorithms that indicate a new trend of co-existence and integration between many scientific communities to solve a specific task.

In this chapter geotechnical aspects of earthquake engineering under a cognitive examination are covered. Geotechnical earthquake engineering, an area that deals with the design and construction of projects in order to resist the effect of earthquakes, requires an understanding of geology, seismology and earthquake engineering. Furthermore, practice of geotechnical earthquake engineering also requires consideration of social, economic and political factors. Via the development of cognitive interpretations of selected topics: i) spatial variation of soil dynamic properties, ii) attenuation laws for rock sites (seismic input), iii) generation of artificial-motion time histories, iv) effects of local site conditions (site effects), and iv) evaluation of liquefaction susceptibility, CC techniques (Neural Networks NNs, Fuzzy Logic FL and Genetic Algorithms GAs) are presented as appealing alternatives for integrated data-driven and theoretical procedures to generate reliable seismic models.

2. Geotechnical earthquake hazards

The author is well aware that standards for geotechnical seismic design are under development worldwide. While there is no need to "reinvent the wheel" there is a requirement to adapt such initiatives to fit the emerging safety philosophy and demands. This investigation also strongly endorses the view that "guidelines" are far more desirable than "codes" or "standards" disseminated all over seismic regions. Flexibility in approach is a key ingredient of geotechnical engineering and the cognitive technology in this area is rapidly advancing. The science and practice of geotechnical earthquake engineering is far from mature and need to be expanded and revised periodically in coming years. It is important that readers and users of the computational models presented here familiarize themselves with the latest advances and amend the recommendations herein appropriately.

This document is not intended to be a detailed treatise of latest research in geotechnical earthquake engineering, but to provide sound guidelines to support rational cognitive approaches. While every effort has been made to make the material useful in a wider range of applications, applicability of the material is a matter for the user to judge. The main aim of this guidance document is to promote consistency of cognitive approach to everyday situations and, thus, improve geotechnical-earthquake aspects of the performance of the built safe-environment.

2.1. A "soft" interpretation of ground motions

After a sudden rupture of the earth's crust (caused by accumulating stresses, elastic strain-energy) a certain amount of energy radiates from the rupture as seismic waves. These waves are attenuated, refracted, and reflected as they travel through the earth, eventually reaching the surface where they cause ground shaking. The principal geotechnical hazards associated with this event are fault rupture, ground shaking, liquefaction and lateral spreading, and landsliding. Ground shaking is one of the principal seismic hazards that causes extensive damage to the built environment and failure of engineering systems over large areas. Earthquake loads and their effects on structures are directly related to the intensity and duration of ground shaking. Similarly, the level of ground deformation, damage to earth structures and ground failures are closely related to the severity of ground shaking.

In engineering evaluations, three characteristics of ground shaking are typically considered: i) the amplitude, ii) frequency content and iii) significant duration of shaking (time over which the ground motion has relatively significant amplitudes).These characteristics of the ground motion at a given site are affected by numerous complex factors such as the source mechanism, earthquake magnitude, rupture directivity, propagation path of seismic waves, source distance and effects of local soil conditions. There are many unknowns and uncertainties associated with these issues which in turn result in significant uncertainties regarding the characteristics of the ground motion and earthquake loads.

If the random nature of response to earthquakes (aleatory uncertainty) cannot be avoided [1,2], it is our limited knowledge about the patterns between seismic events and their manifestations -ground motions- at a site (epistemic uncertainty) that must be improved thorough more scientific seismic analyses. A strategic factor in seismic hazard analysis is the ground motion model or attenuation relation. These attenuation relationships has been developed based on magnitude, distance and site category, however, there is a tendency to incorporate other parameters, which are now known to be significant, as the tectonic environment, style of faulting and the effects of topography, deep basin edges and rupture directivity. These distinctions are recognized in North America, Japan and New Zealand [3-6], but ignored in most other regions of the world [7]. Despite recorded data suggest that ground motions depend, in a significant way, on these aspects, these inclusions did not have had a remarkable effect on the predictions confidence and the geotechnical earthquake engineer prefers the basic and clear-cut approximations on those that demand a *blind* use of coefficients or an intricate determination of soil/fault conditions.

A key practice in current aseismic design is to develop design spectrum compatible time histories. This development entails the modification of a time history so that its response spectrum matches within a prescribed tolerance level, the target design spectrum. In such matching it is important to retain the phase characteristics of the selected ground motion time history. Many of the techniques used to develop compatible motions do not retain the phase [8]. The response spectrum alone does not adequately characterize specific-fault ground motion. Near-fault ground motions must be characterized by a long period pulse of strong motion of a fairly brief duration rather than the stochastic process of long duration that characterizes more distant ground motions. Spectrum compatible with these specific motions will not have these characteristics unless the basic motion being modified to ensure compatibility has these effects included. Spectral compatible motions could match the entire spectrum but the problem arises on finding a "real" earthquake time series that match the specific nature of ground motion. For nonlinear analysis of structures, spectrum compatible motions should also correspond to the particular energy input [9], for this reason, designers should be cautious about using spectrum compatible motions when estimating the displacements of embankment dams and earth structures under strong shaking, if the acceptable performance of these structures is specified by criteria based on tolerable displacements.

Another important seismic phenomenon is the liquefaction. Liquefaction is associated with significant loss of stiffness and strength in the shaken soil and consequent large ground deformation. Particularly damaging for engineering structures are cyclic ground movements during the period of shaking and excessive residual deformations such as settlements of the ground and lateral spreads. Ground surface disruption including surface cracking, dislocation, ground distortion, slumping and permanent deformations, large settlements and lateral spreads are commonly observed at liquefied sites. In sloping ground and backfills behind retaining structures in waterfront areas, liquefaction often results in large permanent ground displacements in the down-slope direction or towards waterways (lateral spreads). Dams, embankments and sloping ground near riverbanks where certain shear strength is required for stability under gravity loads are particularly prone to such failures. Clay soils may also suffer some loss of strength during shaking but are not subject to boils and other "classic" liquefaction phenomena. For intermediate soils, the transition from "sand like" to "clay-like" behavior depends primarily on whether the soil is a matrix of coarse grains with fines contained within the pores or a matrix of plastic fines with coarse grained "filler". Recent papers by Boulanger and Idriss [10, 11] are helpful in clarifying issues surrounding the liquefaction and strain softening of different soil types during strong ground shaking. Engineering judgment based on good quality investigations and data interpretation should be used for classifying such soils as liquefiable or non-liquefiable.

Procedures for evaluating liquefaction, potential and induced lateral spread, have been studied by many engineering committees around the world. The objective has been to review research and field experience on liquefaction and recommended standards for practice. Youd and Idriss [12] findings and the liquefaction-resistance chart proposed by Seed et al. [13] in 1985, stay as standards for practice. They have been slightly modified to adjust new registered input-output conditions and there is a strong tendency to recommend i) the adoption of the cone penetration test CPT, standard penetration test SPT or the shear wave velocities for describing the *in situ* soil conditions [14] and ii) the modification of magnitude factors used to convert the critical stress ratios from the liquefaction assessment charts (usually developed for M7:5) to those appropriate for earthquakes of diverse magnitudes [12, 15].

3. Cognitive Computing

Cognitive Computing CC as a discipline in a narrow sense, is an application of computers to solve a given computational problem by imperative instructions; while in a broad sense, it is a process to implement the instructive intelligence by a system that transfers a set of given information or instructions into expected behaviors. According to theories of cognitive informatics [16-18], computing technologies and systems may be classified into the categories of imperative, autonomic, and cognitive from the bottom up. Imperative computing is a traditional and passive technology based on stored-program controlled behaviors for data processing [19-24]. An autonomic computing is goal-driven and self-decision-driven technologies that do not rely on instructive and procedural information [25-28]. Cognitive computing is more intelligent technologies beyond imperative and autonomic computing, which embodies major natural intelligence behaviors of the brain such as thinking, inference, learning, and perceptions.

Cognitive computing is an emerging paradigm of intelligent computing methodologies and systems, which implements computational intelligence by autonomous inferences and perceptions mimicking the mechanisms of the brain. This section presents a brief description on the theoretical framework and architectural techniques of cognitive computing beyond conventional imperative and autonomic computing technologies. Cognitive models are explored on the basis of the latest advances in applying computational intelligence. These applications of cognitive computing are described from the aspects of cognitive search engines, which demonstrate how machine and computational intelligence technologies can drive us toward autonomous knowledge processing.

3.1. Computational intelligence: Soft Computing technologies

The *computational intelligence* is a synergistic integration of essentially three computing paradigms, viz. neural networks, fuzzy logic and evolutionary computation entailing probabilistic reasoning (belief networks, genetic algorithms and chaotic systems) [29]. This synergism provides a framework for flexible information processing applications designed to operate in the real world and is commonly called *Soft Computing SC* [30]. Soft computing technologies are robust by design, and operate by trading off precision for tractability. Since they can handle uncertainty with ease, they conform better to real world situations and provide lower cost solutions.

The three components of soft computing differ from one another in more than one way. Neural networks operate in a numeric framework, and are well known for their learning and generalization capabilities. Fuzzy systems [31] operate in a linguistic framework, and their strength lies in their capability to handle linguistic information and perform approximate reasoning. The evolutionary computation techniques provide powerful search and optimization methodologies. All the three facets of soft computing differ from one another in their time scales of operation and in the extent to which they embed *a priori* knowledge.

Figure 1 shows a general structure of Soft Computing technology. The following main components of SC are known by now: fuzzy logic FL, neural networks NN, probabilistic reasoning PR, genetic algorithms GA, and chaos theory ChT (Figure 1). In SC FL is mainly concerned with imprecision and approximate reasoning, NN with learning, PR with uncertainty and propagation of belief, GA with global optimization and search and ChT with nonlinear dynamics. Each of these computational paradigms (emerging reasoning technologies) provides us with complementary reasoning and searching methods to solve complex, real-world problems. In large scope, FL, NN, PR, and GA are complementary rather that competitive [32-34]. The interrelations between the components of SC, shown in Figure 1, make the theoretical foundation of Hybrid Intelligent Systems. As noted by L. Zadeh: "... the term hybrid intelligent systems is gaining currency as a descriptor of systems in which FL, NC, and PR are used in combination. In my view, hybrid intelligent systems are the wave of the future" [35]. The use of Hybrid Intelligent Systems are leading to the development of numerous manufacturing system, multimedia system, intelligent robots, trading systems, which exhibits a high level of MIQ (machine intelligence quotient).

3.1.1. Comparative characteristics of SC tools

The constituents of SC can be used independently (fuzzy computing, neural computing, evolutionary computing etc.), and more often in combination [36, 37, 38- 40, 41]. Based on independent use of the constituents of Soft Computing, fuzzy technology, neural technology, chaos technology and others have been recently applied as emerging technologies to both industrial and non-industrial areas.

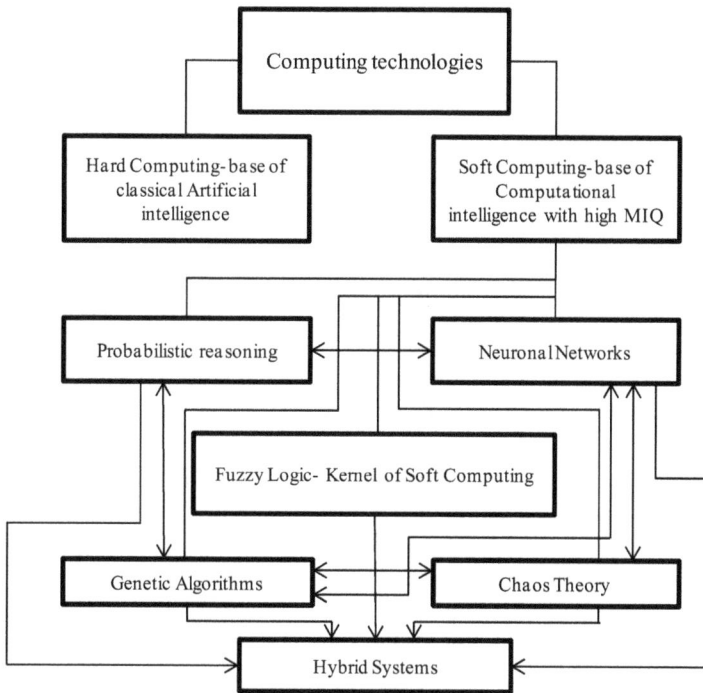

Figure 1. Soft Computing Components

Fuzzy logic is the leading constituent of Soft Computing. In Soft Computing, fuzzy logic plays a unique role. FL serves to provide a methodology for computing [36]. It has been successfully applied to many industrial spheres, robotics, complex decision making and diagnosis, data compression, and many other areas. To design a system processor for handling knowledge represented in a linguistic or uncertain numerical form we need a fuzzy model of the system. Fuzzy sets can be used as a universal approximator, which is very important for modeling unknown objects. If an operator cannot tell linguistically what kind of action he or she takes in a specific situation, then it is quite useful to model his/her control actions using numerical data. However, fuzzy logic in its so called *pure form* is not always useful for easily constructing intelligent systems. For example, when a designer does not have sufficient prior information (knowledge) about the system, development of acceptable fuzzy rule base becomes impossible. As the complexity of the system increases, it becomes difficult to specify a correct set of rules and membership functions for describing adequately the behavior of the system. Fuzzy systems also have the disadvantage of not being able to extract additional knowledge from the experience and correcting the fuzzy rules for improving the performance of the system.

Another important component of Soft Computing is neural networks. Neural networks NN viewed as parallel computational models, are parallel fine-grained implementation of non-linear static or dynamic systems. A very important feature of these networks is their adaptive nature, where "learning by example" replaces traditional "programming" in problems solving. Another key feature is the intrinsic parallelism that allows fast computations. Neural networks are viable computational models for a wide variety of problems including pattern classification, speech synthesis and recognition, curve fitting, approximation capability, image data compression, associative memory, and modeling and control of non-linear unknown systems [42, 43]. NN are favorably distinguished for efficiency of their computations and hardware implementations. Another advantage of NN is generalization ability, which is the ability to classify correctly new patterns. A significant disadvantage of NN is their poor interpretability. One of the main criticisms addressed to neural networks concerns their black box nature [35].

Evolutionary Computing EC is a revolutionary approach to optimization. One part of EC — genetic algorithms — are algorithms for global optimization. Genetic algorithms GAs are based on the mechanisms of natural selection and genetics [44]. One advantage of genetic algorithms is that they effectively implement parallel multi-criteria search. The mechanism of genetic algorithms is simple. Simplicity of operations and powerful computational effect are the two main advantages of genetic algorithms. The disadvantages are the problem of convergence and the absence of strong theoretical foundation. The requirement of coding the domain of the real variables' into bit strings also seems to be a drawback of genetic algorithms. It should be also noted that the computational speed of genetic algorithms is low.

Because in this investigation PR and ChT are not exploited, they are not going to be explained. For the interested reader [41] is recommended. Table 1 presents the comparative characteristics of the components of Soft Computing. For each component of Soft Computing there is a specific class of problems, where the use of other components is inadequate.

3.1.2. Intelligent Combinations of SC

As it was shown above, the components of SC complement each other, rather than compete. It becomes clear that FL, NC and GA are more effective when used in combinations. Lack of interpretability of neural networks and poor learning capability of fuzzy systems are similar problems that limit the application of these tools. Neurofuzzy systems are hybrid systems which try to solve this problem by combining the learning capability of connectionist models with the interpretability property of fuzzy systems. As it was noted above, in case of dynamic work environment, the automatic knowledge base correction in fuzzy systems becomes necessary. On the other hand, artificial neural networks are successfully used in problems connected to knowledge acquisition using learning by examples with the required degree of precision.

Incorporating neural networks in fuzzy systems for fuzzification, construction of fuzzy rules, optimization and adaptation of fuzzy knowledge base and implementation of fuzzy reasoning is the essence of the Neurofuzzy approach.

	Fuzzy Sets	Artificial Neural Networks	Evolutionary Computing, GA	Probabilistic Reasoning	Chaotic computing
Weaknesses	•Knowledge acquisition •Learning	•Black Box interpretability	•Coding •Computational speed	•Limitation of the axioms of Probability Theory •Lack of complete knowledge •Copmputational complexity	•Computational complexity •Chaos identification complexity
Strengths	•Interpretability •Transparency •Plausibility •Graduality •Modeling •Reasoning •Tolerance to imprecision	•Learning •Adaptation •Fault tolerance •Curve fiting •Generalization ability •Approximation ability	Computational efficiency •Global optimization	•Rigorous framework •Well understanding	•Nonlinear dynamics simulation •Discovering chaos in observed data (with noise) •Determinig the predictability •Prediction startegies formulation

Table 1. Central characteristics of Soft Computing technologies

The combination of genetic algorithms with neural networks yields promising results as well. It is known that one of main problems in development of artificial neural systems is selection of a suitable learning method for tuning the parameters of a neural network (weights, thresholds, and structure). The most known algorithm is the "error back propagation" algorithm. Unfortunately, there are some difficulties with "back propagation". First, the effectiveness of the learning considerably depends on initial set of weights, which are generated randomly. Second, the "back propagation", like any other gradient-based method, does not avoid local minima. Third, if the learning rate is too slow, it requires too much time to find the solution. If, on the other hand, the learning rate is too high it can generate oscillations around the desired point in the weight space. Fourth, "back propagation" requires the activation functions to be differentiable. This condition does not hold for many types of neural networks. Genetic algorithms used for solving many optimization problems when the "strong" methods fail to find appropriate solution, can be successfully applied for learning neural networks, because they are free of the above drawbacks.

The models of artificial neurons, which use linear, threshold, sigmoidal and other transfer functions, are effective for neural computing. However, it should be noted that such models are very simplified. For example, reaction of a biological axon is chaotic even if the input is periodical. In this aspect the more adequate model of neurons seems to be chaotic. Model of a chaotic neuron can be used as an element of chaotic neural networks. The more adequate results can be obtained if using fuzzy chaotic neural networks, which are closer to biological computation. Fuzzy systems with If-Then rules can model non-linear dynamic systems and capture chaotic attractors easily and accurately. Combination of Fuzzy Logic and Chaos Theory gives us useful tool for building system's chaotic behavior into rule structure. Identification of chaos allows us to determine predicting strategies. If we use a Neural Network Predictor for predicting the system's behavior, the parameters of the strange attractor (in particular fractal dimension) tell us how much data are necessary to train the neural network. The combination of Neurocomputing and Chaotic computing technologies can be very helpful for prediction and control.

The cooperation between these formalisms gives a useful tool for modeling and reasoning under uncertainty in complicated real-world problems. Such cooperation is of particular importance for constructing perception-based intelligent information systems. We hope that the mentioned intelligent combinations will develop further, and the new ones will be proposed. These SC paradigms will form the basis for creation and development of Computational Intelligence.

4. Cognitive models of ground motions

The existence of numerous databases in the field of civil engineering, and in particular in the field of geotechnical earthquake, has opened new research lines through the introduction of analysis based on soft computing. Three methods are mainly applied in this emerging field: the ones based on the Neural Networks NN, the ones created using Fuzzy Sets FS theory and the ones developed from the Evolutionary Computation [45].

The SC hybrids used in this investigation are directed to tasks of prediction (classification and/or regression). The central objective is obtaining numerical and/or categorical values that mimic input-output conditions from experimentation and in situ measurements and then, through the recorded data and accumulated experience, predict future behaviors. The examples presented herein have been developed by an engineering committee that works for generating useful guidance to geotechnical practitioners with geotechnical seismic design. This effort could help to minimize the perceived significant and undesirable variability within geotechnical earthquake practice. Some urgency in producing the alternative guidelines was seen, after the most recent earthquakes disasters, as being necessary with a desire to avoid a long and protracted process. To this end, a two stage approach was suggested with the first stage being a cognitive interpretation of well-known procedures with appropriate factors for geotechnical design, and a posterior step identifying the relevant philosophy for a new geotechnical seismic design.

4.1. Spatial variation of soil dynamic properties

The spatial variability of subsoil properties constitutes a major challenge in both the design and construction phases of most geo-engineering projects. Subsoil investigation is an imperative step in any civil engineering project. The purpose of an exploratory investigation is to infer accurate information about actual soil and rock conditions at the site. Soil exploration, testing, evaluation, and field observation are well-established and routine procedures that, if carried out conscientiously, will invariably lead to good engineering design and construction. It is impossible to determine the optimum spacing of borings before an investigation begins because the spacing depends not only on type of structure but also on uniformity or regularity of encountered soil deposits. Even the most detail soil maps are not efficient enough for predicting a specific soil property because it changes from place to place, even for the same soil type. Consequently interpolation techniques have been extensively exploited. The most commonly used methods are kriging and co-kriging but for better estimations they require a great number of measurements available for each soil type, what is generally impossible.

Based on the high cost of collecting soil attribute data at many locations across landscape, new interpolation methods must be tested in order to improve the estimation of soil properties. The integration of GIS and Soft Computing SC offers a potential mechanism to lower the cost of analysis of geotechnical information by reducing the amount of time spent understanding data. Applying GIS to large sites, where historical data can be organized to develop multiple databases for analytical and stratigraphic interpretation, originates the establishment of spatial/chronological efficient methodologies for interpreting properties (soil exploration) and behaviors (in situ measured). GIS-SC modeling/simulation of natural systems represents a new methodology for building predictive models, in this investigation NN and GAs, nonparametric cognitive methods, are used to analyze physical, mechanical and geometrical parameters in a geographical context. This kind of spatial analysis can handle uncertain, vague and incomplete/redundant data when modeling intricate relationships between multiple variables. This means that a NN has not constraints about the spacing (minimum distance) between the drill holes used for building (training) the SC model. The NNs-GAs acts as computerized architectures that can approximate nonlinear functions of several variables, this scheme represent the relations between the spatial patterns of the stratigraphy without restrictive assumptions or excessive geometrical and physical simplifications.

The geotechnical data requirements (geo-referenced properties) for an easy integration of the SC technologies are explained through an application example: a geo-referenced three-dimensional model of the soils underlying Mexico City. The classification/prediction criterion for this very complex urban area is established according to two variables: the cone penetration resistance q_c (mechanical property) and the shear wave velocity V_s (dynamic property). The expected result is a 3D-model of the soils underlying the city area that would eventually be improved for a more complex and comprehensive model adding others mechanical, physical or geometrical geo-referenced parameters.

Cone-tip penetration resistances and shear wave velocities have been measured along 16 bore holes spreaded throughout the clay deposits of Mexico City (Figure 2). This information was used as the set of examples inputs (latitude, longitude and depth) → output (q_c / V_s). The analysis was carried out in an approximate area of 125 km^2 of Mexico City downtown. It is important to point out that 20% of these patterns (sample points and complete variables information) are not used in the training stage; they will be presented for testing the generalization capabilities of the closed system components (once the training is stopped).

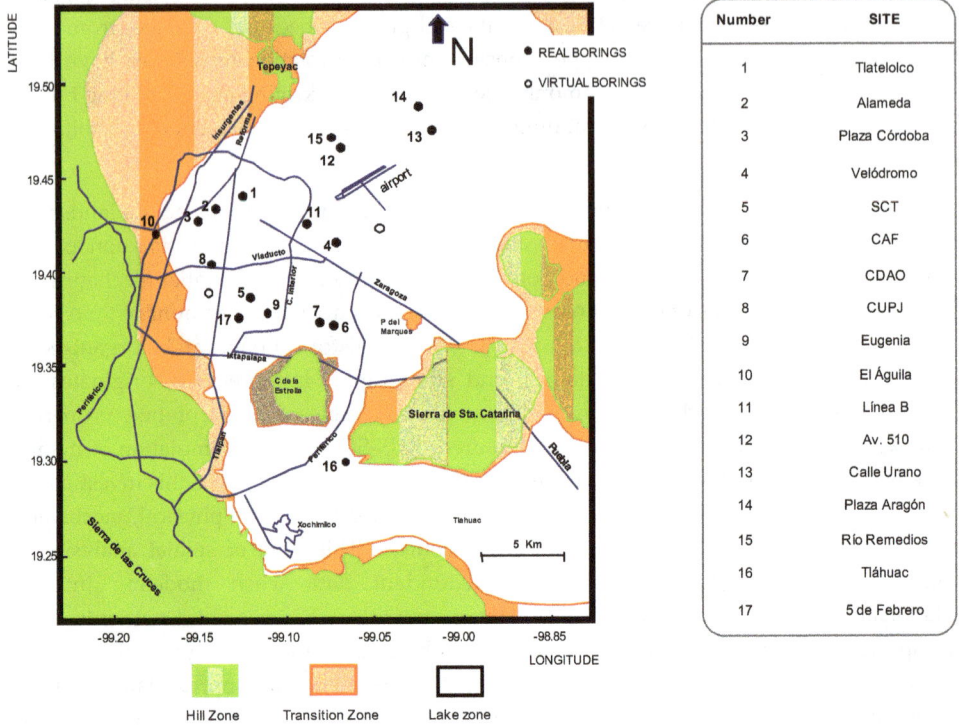

Number	SITE
1	Tlatelolco
2	Alameda
3	Plaza Córdoba
4	Velódromo
5	SCT
6	CAF
7	CDAO
8	CUPJ
9	Eugenia
10	El Águila
11	Línea B
12	Av. 510
13	Calle Urano
14	Plaza Aragón
15	Río Remedios
16	Tláhuac
17	5 de Febrero

Figure 2. Mexico City Zonation

In the 3D-neurogenetic analysis, the functions $q_c = \{q_c\ (X,Y,Z)\}/V_s = \{V_s\ (X,Y,Z)\}$ are to be approximated using the procedure outlined below:

1. Generate the database including identification of the site [borings or stations] (X,Y – geographical coordinates, Z –depth, and a CODE –ID number), elevation reference (meters above de sea level, m.a.s.l.), thickness of predetermined structures (layers), and additional information related to geotechnical zoning that could be useful for results interpretation.

2. Use the database to train an initial neural topology whose weights and layers are tuned by an evolutive algorithm (see [46] for details), until the minimum error between calculated and measured values $q_c = fNN\ (X,Y,Z)\}/V_s = \{fNN\ (X,Y,Z)\}$ is achieved (Figure 3a). The generalization capabilities of the optimal 3D neural model are tested presenting real work cases (information from borings not included in the training set) to the net. Figure 3b presents the comparison between the measured q_c, V_s values and the NN calculations for testing cases. Through the neurogenetic results for unseen situations we can conclude that the procedure works extremely well in identifying the general trend in materials resistance (stiffness). The correlation between NN calculations and "real" values is over 0.9.

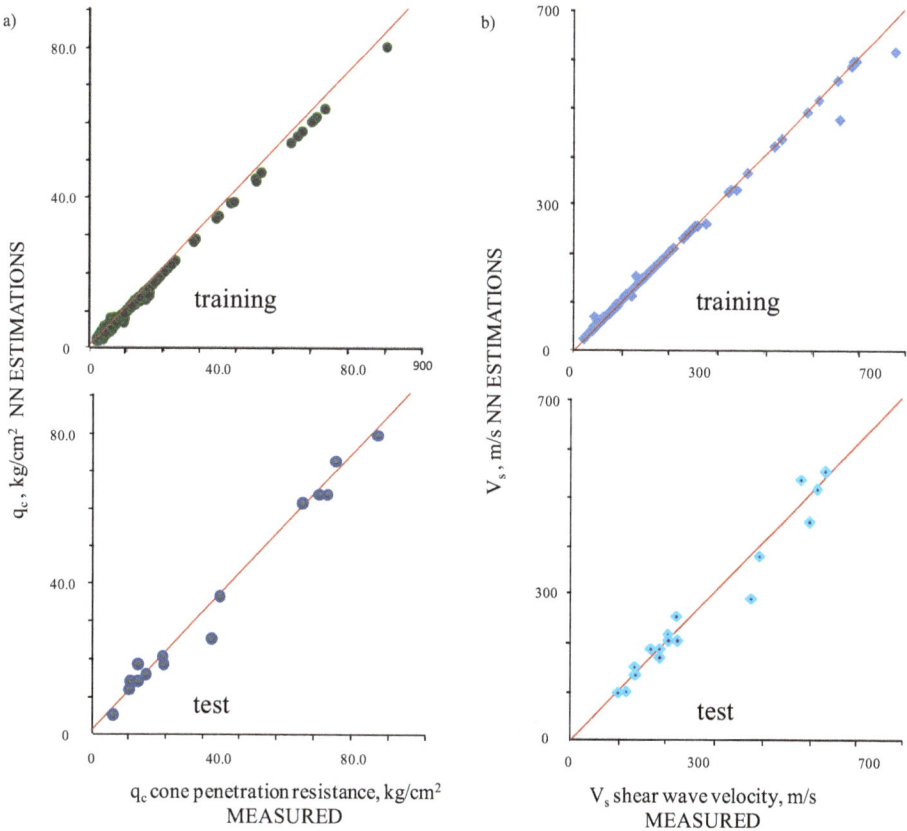

Figure 3. Neural estimations of mechanical and dynamic parameters

3. For visual environment requirements a grid is constructed using raw information and neurogenetic estimations for defining the spatial variation of properties (Figure 4). The 3D view of the studied zone represents an easier and more understandable engineering system. The 3D neurogenetic-database also permits to display property-contour lines for specific depths. Using the neurogenetic contour maps, the spatial distribution of the mechanical/dynamic variables can be visually appreciated. The 3D model is able to reflect the stratigraphical patterns (Figure 5), indicating that the proposed networks are effective in site characterization with remarkable advantages if comparing with geostatistical approximations: it is easier to use, to understand and to develop graphical user interfaces. The confidence and practical advantages of the defined neurogenetic layers is evident. Precision of predictions depends on neighborhood structure, grid size, and variance response, but based on the results we can conclude that despite of the grid cell (size) is not too small the spatial correlation extends beyond the training neighborhood, but the higher confidence is obviously only within.

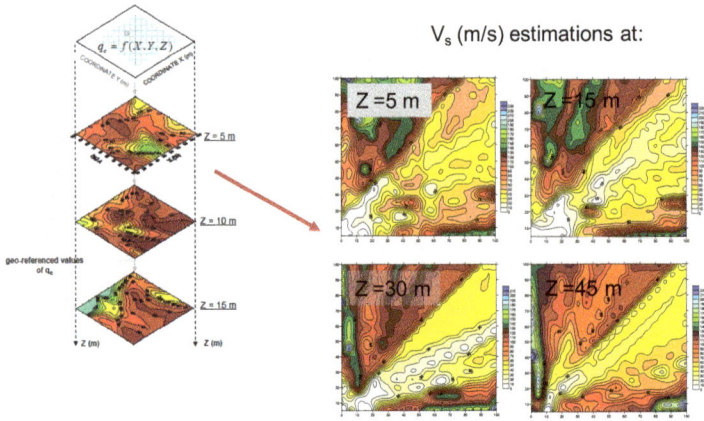

Figure 4. 3D Neural response

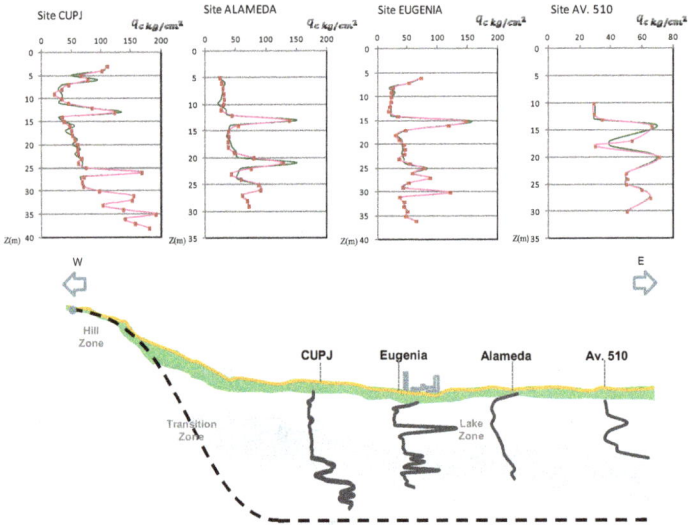

Figure 5. Stratigraphy sequence obtained using the 3D Neural estimations

4.2. Attenuation laws for rock site (outcropping motions)

Source, path, and local site response are factors that should be considered in seismic hazard analyses when using attenuation relations. These relations, obtained from statistical regression, are derived from strong motion recordings to define the occurrence of an earthquake with a specific magnitude at a particular distance from the site. Because of the uncertainties inherent in the variables describing the source (e.g. magnitude, epicentral distance, focal depth and fault rupture dimension), the difficulty to define broad categories to classify the site (e.g. rock or soil) and our lack of understanding regarding wave propagation processes and the ray path characteristics from source to site, commonly the predictions from attenuation regression analyses are inaccurate. As an effort to recognize these aspects, multiparametric attenuation relations have been proposed by several researchers [47-53]. However, most of these authors have concluded that the governing parameters are still source, ray path, and site conditions. In this section an empirical NN formulation that uses the minimal information about magnitude, epicentral distance, and focal depth for subduction-zone earthquakes is developed to predict the peak ground acceleration PGA and spectral accelerations S_a at a rock-like site in Mexico City.

The NN model was training from existing information compiled in the Mexican strong motion database. The NN uses earthquake moment magnitude M_w, epicentral distance E_D, and focal depth F_D from hundreds of events recorded during Mexican subduction earthquakes (Figure 6) from 1964 to 2007. To test the predicting capabilities of the neuronal model, 186 records were excluded from the data set used in the learning phase. Epicentral distance E_D is considered to be the length from the point where fault-rupture starts to the recording site, and the focal depth F_D is not declared as mechanism classes, the NN should identify the event type through the F_D crisp value coupled with the others input parameters [54, 47, 55], The interval of M_w goes from 3 to 8.1 approximately and the events were recorded at near (a few km) and far field stations (about 690 km). The depth of the zone of energy release ranged from very shallow to about 360 km.

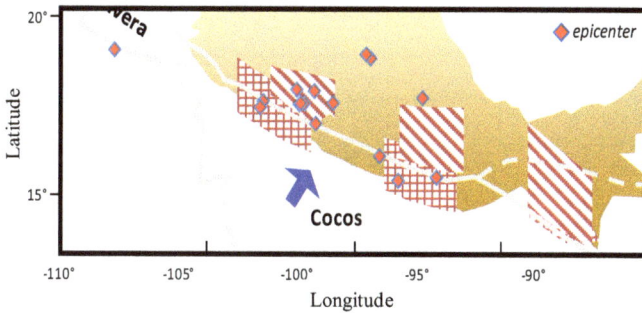

EVENT SUMMARY	
Events	80 from 1964 to 2007
Epicentral coordinates	Latitude from 13.98° to 18.74°
	Longitude from 92.79° to 104.67°
Magnitudes	from 3.9 to 8.1
Focal depth	from <3 to 360 km
Epicentral distance	from 112 to 690 km

Figure 6. Earthquakes characteristics

Modeling of the data base has been performed using backpropagation learning algorithm. Horizontal (mutually orthogonal PGA_{h1}, N-S component, and PGA_{h2}, E-W component) and vertical components (PGA_v) are included as outputs for neural mapping. After trying many topologies, the best horizontal and vertical modules with quite acceptable approximations were the simpler alternatives (BP backpropagation, 2 hidden layers/15 units or nodes each). The neuronal attenuation model for $\{M_w, E_D, F_D\} \rightarrow \{PGA_{h1}, PGA_{h2}, PGA_v\}$ was evaluated by performing testing analyses. The predictive capabilities of the NNs were verified by comparing the estimated PGA's to those induced by the 186 events excluded from the original database (data for training stage). In Figure 7 are compared the computed PGA's during training and testing stages to the measured values. The relative correlation factors ($R^2 \approx 0.97$), obtained in the training phase, indicate that those topologies selected as optimal behave consistently within the full range of intensity, distances and focal depths depicted by the patterns. Once the networks converge to the selected stop criterion, learning is finished and each of these black-boxes become a nonlinear multidimensional functional. Following this procedure 20 NN are trained to evaluate de S_a at different response spectra periods (from T= 0.1 s to T= 5.0 s with DT=0.25 s). Forecasting of the spectral components is reliable enough for practical applications.

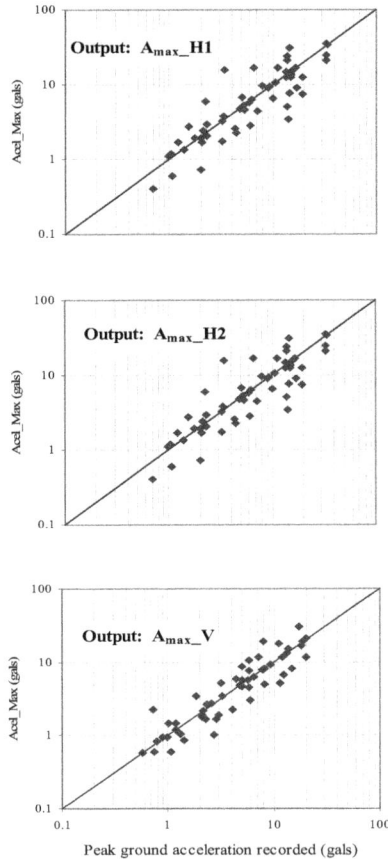

Figure 7. Some examples of measured and NN-estimated PGA values

In Figure 8 two case histories correspond to large and medium size events are shown, the estimated values obtained for these events using the relationships proposed by Gómez, Ordaz &Tena [56], Youngs et al. [47], Atkinson and Boore [55] –proposed for rock sites– and Crouse et al. [51] –proposed for stiff soil sites– and the predictions obtained with the PGA_{h1-h2} modules are shown. It can be seen that the estimation obtained with Gómez, Ordaz y Tena [56] seems to underestimate the response for the large magnitude event. However, for the lower magnitude event follows closely both the measured responses and NN predictions. Youngs et al. [47] attenuation relationship follows closely the overall trend but tends to fall sharply for long epicentral distances.

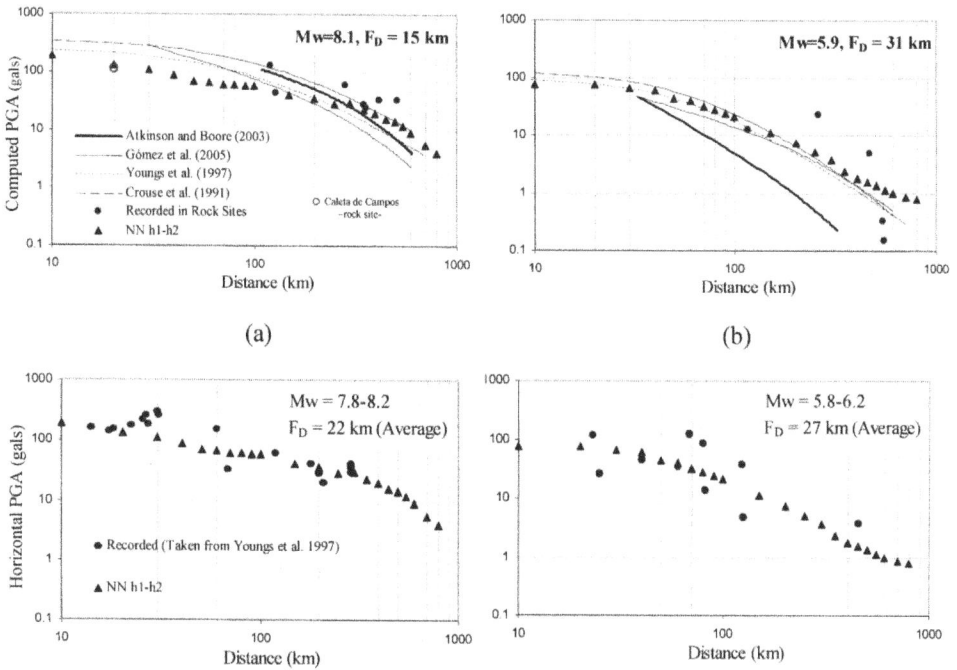

Figure 8. Attenuation laws comparisons

Furthermore, it should be stressed the fact that, as can be seen in Figure 9 the neural attenuation model is capable to follow the general behavior of the measure data expressed as spectra while the traditional functional approaches are not able to reproduce. A neural sensitivity study for the input variables was conducted for the neuronal modules. The results are strictly valid only for the data base utilized, nevertheless, after several sensitivity analyses conducted changing the database composition, it was found that the following trend prevails; the M_w would be the most relevant parameter then would follow E_D coupled with F_D. However, for near site events the epicentral distance could become as relevant as the magnitude, particularly, for the vertical component and for minor earthquakes (M low) the F_D becomes very transcendental.

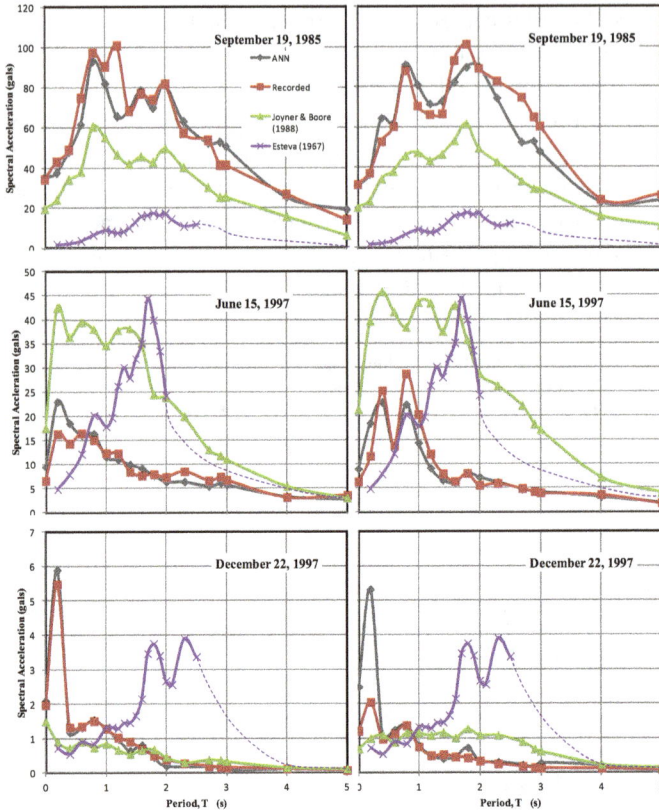

Figure 9. Response Spectra: NN-calculated vs traditional functions

Through $\{M_w, E_D, F_D\} \rightarrow \{PGA_{hi}, S_a\}$ mapping, this neuronal approach offers the flexibility to fit arbitrarily complex trends in magnitude and distance dependence and to recognize and select among the tradeoffs that are present in fitting the observed parameters within the range of magnitudes and distances present in data. This approach seems to be a promising alternative to describe earthquake phenomena despite of the limited observations and qualitative knowledge of the recording stations geotechnical site conditions, which leads to a reasoning of a partially defined behavior.

4.3. Generation of artificial time series: Accelerograms application

For nonlinear seismic response analysis, where the superposition techniques do not apply, earthquake acceleration time histories are required as inputs. Virtually all seismic design codes and guidelines require scaling of selected ground motion time histories so that they match or exceed the controlling design spectrum within a period range of interest. Considerable variability in the characteristics of the recorded strong-motions under similar conditions may still require a characterization of future shaking in terms of an ensemble of accelerograms rather than in terms of just one or two "typical" records. This situation has thus created a need for the generation of synthetic (artificial) strong-motion time histories that simulate realistic ground motions from different points of views and/or with different degrees of sophistication. To provide the ground motions for analysis and design, various methods have been developed: i) frequency-domain methods where the frequency content of recorded signals is manipulated [57-60] and ii) time-domain methods where the recorded ground motions amplitude is controlled [61, 62]. Regardless of the method, first, one or more time histories are selected subjectively, and then scaling mechanisms for spectrum matching are applied. This is a trial and error procedure that leads artificial signals very far from real-earthquake time series.

In this investigation a Genetic Generator of Signals is presented. This genetic generator is a tool for finding the coefficients of a pre-specified functional form, which fit a given sampling of values of the dependent variable associated with particular given values of the independent variable(s). When the genetic generator is applied to synthetic accelerograms construction, the proposed tool is capable of i) searching, under specific soil and seismic conditions (within thousands of earthquake records) and recommending a desired subset that better match a target design spectrum, and ii) through processes that mimic mating, natural selection, and mutation, producing new generations of accelerograms until an optimum individual is obtained. The procedure is fast and reliable and results in time series that match any type of target spectrum with minimal tampering and deviation from recorded earthquakes characteristics.

The objective of the genetic generator, when applied to synthetic earthquakes construction, is to produce compatible artificial signals with specific design spectra. In this model specific seismic (fault rupture, magnitude, distance, focal depth) and site characteristics (soil/ rock) are the first set of inputs. They are included to take into consideration that a typical strong motion record consists of a variety of waves whose contribution depends on the earthquake source mechanism (wave path) and its particular characteristics are influenced by the distance between the source and the site, some measure of the size of the earthquake, and the surrounding geology and site conditions; and that the design spectra can be an envelope or integration of many expected ground motions that are possible to occur in certain period of time, or the result of a formulation that involves earthquake magnitude, distance and soil conditions. The second set of inputs consist of the target spectrum, the period range for the matching, lower- and upper-bound acceptable values for scaling signal shape, and a collection of GAs parameters (a population size, number of generations, crossover ratio, and mutation ratio). The output is the more success individual with a chromosome array generated from "real" accelerograms parents (a set of).

The algorithm (see Figure 10) is started with a set of solutions (each solution is called a chromosome). A solution is composed of thousands of components or genes (accelerations recorded at the time), each one encoding a particular trait. The initial solutions (original population) are selected based on the seismic parameters at a site (defined previously by the user): fault mechanism, moment magnitude, epicentral distance, focal depth, geotechnical and geological site classification, depth of sediments. If the user does not have a priori seismic/site knowledge, the genetic generator could select the initial population randomly (Figure 11). Once the model has found the seed-accelerogram(s) or chromosome(s), the space of all feasible solutions can be called accelerograms space (state space). Each point in this search space represents one feasible solution and can be "marked" by its value or fitness for the problem. The looking for a solution is then equal to a looking for some extreme (minimum or maximum) in the space.

According to the individuals' fitness, expressed by difference between the target design spectrum and the chromosome response spectrum, the problem is formulated as the minimization of the error function, Z, between the actual and the target spectrum in a certain period range. Solutions with highest fitness are selected to form new solutions (offspring). During reproduction, the recombination (or crossover) and mutation permits to change the genes (accelerations) from parents (earthquake signals) in some way that the whole new chromosome (synthetic signal) contains the older organisms attributes that assure success. This is repeated until some user's condition (for example number of populations or improvement of the best solution) is satisfied (Figure 12).

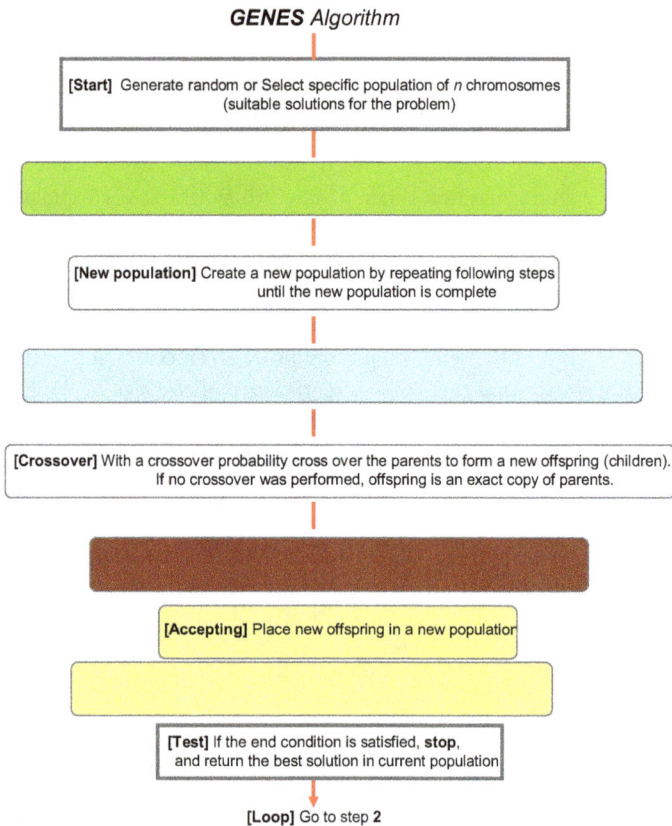

GENES *Algorithm*

[Start] Generate random or Select specific population of *n* chromosomes (suitable solutions for the problem)

[New population] Create a new population by repeating following steps until the new population is complete

[Crossover] With a crossover probability cross over the parents to form a new offspring (children). If no crossover was performed, offspring is an exact copy of parents.

[Accepting] Place new offspring in a new population

[Test] If the end condition is satisfied, **stop**, and return the best solution in current population

[Loop] Go to step **2**

Figure 10. Genetic Generator: flow diagram

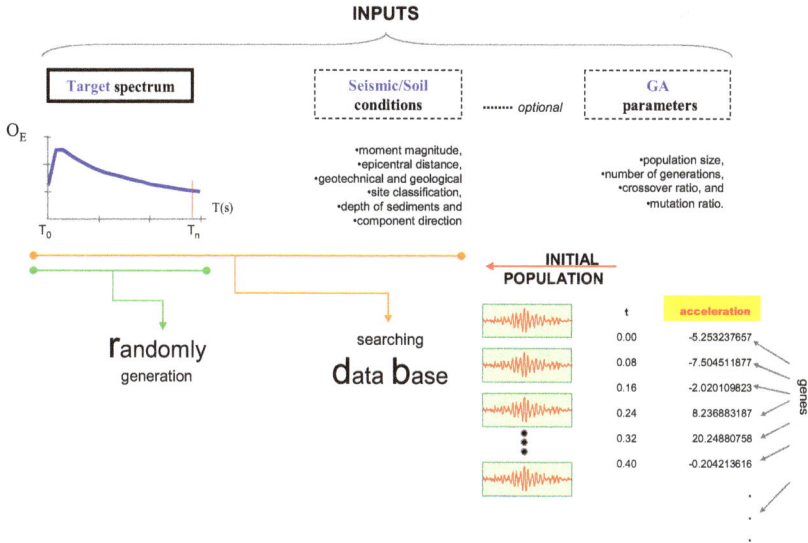

Figure 11. Genetic Generator: working phase diagram

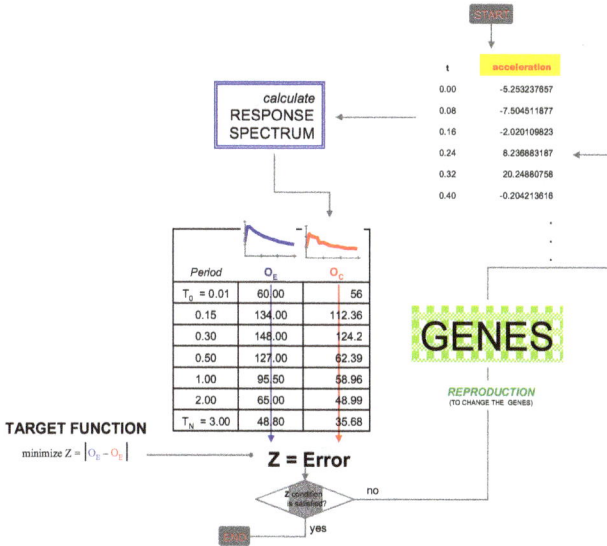

Figure 12. Iteration process of the Genetic Generator

One of the genetic advantages is the possibility of modifying on line the image of the expected earthquake. While the genetic model is running the user interface shows the chromosome per epoch and its response spectra in the same window, if the duration time, the highest intensities interval or the Δt are not convenient for the user's interests, these values can be modified without retraining or a change on model' structure.

In Figure 13 are shown three examples of signals recovered following this methodology. The examples illustrate the application of the genetic methodology to select any number of records to match a given target spectrum (only the more successful individuals for each target are shown in the figure). It can be noticed the stability of the genetic algorithm in adapting itself to smooth, code or scarped spectrum shapes. The procedure is fast and reliable as results in records match the target spectrum with minimal deviation. The genetic procedure has been applied successfully to generate synthetic ground motions having different amplitudes, duration and combinations of moment magnitude and epicentral distance. Although the variations in the target spectra, the genetic signals maintain the nonlinear and nonstationary characteristics of real earthquakes. It is still under development an additional toolbox that will permit to use advanced signal analysis instruments because, as it has been demonstrated [63] [64], studying nonstationary signals through Fourier or response spectra is not convenient for all applications.

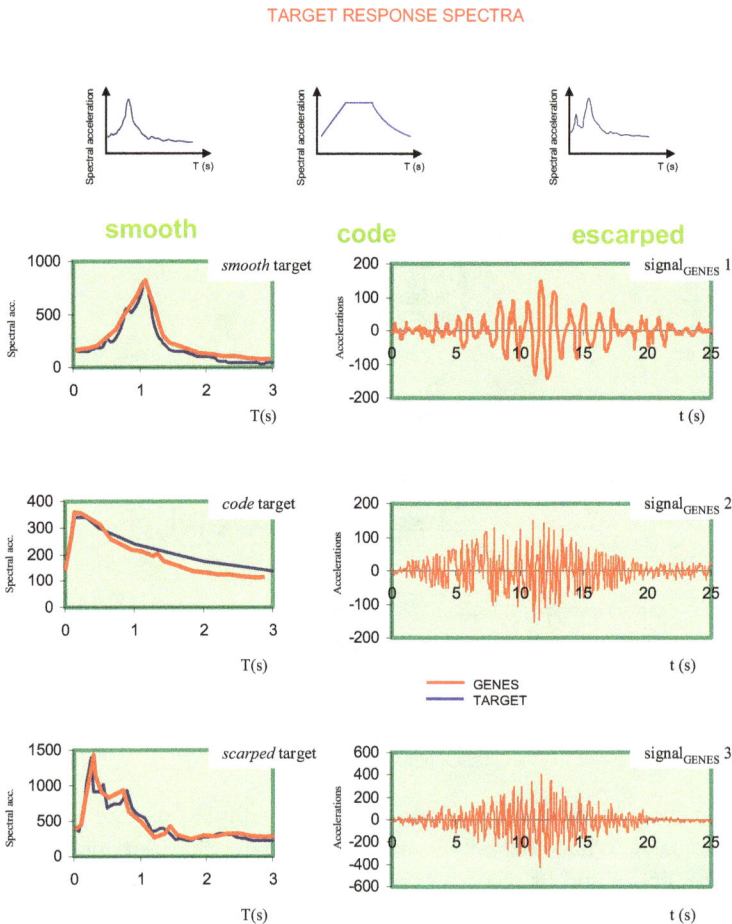

Figure 13. Some Generator results: accelerograms application

4.4. Effects of local site conditions on ground motions

Geotechnical and structural engineers must take into account two fundamental characteristics of earthquake shaking: 1) how ground shaking propagates through the Earth, especially near the surface (site effects), and 2) how buildings respond to this ground motion. Because neither characteristic is completely understood, the seismic phenomenon is still a challenging research area.

Site effects play a very important role in forecasting seismic ground responses because they may strongly amplify (or deamplify) seismic motions at the last moment just before reaching the surface of the ground or the basement of man-made structures. For much of the history of seismological research, site effects have received much less attention than they should, with the exception of Japan, where they have been well recognized through pioneering work by Sezawa and Ishimoto as early as the 1930's [65]. The situation was drastically changed by the catastrophic disaster in Mexico City during the Michoacan, Mexico earthquake of 1985, in which strong amplification due to extremely soft clay layers caused many high-rise buildings to collapse despite their long distance from the source. The cause of the astounding intensity and long duration of shaking during this earthquake is not well resolved yet even though considerable research has been conducted since then, however, there is no room for doubt that the primary cause of the large amplitude of strong motions in the soft soil (lakebed) zone relative to those in the hill zone is a site effect of these soft layers.

The traditional data-analysis methods to study site effects are all based on linear and stationary assumptions. Unfortunately, in most soil systems, natural or manmade ones, the data are most likely to be both nonlinear and nonstationary. Discrepancies between calculated responses (using code site amplification factors) and recent strong motion evidence point out serious inaccuracies may be committed when analyzing amplification phenomena. The problem might be due partly because of the lack of understanding regarding the fundamental causes in soil response but also a consequence of the distorted soil amplification quantification and the incomplete characterization of nonlinearity-induced nonstationary features exposed in motion recordings [66]. The objective of this investigation is to illustrate a manner in which site effects can be dealt with for the case of Mexico City soils, making use of response spectra calculated from the motions recorded at different sites during extreme and minor events (see Figure 6). The variations in the spectral shapes, related to local site conditions, are used to feed a multilayer neural network that represent a very advantageous nonlinear-amplification relation. The database is composed by registered information earthquakes affecting Mexico City originated by different source mechanisms.

The most damaging shocks, however, are associated to the subduction of the Cocos Plate into the Continental Plate, off the Mexican Pacific Coast. Even though epicentral distances are rather large, these earthquakes have recurrently damaged structures and produced severe losses in Mexico City. The singular geotechnical environment that prevails in Mexico

City is the one most important factor to be accounted for in explaining the huge amplification of seismic movements [67-70]. The soils in Mexico City were formed by the deposition into a lacustrine basin of air and water transported materials. From the view point of geotechnical engineering, the relevant strata extend down to depths of 50 m to 80 m, approximately. The superficial layers formed the bed of a lake system that has been subjected to dessication for the last 350 years. Three types of soils may be broadly distinguished: in Zone I, firm soils and rock-like materials prevail; in Zone III, very soft clay formations with large amounts of microorganisms interbedded by thin seams of silty sands, fly ash and volcanic glass are found; and in Zone II, which is a transition between zones I and III, sequences of clay layers an coarse material strata are present (Figure 14).

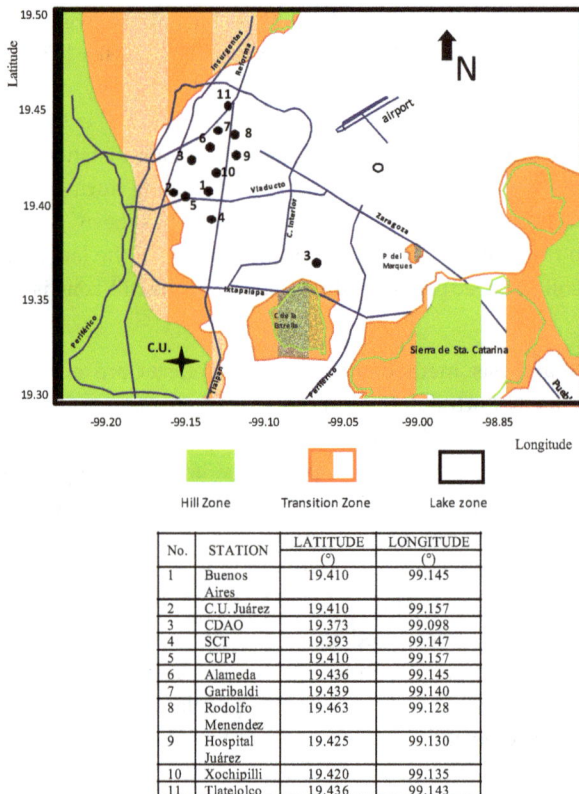

No.	STATION	LATITUDE (°)	LONGITUDE (°)
1	Buenos Aires	19.410	99.145
2	C.U. Juárez	19.410	99.157
3	CDAO	19.373	99.098
4	SCT	19.393	99.147
5	CUPJ	19.410	99.157
6	Alameda	19.436	99.145
7	Garibaldi	19.439	99.140
8	Rodolfo Menendez	19.463	99.128
9	Hospital Juárez	19.425	99.130
10	Xochipilli	19.420	99.135
11	Tlatelolco	19.436	99.143

Figure 14. Accelerographic stations used in this study

Due to space limitations, reference is made only to two seismic events: the June 15, 1999 and the October 9, 1995. This module was developed based in a previous study (see section 4.2 of this Chapter) where the effect of the parameters E_D, F_D and M_w on the ground motion attenuation from epicentre to the site, were found to be the most significant [71]. The recent disaster experience showed that the imprecision that is inherent to most variables measurements or estimations makes crucial the consideration of subjectivity to evaluate and to derive numerical conclusions according to the phenomena behavior. The neuronal training process starts with the training of four input variables booked: E_D, F_D and M_w. The output linguistic variables are PGA_{h1} (peak ground acceleration horizontal, component 1) and PGA_{h2} (peak ground acceleration, horizontal component 2) registered in a rock-like site in Zone I .The second training process is linked *feed-forward* with the previous module (PGA for rock-like site) and the new seismic inputs are Seismogenic Zone and PGArock and the Latitude and Longitude coordinates are the geo-referenced position needed to draw the deposition variation into the basin. This neuro-training runs one step after the first training phase and until the minimum difference between the S_a and the neuronal calculations is attained. In Figure 15 some results from training and testing modes are shown.

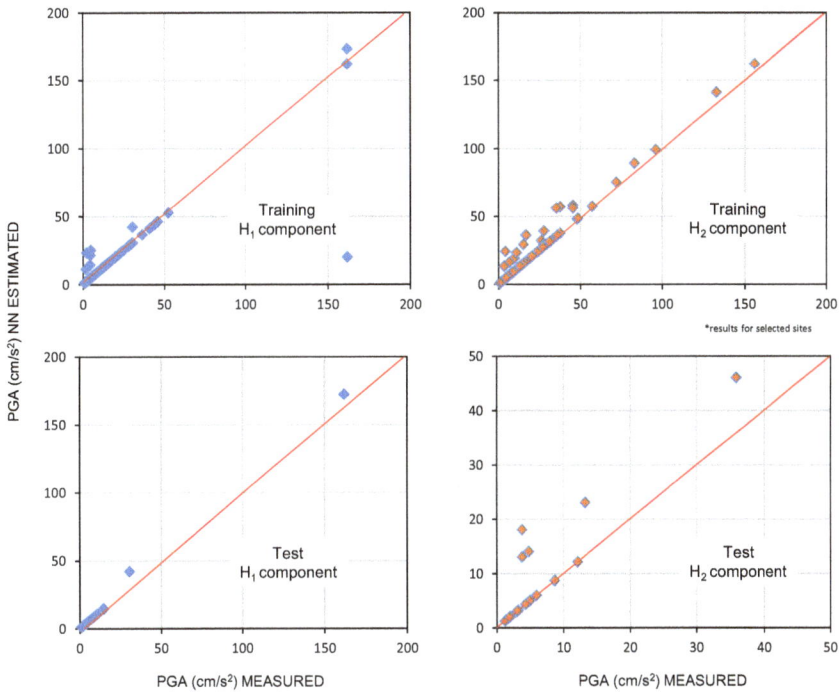

Figure 15. Neural estimations for PGA in Lake Zone sites

This second NN represents the geo-referenced amplification ratio that take into consideration the topographical, geotechnical and geographical conditions, implicit in the recorded accelerograms. The results of these two NNs are summarized in Figure 16. These graphs show the predicting capabilities of the neural system comparing the measured values with those obtained in neural-working phase. It can be observed a good correspondence throughout the full distance and magnitude range for the seismogenic zones considered in this study for the whole studied area (Lake Zone).

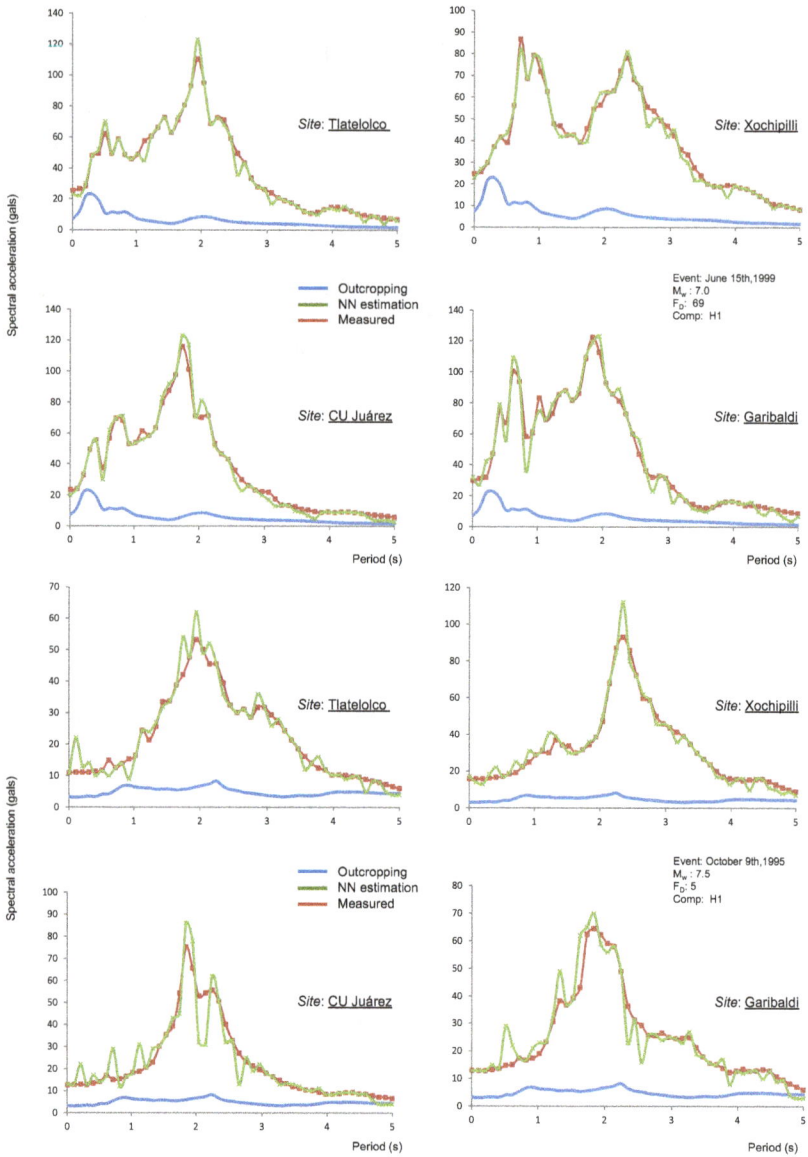

Figure 16. Spectral accelerations in some Lake-Zone sites: measured vs NN

4.5. Liquefaction phenomena: potential assessment and lateral displacements estimation

Over the past forty years, scientists have conducted extensive research and have proposed many methods to predict the occurrence of liquefaction. In the beginning, undrained cyclic loading laboratory tests had been used to evaluate the liquefaction potential of a soil [72] but due to difficulties in obtaining undisturbed samples of loose sandy soils, many researchers have preferred to use *in situ* tests [73]. In a semi-empirical approach the theoretical considerations and experimental findings provides the ability to make sense out of the field observations, tying them together, and thereby having more confidence in the validity of the approach as it is used to interpolate or extrapolate to areas with insufficient field data to constrain a purely empirical solution. Empirical field-based procedures for determining liquefaction potential have two critical constituents: i) the analytical framework to organize past experiences, and ii) an appropriate *in situ* index to represent soil liquefaction characteristics. The original simplified procedure [74] for estimating earthquake-induced cyclic shear stresses continues to be an essential component of the analysis framework. The refinements to the various elements of this context include improvements in the in-situ index tests (e.g., SPT, CPT, BPT, V_s), and the compilation of liquefaction/no-liquefaction cases.

The objective of the present study is to produce an empirical machine learning ML method for evaluating liquefaction potential. ML is a scientific discipline concerned with the design and development of algorithms that allow computers to evolve behaviours based on empirical data, such as from sensor data or databases. Data can be seen as examples that illustrate relations between observed variables. A major focus of ML research is to automatically learn to recognize complex patterns and make intelligent decisions based on data; the difficulty lies in the fact that the set of all possible behaviours given all possible inputs is too large to be covered by the set of observed examples (training data). Hence the learner must generalize from the given examples, so as to be able to produce a useful output in new cases. In the following two ML tools, Neural Networks NN and Classification Trees CTs, are used to evaluate liquefaction potential and to find out the liquefaction control parameters, including earthquake and soil conditions. For each of these parameters, the emphasis has been on developing relations that capture the essential physics while being as simplified as possible. The proposed cognitive environment permits an improved definition of i) seismic loading or cyclic stress ratio CSR, and ii) the *resistance* of the soil to triggering of liquefaction or cyclic resistance ratio CRR.

The factor of safety FS against the initiation of liquefaction of a soil under a given seismic loading is commonly described as the ratio of cyclic resistance ratio (CRR), which is a measure of liquefaction resistance, over cyclic stress ratio (CSR), which is a representation of seismic loading that causes liquefaction, symbolically, $FS = CRR / CSR$. The reader is referred to Seed and Idriss [74], Youd et al. [75], and Idriss and Boulanger [76] for a historical perspective of this approach. The term CSR $CSR = f\left(0.65, \sigma_{vo}, a_{max}, \sigma'_{vo}, r_d, MSF\right)$ is function of the vertical total stress of the soil σ_{vo} at the depth considered, the vertical effective stress σ'_{vo}, the peak horizontal ground surface acceleration a_{max}, a depth-dependent shear stress reduction factor r_d (dimensionless), a magnitude scaling factor MSF (dimensionless). For CRR, different in situ-resistance measurements and overburden correction factors are included in its determination; both terms operate depending of the geotechnical conditions. Details about the theory behind this topic in Idriss and Boulanger, [76] and Youd et al. [75].

Many correction/adjustment factors have been included in the conventional analytical frameworks to organize and to interpret the historical data. The correction factors improve the consistency between the geotechnical/seismological parameters and the observed liquefaction behavior, but they are a consequence of a constrained analysis space: a 2D plot [CSR *vs.* CRR] where regression formulas (simple equations) intend to relate complicated nonlinear/multidimensional information. In this investigation the ML methods are applied to discover unknown, valid patterns and relationships between geotechnical, seismological and engineering descriptions using the relevant available information of liquefaction phenomena (expressed as empirical prior knowledge and/or input-output data). These ML techniques "work" and "produce" accurate predictions based on few logical conditions and they are not restricted for the mathematical/analytical environment. The ML techniques establish a *natural* connection between experimental and theoretical findings.

Following the format of the simplified method pioneered by Seed and Idriss [74], in this investigation a nonlinear and adaptative *limit state* (a fuzzy-boundary that separates liquefied cases from nonliquefied cases) is proposed (Figure 17). The database used in the present study was constructed using the information included in Table 3 and it was compiled by Agdha et al., [77], Juang et al., [78], Juang [79], Baziar, [80] and Chern and Lee [81]. The cases are derived from cone penetration tests CPT, and shear wave velocities V_s measurements and different world seismic conditions (U.S., China, Taiwan, Romania, Canada and Japan). The soils types ranges from clean sand and silty sand to silt mixtures (sandy and clayey silt). Diverse geological and geomorphological characteristics are included. The reader is referred to the citations in Table 3 for details.

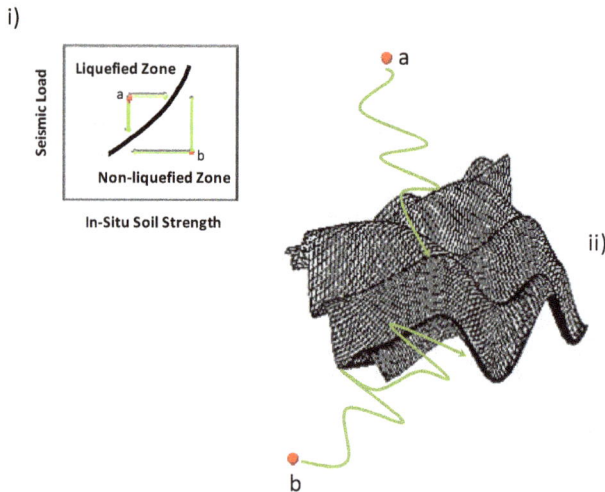

Figure 17. An schematic view of the nonlinear-liquefaction boundary

The ML formulation uses Geotechnical (q_c, V_s, Unit weight, Soil Type, Total vertical stresses, Effective vertical stresses, Geometrical (Layer thickness, Water Level Depth, Top Layer Depth) and Seismological (Magnitude, PGA) input parameters and the output variable is "Liquefaction?" and it can take the values "YES/NO" (Figure 17). Once the NN is trained the number of cases that was correctly evaluated was 100% and applied to "unseen" cases (separated for testing) less than 10% of these examples were not fitted. The CT has a minor efficiency during the training showing 85% of cases correctly predicted, but when the CT runs on the unseen patterns its capability is not diminished and it asserts the same proportion. From these findings it is concluded that the neuro system is capable of predicting the in situ measurements with a high degree of accuracy but if improvement of knowledge is necessary or there are missed, vague even contradictory values in the analyzed case, the CT is a better option.

Set	Input Parameters	Number .of Patterns	Ref.
A	Z, Z_{NAF}; H, Soil Class, Geomorphological units, Geological units, Site amplification, a_{max}	56	Fatemi-Agdha et al., 1988
B	Z, q_c, F_s, σ_0, σ_0', a_{max}, M	21	Juang et al., 1999
C	Z, q_c, F_s, σ_0, σ_0', a_{max}, M	242	Juang, 2003
D	D_{50}, a_{max}, σ_0', σ_0, M, F_s, q_c, SPT, Z	170	Baziar, 2003
E	M, σ_0, σ_0', q_c, a_{max}	466	Chern and Lee, 2009
F	Z_{NAF}, Z, H, σ_0, σ_0', Soil Class, V_s	80	Andrus and Stokoe, 1997; 2000
	Total:	**1035**	

Table 2. Database for liquefaction analysis

Figure 18 shows the pruned liquefaction trees (two, one runs using q_c values and the other through the V_s measurements) with YES/NO as terminal nodes. In the Figure 19, some examples of tree reading are presented. The trees incorporate soil type dependence through the resistance values (q_c, and V_s) and fine content, and it is not necessary to label the material as "sand" or "silt". The most general geometrical branches that split the behaviors are the Water table depth and the Layer thickness but only when the soil description is based on V_s, when q_c, serves as rigidity parameter this geometrical inputs are not explicit exploited. This finding can be related to the nature of the measurement: the cone penetration value contains the effect of the saturated material while the shear wave velocities need the inclusion of this situation explicitly. Without potentially confusing regression strategies, the liquefaction trees results can be seen as an indication of how effectively the ML model maps the assigned predictor variables to the response parameter. Using data from all regions and wide parameters ranges, the prediction capabilities of the neural network and classification trees are superior to many other approximations used in common practice, but the most important remark is the generation of meaningful clues about the reliability of physical parameters, measurement and calculation process and practice recommendations.

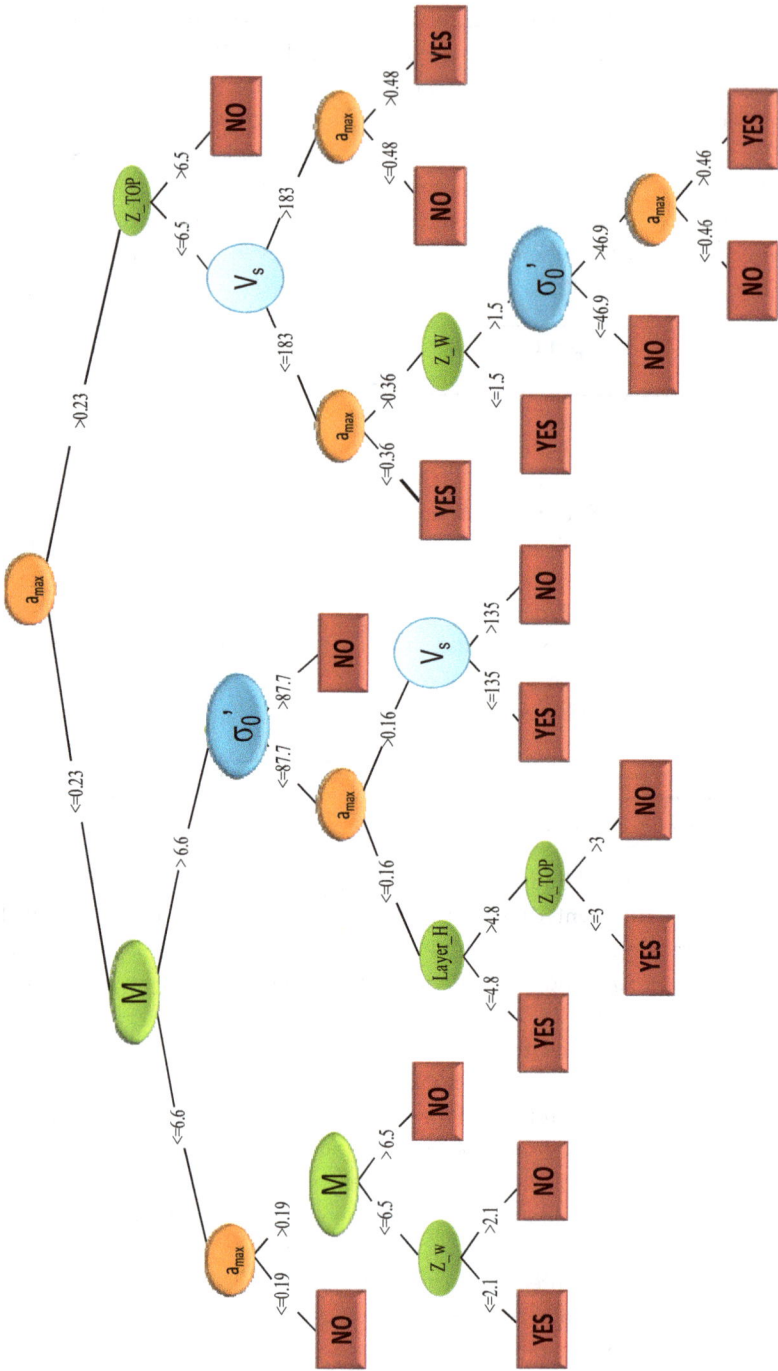

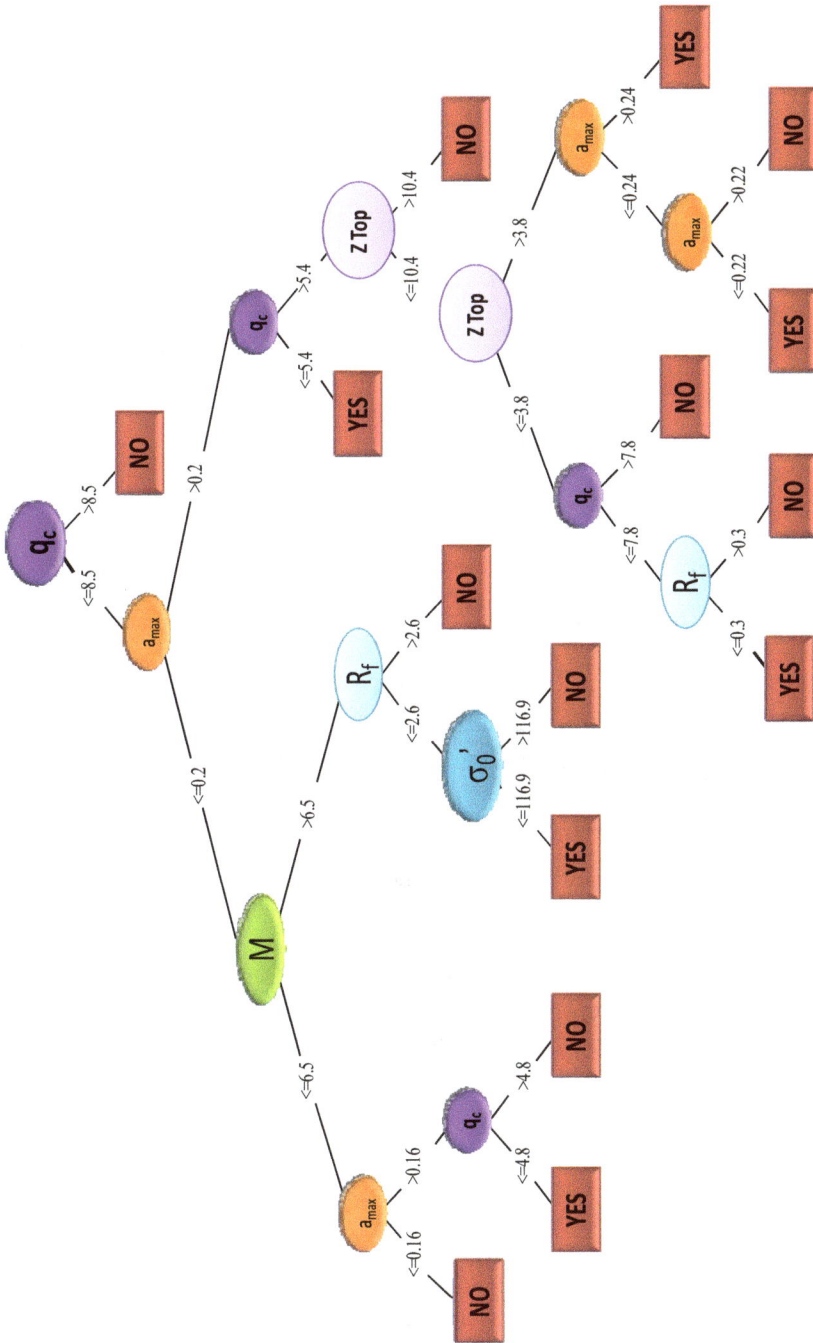

Figure 18. Classification tree for liquefaction potential assessment

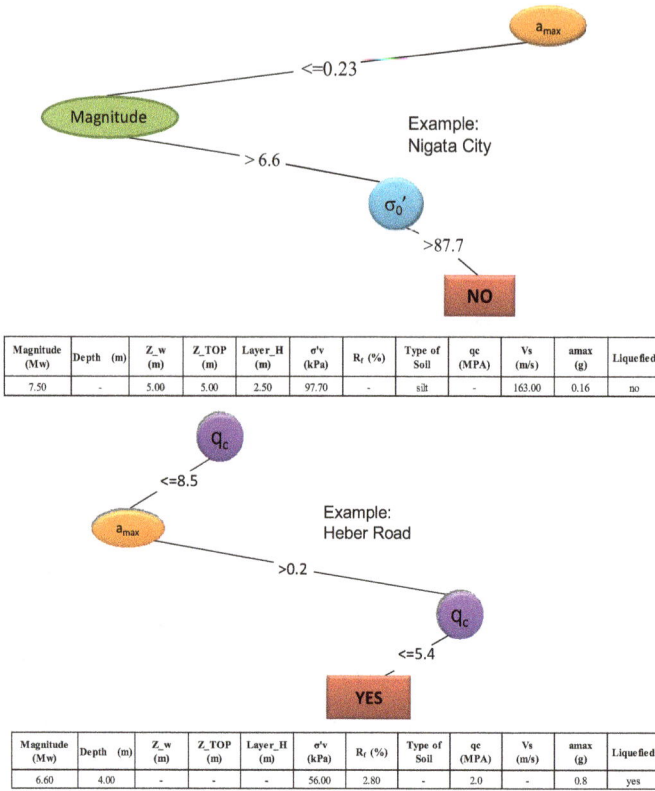

Figure 19. CT classification examples

The intricacy and nonlinearity of the phenomena, an inconsistent and contradictory database, and many subjective interpretations about the observed behavior, make SC an attractive alternative for estimation of liquefaction induced lateral spread. NEFLAS [82], NEuroFuzzy estimation of liquefaction induced LAteral Spread, profits from fuzzy and neural paradigms through an architecture that uses a fuzzy system to represent knowledge in an interpretable manner and proceeds from the learning ability of a neural network to optimize its parameters. This blending can constitute an interpretable model that is capable of learning the problem-specific prior knowledge.

NEFLAS is based on the Takagi-Sugeno model structure and it was constructed according the information compiled by Bartlett and Youd [83] and extended later by Youd et al. [84]. The output considered in NEFLAS is the horizontal displacements due to liquefaction, dependent of moment magnitude, the PGA, the nearest distance from the source in kilometers; the free face ratio, the gradient of the surface topography or the slope of the liquefied layer base, the cumulative thickness of saturated cohesionless sediments with number of blows (modified by overburden and energy delivered to the standard penetration probe, in this case 60%) , the average of fines content, and the mean grain size.

One of the most important NEFLAS advantages is its capability of dealing with the imprecision, inherent in geoseismic engineering, to evaluate concepts and derive conclusions. It is well known that engineers use words to classify qualities ("strong earthquake", "poor graduated soil" or "soft clay" for example), to predict and to validate "first principle" theories, to enumerate phenomena, to suggest new hypothesis and to point the limits of knowledge. NEFLAS mimics this method. See the technical quantity "magnitude" (earthquake input) depicted in Figure 20. The degree to which a crisp magnitude belongs to LOW, MEDIUM or HIGH linguistic label is called the degree of membership. Based on the figure, the expression, "the magnitude is LOW" would be true to the degree of 0.5 for a M_w of 5.7. Here, the degree of membership in a set becomes the degree of truth of a statement.

On the other hand, the human logic in engineering solutions generates sets of behavior rules defined for particular cases (parametric conditions) and supported on numerical analysis. In the neurofuzzy methods the human concepts are re-defined through a flexible computational process (training) putting (empirical or analytical) knowledge into simple "if-then" relations (Figure 20). The fuzzy system uses 1) variables composing the antecedents (premises) of implications; 2) membership functions of the fuzzy sets in the premises, and 3) parameters in consequences for finding simpler solutions with less design time.

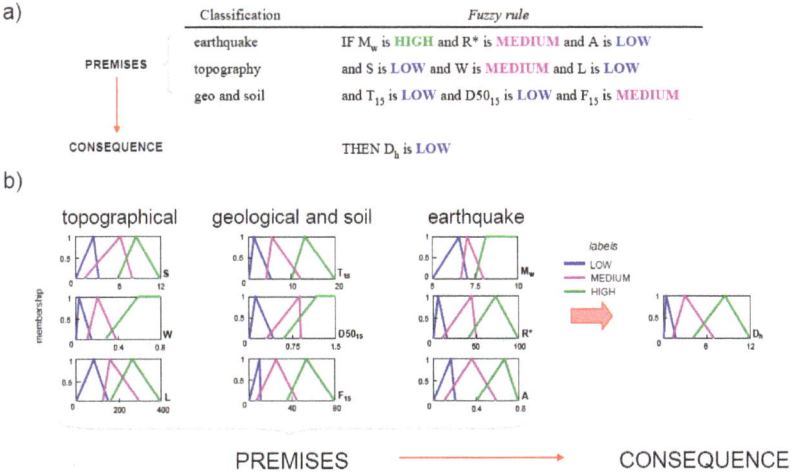

Figure 20. Neurofuzzy estimation of lateral spread

NEFLAS considers the character of the earthquake, topographical, regional and geological components that influence lateral spreading and works through three modules: Reg-NEFLAS, appropriate for predicting horizontal displacements in geographic regions where seismic hazard surveys have been identified; Site- NEFLAS, proper for predictions of horizontal displacements for site-specific studies with minimal data on geotechnical conditions and Geotech-NEFLAS allows more refined predictions of horizontal displacements when additional data is available from geotechnical soil borings. The

NEFLAS execution on cases not included in the database (Figure 21.b and Figure 21.c) and its higher values of correlation when they are compared with evaluations obtained from empirical procedures permit to assert that NEFLAS is a powerful tool, capable of predicting lateral spreads with high degree of confidence.

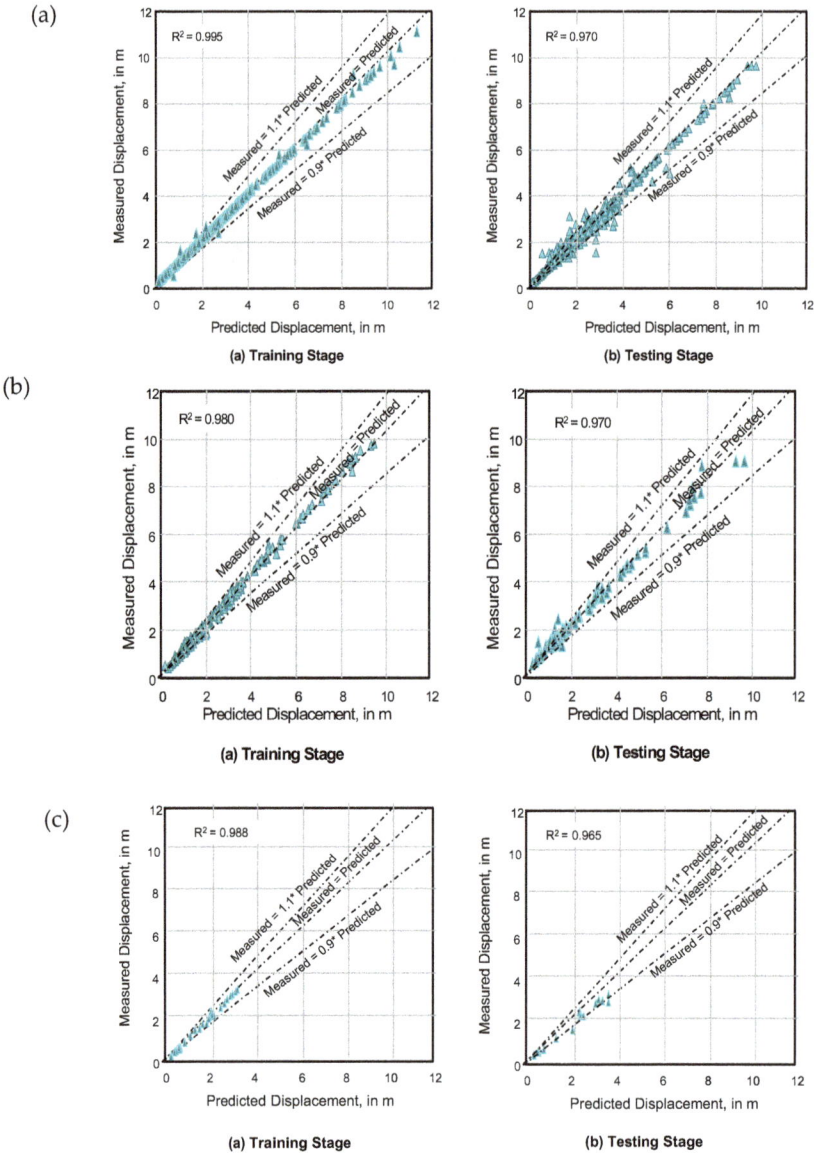

Figure 21. NN estimations vs measured displacements for a) the whole data set, b)Niigata Japan, c) San Francisco USA cases

5. Conclusions

Based on the results of the studies discussed in this paper, it is evident that cognitive techniques perform better than, or as well as, the conventional methods used for modeling complex and not well understood geotechnical earthquake problems. Cognitive tools are having an impact on many geotechnical and seismological operations, from predictive modeling to diagnosis and control.

The hybrid *soft* systems leverage the tolerance for imprecision, uncertainty, and incompleteness, which is intrinsic to the problems to be solved, and generate tractable, low-cost, robust solutions to such problems. The synergy derived from these hybrid systems stems from the relative ease with which we can translate problem domain knowledge into initial model structures whose parameters are further tuned by local or global search methods. This is a form of methods that do not try to solve the same problem in parallel but they do it in a mutually complementary fashion. The push for low-cost solutions combined with the need for intelligent tools will result in the deployment of hybrid systems that efficiently integrate reasoning and search techniques.

Traditional earthquake geotechnical modeling, as physically-based (or knowledge-driven) can be improved using soft technologies because the underlying systems will be explained also based on data (CC data-driven models). Through the applications depicted here it is sustained that cognitive tools are able to make abstractions and generalizations of the process and can play a complementary role to physically-based models.

Author details

Silvia Garcia

Geotechnical Department, Institute of Engineering, National University of Mexico, Mexico

6. References

[1] W.D.L. Finn State of the art of geotechnical earthquake engineering practice Soil Dynamics and Earthquake Engineering 20 (2000) Elsevier

[2] Abrahamson NA, R R, Youngs. A stable algorithm for regression analysis using the random effects model. Bull Seismol Soc Am 1992;82:505±10.

[3] Somerville PG, Sato T. Correlation of rise time with the style-offaulting factor in strong ground motions. Seismol Res Lett 1998:153pp (abstract).

[4] Somerville PG, Greaves RL. Strong ground motions of the Kobe, Japan earthquake of January 17, 1995, and development of a model of forward rupture directivity applicable in California. Proc. Western Regional Tech. Seminar on Earthquake Eng. for Dams. Assoc. of State Dam Safety Oficials, Sacramento, CA, April 11±12, 1996.

[5] Abrahamson NA, Silva WJ. Empirical duration relations for shallow crustal earthquakes. Written communication, 1997.

[6] Graves RW, Pitarka A, Somerville PG. Ground motion amplication in the Santa Monica area: effects of shallow basin edge structure. Submitted for publication.

[7] Somerville PG. Emerging art: earthquake ground motion. In: Dakoulas P, Yegian M, Holtz RD, editors. Proc Geotechnical Earthquake Engineering in Soil Dynamics III, Geotechnical Special Publication No. 75, vol. 1. ASCE, 1998. p. 1±38.

[8] Abrahamson NA. Spatial variation of multiple support inputs. Proc. First US Symp. Seism. Eval. Retrofit Steel Bridges, UC, Berkeley, October 18, 1993.

[9] Naeim F, Lew M. On the use of design spectrum compatible motions. Earthquake Spec 1995;II(1):111±28.

[10] Boulanger R.W. and Idriss, I.M. (2006). "Liquefaction Susceptibility Criteria for Silts and Clays," Journal of Geotechnical and Geoenvironmental Engineering, 132 (11), 1413–1426pp.

[11] Boulanger R.W. and Idriss, I.M. (2007). "Evaluation of Cyclic Softening in Silts and Clays", Journal of Geotechnical and Geoenvironmental Engineering, 133 (6), 641–652pp.

[12] Youd and Idriss, NCEER. Proceedings, Workshop on Evaluation of Liquefaction Resistance of Soils. Technical Report No. NCCER-97-0022. National Center for Earthquake Engineering Research, University of Buffalo, Buffalo, New York, 1997.

[13] Seed HB, Tokimatsu K, Harder LF, Chung RM. Influence of SPT procedures in soil liquefaction resistance evaluations. J Geotech Engng 1985;111(12):1425±45.

[14] Robertson PK, Fear CE. Liquefaction of sands and its evaluation. Proceedings, 1st Int. Conf. on Earthquake Geotechnical Engineering, Tokyo, Japan, 1995.

[15] Ambraseys NN. Engineering seismology. Earthquake Engng Struct Dynam 1988;17:1±105.

[16] Wang, Y. (2008a). On Contemporary Denotational Mathematics for Computational Intelligence. In Transactions of Computational Science (Vol. 2, pp. 6-29). New York: Springer.

[17] Wang, Y. (2009a). On Abstract Intelligence: Toward a Unified Theory of Natural, Artificial, Machinable, and Computational Intelligence. International Journal of Software Science and Computational Intelligence, 1(1), 1–18.

[18] Wang, Y. (2009b). On Cognitive Computing. International Journal of Software Science and Computational Intelligence, 1(3), 1–15.

[19] Turing, A.M. (1950). Computing Machinery and Intelligence. Mind, 59, 433-460.

[20] Von Neumann, J. (1946). The Principles of Large-Scale Computing Machines. Reprinted in Annals of History of Computers, 3(3), 263-273.

[21] Von Neumann, J. (1958). The Computer and the Brain. New Haven: Yale Univ. Press.

[22] Gersting, J.L. (1982). Mathematical Structures for Computer Science. San Francisco: W. H. Freeman & Co.

[23] Mandrioli, D., & Ghezzi, C. (1987). Theoretical Foundations of Computer Science. New York: John Wiley & Sons.

[24] Lewis, H.R., & Papadimitriou, C.H. (1998). Elements of the Theory of Computation, 2nd ed. Englewood Cliffs, NJ: Prentice Hall International.

[25] Kephart, J., & Chess, D. (2003). The Vision of Autonomic Computing. IEEE Computer, 26(1), 41-50.

[26] IBM (2006). Autonomic Computing White Paper. An Architectural Blueprint for Autonomic Computing, 4th ed., June, (pp. 1-37).

[27] Wang, Y. (2004). Keynote: On Autonomic Computing and Cognitive Processes. Proc. 3rd IEEE International Conference on Cognitive Informatics (ICCI'04), Victoria, Canada, IEEE CS Press, (pp. 3-4).

[28] Wang, Y. (2007a, July). Software Engineering Foundations: A Software Science Perspective. CRC Book Series in Software Engineering, Vol. II, Auerbach Publications, NY.

[29] Bouchon-Meunier, B. Yager, R. Zadeh, L. 1995. Fuzzy Logic and SoftComputing. World Scientific, Singapore.

[30] Bezdek, J.C.: What is a computational intelligence? In: Zurada, J.M., Marks II, R.J., Robinson, C.J. (eds.) Computational Intelligence: Imitating Life, pp. 1–12. IEEE Press, Los Alamitos (1994)

[31] Zadeh, L.A.: Fuzzy Sets. Information and Control 8, 338–353 (1965)

[32] Zadeh L.A. The roles of fuzzy logic and soft computing in the conception, design and deployment of intelligent systems. BT Technol J. 14(4): 32-36, 1994.

[33] Aliev R.A. and Aliev R.R., Soft Computing, volumes I, II, III. Baku: ASOA Press, 1997-1998 (in Russian).

[34] Aliev R., Bonfig K., and Aliew F., Soft Computing. Berlin: Verlag Technic, 2000.

[35] Zadeh L.A. Foreword. In Proc. First European Congress on Intelligent Techniques and Soft Computing - EUFIT'95, page VII, 1995.

[36] Zadeh L. A. Soft Computing and Fuzzy Logic. IEEE Software 11 (6): 48-58, 1994.

[37] Aliev R.A. Fuzzy Expert Systems. In Aminzadeh F. and Jamshidi M.(eds) SOFT COMPUTING: Fuzzy Logic, Neural Networks and Distributed Artificial Intelligence.pages 99-108. NJ: PTR Prentice Hall, 1994.

[38] Zadeh L.A. Fuzzy logic, Neural Networks and Soft Computing . Comm of ACM 37(3): 77-84, 1994.

[39] Welstead S.T.(ed) Neural Networks and Fuzzy Logic Applications in C/C++, Professional Computing. NY: John Wiley, 1994.

[40] Yager R.R. and Zadeh L.A.(eds) Fuzzy sets, neural networks and Soft Computing. NY: VAN Nostrand Reinhold , 1994.

[41] Nauck D., Klawonn F., and Kruse R., Foundations of Neuro-Fuzzy Systems.NY: John Wiley and Sons, 1997.

[42] Mohamad H.Hassoun, Fundamentals of artificial neural networks. Cambridge: MIT Press, 1995.

[43] Haykin S., Neural Networks: A Comprehensive Foundation. Marmillau and IEEE Computer Society, 1994.

[44] Goldberg D.E., Genetic algorithms in search, optimization and machine learning. Reading, MA: Addison-Wesley, 1989.

[45] Arciszewski T. and De Jong K.A. (2001). Evolutionary computation in civil engineering: research frontiers. Eight International Conference on Civil and Structural Engineering Computing, (Topping B. H. V., ed.), Saxe-Coburg Publications, Eisenstadt, Vienna, Austria.

[46] Miettinen, K.; Neittaanmaki, P.and Periaux, J. 1999, Evolutionary Algorithms in Engineering and Computer Science : Recent Advances in Genetic Algorithms, Evolution Strategies, Evolutionary Programming, John Wiley & Sons Ltd., pps. 483. ISBN 0471999024

[47] Youngs, R. R., S. J. Chiou, W. J. Silva and J. R. Humphrey, 1997. Strong ground motion attenuation relationships for subduction zone earthquakes. Seismol. Res. Lett., (68) 1, 58-75.

[48] Anderson, J. G., 1997. Nonparametric description of peak acceleration above a subduction thrust. Seismol. Res. Lett., (68) 1, 86-94.

[49] Crouse, C. B., 1991. Ground motion attenuation equations for earthquakes on the Cascadia subduction zone. Earth. Spectra, 7, 210-236.

[50] Singh, S. K., M. Ordaz, M. Rodríguez, R. Quaas, V. Mena, M. Ottaviani, J. G. Anderson and D. Almora, 1989. Analysis of near-source strong motion recordings along the Mexican subduction zone. Bull. Seism. Soc. Am., 79, 1697-1717.

[51] Crouse, C. B., Y. K. Vyas and B. A. Schell, 1988. Ground motions from subduction-zone earthquakes. Bull. Seism. Soc. Am.,78, 1-25.

[52] Singh, S. K., E. Mena, R. Castro and C. Carmona, 1987. Empirical prediction of ground motion in Mexico City from coastal earthquakes. Bull. Seism. Soc. Am., 77, 1862-1867.

[53] Sadigh, K., 1979. Ground motion characteristics for earthquakes originating in subduction zones and in the western United States. Proc. Sixth Pan Amer. Conf., Lima, Peru.

[54] Tichelaar, B.F., and L. J. Ruff, 1993. Depth of seismic coupling along subduction zones. J. Geophys. Res., 98, 2017-2037.

[55] Atkinson, G. M. and D. M. Boore, 2003. Empirical ground-motion Relations for Subduction-Zone Earthquakes and Their Applications to Cascadia and other regions. Bull. Seism. Soc. Am., 93, 4, 1703-1729

[56] Gómez, S. C., M. Ordaz and C. Tena, 2005. Leyes de atenuación en desplazamiento y aceleración para el diseño sísmico de estructuras con aislamiento en la costa del Pacífico. Memorias del XV Congreso Nacional de Ingeniería Sísimica, México, Nov. A-II-02

[57] Gasparini, D., and Vanmarcke, E. H. 1976. SIMQKE: A Program for Artificial Motion Generation, Department of Civil Engineering, Massachusetts Institute of Technology, Cambridge, MA.

[58] Silva, W.J., and Lee, K. (1987). "WES RASCAL code for synthesizing earthquake ground motions." State-of-the-Art for Assessing Earthquake Hazards in the United States, Report 24, U.S. Army Engineers Waterways Experiment Station, Misc. Paper S-73-1.

[59] Bolt, B. A., and Gregor, N. J. 1993. "Synthesized Strong Ground Motions for the Seismic Condition Assessment of the Eastern Portion of the San Francisco Bay Bridge", Report UCB /EERC-93/12, University of California, Earthquake Engineering Research Center, Berkeley, CA.

[60] Carballo, J E, y C A Cornell (2000), "Probabilistic seismic demand analysis: spectrum matching and design", Department of Civil and Environmental Engineering, Stanford University, Report No. RMS-41.

[61] Kircher, C., 1993. Personal communication with Farzad Naeim and Marshall Lew.

[62] Naeim, F; J. Kelly. 1999. Design of Seismic Isolated Structures from Theory to Practice. New York, John Wiley & Sons. 289p.

[63] Huang, N. E., Zheng, S., Long, S. R., Wu, M. C., Shih, H. H., Zheng, Q., Yen, N.-C., Tung, C. C., and Liu, M. H., (1998). "The empirical mode decomposition and Hilbert spectrum for nonlinear and nonstationary time series analysis", Proc. R. Soc. London, Ser. A 454, 903–995.

[64] Roulle, A., and F. J. Chavez-Garcia (2006). The strong ground motion in Mexico City: Analysis of data recorded by a 3D array, Soil. Dyn. Eq. Eng. 26 71-89.

[65] Kawase, H. and K. Aki (1989). A study of the response of a soft bas in for incident S, P and Rayleigh waves with special reference to the long duration observed in Mexico City, Bull Seism. Soc. Am. 79, 1361-1382.

[66] Yoshida, N. & Iai, S. (1998). "Nonlinear site response and its evaluation and prediction," IN Irikura, K., Kudo, K., Okada, K. & Sasatani, T. (Eds.) The Second International Symposium on the Effects of Surface Geology on Seismic Motion, Yokohama, Japan, A.A.Balkema, 71-90.

[67] Herrera, I. y Rosenblueth, E. "Response Spectra on Stratified Soil". Proc. 3rd. World Conference on Earthquake Engineering. Nueva Zelandia, pp. 44-56, 1965.

[68] Romo, M. P. and Jaime, A. (1986). "Dynamic characteristics of some clays of the Mexico Valley and seismic response of the ground". Technical Report, Apr., Instituto de Ingenieria, Mexico City, Mexico (in Spanish).

[69] Jaime, A., Romo, M. P., and Reséndiz, D. (1988). "Comportamiento de pilotes de fricción en arcilla del valle de México." Series of the Instituto de Ingeniería, Mexico City, Mexico

[70] Romo, M. P., and Seed (1986). "Analytical modelling of dynamic soil response in the Mexico Earthquake of September 19, 1985". Proc. ASCE Int. Conf. on the Mexico Earthquakes-1985, 148-162

[71] García S R, Romo M P and Mayoral J, (2007), "Estimation of Peak Ground Accelerations for Mexican Subduction Zone Earthquakes using Neural Networks", Geofísica Internacional, Vol 46-1, pp 51-63, enero-marzo

[72] Castro, G., Poulos, S.J., France, JW., Enos, J.L. 1982. Liquefaction induced by cyclic loading. Winchester, Mass: Geotechnical Engineers Inc.

[73] Seed, H. B., Idriss, I. M., and Arango, I., "Evaluation of Liquefaction Potential Using Field Performance Data," Journal of the Geotechnical Engineering Division, ASCE, Vol. 109, No. GT3, 1983. Seed et al., 1983

[74] Seed, H. B. and Idriss, I. M.1971. Simplified procedure for evaluation soil liquefaction potential. Journal of the Soil Mechanics and Foundations ASCE, 97 (9), 1249-1273.

[75] Youd, T.L., Idriss, I.M. , Andrus, R.D., Arango, I., Castro, G., Christian, J.T., Dobry, R., Liam F., Harder, L.F., Hynes M.E., Ishihara, K., Koester, J.P., Liao,S.S.C., Marcuson III, W.F., Martin, G.R., Mitchell, J.K., Moriwaki, Y., Power, M.S., Robertson, P.K., Seed, R.B.,

and Stokoe, K.H. 2001. Liquefaction resistance of soils. Summary report from the 1996 NCEER and 1998 NCEER/NSF workshops on evaluation of liquefaction resistance of soils. J. Geotech. Geoenviron. Eng., 127(10), 817–833.

[76] Boulanger, R. and Idriss, I.M. 2004. State normalization of penetration resistance and the effect of overburden stress on liquefaction resistance. Proc. 11th International Conf. on Soil Dynamics and Earthquake Engineering and 3rd International Conference on Earthquake Geotechnical Engineering, Univ. of California, Berkeley, CA.

[77] Fatemi-Agdha, S.M., Teshnehlab, M., Suzuki, A., Akiyoshi, T., and Kitazono, Y. 1998. Liquefaction potential assesment using multilayer artificial neural network. J. Sci. I.R. Iran, 9(3).

[78] Juang, C. H., Chen, C. J., and Tien, Y. M. 1999. Appraising cone penetration test based liquefaction resistance evaluation methods: Artificial neural networks approach. Canadian Geotechnical Journal, 36(3) 443-454.

[79] Juang, C. H., Yuan, H. M., Lee, D. H., and Lin 2003, P. S., "Simplified cone penetration test-based method for evaluating liquefaction resistance of soils," Journal of Geotechnical and Geoenvironmental Engineering, Vol. 129, No. 1, pp. 66-80.

[80] Baziar, M. H. and Nilipour, N. 2003. Evaluation of liquefaction potential using neural-networks and CPT results. Soil Dynamics and Earthquake Engineering, 23(7) 631-636.

[81] Chern, S. and Lee, C. 2009. CPT-based simplified liquefaction assessment by using fuzzy-neural network. Journal of Marine Science and Technology, 17(4) 326-331

[82] García S.R. and Romo M.P. 2007. GENES: Genetic Generator of Signals, a Synthetic Accelerograms Application. , Proc. of the SEE5, SM-80.

[83] Bartlett, S. F. and Youd, T. L., 1992, "Empirical Analysis of Horizontal Ground Displacement Generated by Liquefaction Induced Lateral Spreads", Tech. Rept. NCEER 92 - 0021, National Center for Earthquake Engineering Research, SUNY - Buffalo, Buffalo, NY.

[84] Youd T.L., Hansen C.M. and Bartlett S.F. (2002) 'Revised Multilinear Regression Equations for Prediction of Lateral Spread Displacement' Journal of Geotechnical and Geoenvironmental Engineering, Vol. 128, No. 12, pp. 1007-1017.

Chapter 4

Three-Dimensional Wavefield Simulation in Heterogeneous Transversely Isotropic Medium with Irregular Free Surface

Haiqiang Lan and Zhongjie Zhang

Additional information is available at the end of the chapter

1. Introduction

Rough topography is very common and we have to deal with it during the acquisition, processing and interpretation of seismic data. For example, in the context of the deep seismic soundings to explore the crustal structure, seismic experiments are usually carried out across: (a) orogenic belts for understanding the mechanisms; (b) basins to understand the formation mechanisms; (c) transition zones for the study of its interaction (Al-Shukri et al., 1995; Ashford et al., 1997; Boore, 1972; Jih et al., 1988; Levander, 1990; Robertsson, 1996; Zhang et al., 2010). In oil/gas seismic exploration, seismologists also have a similar problem with the undulating topography along the survey line.

In the last two decades, several approaches have been proposed to simulate wave propagation in heterogeneous medium with irregular topography. These schemes include finite element method (Rial et al., 1992; Toshinawa and Ohmachi, 1992), spectral element method (Komatitsch and Tromp, 1999, 2002), pseudo-spectral method (Nielsen et al., 1994; Tessmer et al., 1992; Tessmer and Kosloff, 1994), boundary element method (Bouchon et al., 1989; Campillo and Bouchon, 1985; Sánchez-Sesma and Campillo, 1993; Sánchez-Sesma et al., 2006), finite difference method (Frankel and Vidale, 1992; Gao and Zhang, 2006; Hestholm and Ruud, 1994, 1998; Jih et al., 1988; Lombard et al., 2008; Robertsson, 1996; Zhang and Chen, 2006), and also a hybrid approach which combines the staggered-grid finite difference scheme with the finite element method (Galis et al., 2008; Moczo et al., 1997). Both the spectral element and the finite element methods satisfy boundary conditions on the free surface naturally. 3D surface and interface topographies can be modeled using curved piecewise elements. However, the classical finite element method suffers from a high computational cost, and, on the other hand, a smaller spectral element than the one required by numerical dispersion is required to describe a highly curved topography, as demonstrated in seismic modeling of a hemispherical crater (Komatitsc and Tromp, 1999). The pseudo-spectral method is limited to a free surface with smoothly varying topography and leads to inaccuracies for models with strong heterogeneity or sharp boundaries (Tessmer et al., 1992). The boundary integral equation and boundary element methods are not suitable for near-surface regions with large velocity contrasts (Bouchon et al., 1995). The finite difference method is one of the most popular numerical methods used in computational seismology. In comparison to other methods, the finite difference method is simpler and more flexible, although it has some difficulty in dealing with surface topography. The situation has improved recently. For rectangular domains, a stable and explicit discretization of the free surface boundary conditions has been presented by Nilsson et al. (2007). By using boundary-modified difference operators, Nilsson et al. (2007) introduce a discretization of the mixed derivatives in the governing equations; they also show that the method is second order accurate for problems with smoothly varying material properties and stable under standard Courant-Friedrichs-Lewy (CFL) constraints, for arbitrarily varying material properties. We have investigated 6 free-surface boundary condition approximate schemes in seismic wavefield modelling and evaluated their stability and applicability by comparing with corresponding analytical solutions, the results reveal that Nilsson et al.'s method is more effective than others (Lan & Zhang, 2011a). Recently, Appelo and Petersson (2009) have generalized the results of Nilsson et al. (2007) to curvilinear coordinate systems, allowing for simulations on non-rectangular domains. They construct a stable discretization of the free surface boundary conditions on curvilinear grids, and they prove that the strengths of the proposed method are its ease of implementation, efficiency (relative to low-order unstructured grid methods), geometric flexibility, and, most importantly, the "bullet-proof" stability (Appelo and Petersson, 2009), even though they deal with 2D isotropic medium.

Nevertheless, the earth is often seismically anisotropic resulting from fractured rocks, fluid-filled cracks (Crampin, 1981; Hudson, 1981; Liu et al., 1993; Schoenberg and Muir, 1989; Zhang et al., 1999), thin isotropic layering (Backus, 1962; Helbig, 1984), lack of homogeneity (Grechka and McMechan, 2005), or even preferential orientation of olivine (Dziewonski and Anderson, 1981; Forsyth, 1975). Here, we give an introduction of the method in 3D case with the purpose of simulating seismic wave propagation in 3D heterogeneous anisotropic medium with non-flat surface topography. The chapter is organized as follows: firstly, we give a brief introduction on the boundary-conforming grid and the transformation between curvilinear coordinates and Cartesian coordinates; then we write the wave equations and free boundary conditions in these two coordinate systems; after that we introduce a numerical method to discretize both the wave equations and the free surface boundary conditions. Finally, several numerical examples are presented to demonstrate the accuracy and efficiency of the method.

2. Transformation between curvilinear and Cartesian coordinates

As to the topographic surface, the discrete grid must conform to the free surface to suppress artificial scattered waves. Such grid is named boundary-conforming grid (Hvid, 1994;

Thompson et al., 1985), and it was early used by Fornberg (1988) in seismic wave simulation with the pseudo-spectral method. A grid of this type is achieved by carrying out a transformation between the (curvilinear) computational space and the (Cartesian) physical space as illustrated in Figure 1. By means of this transformation, the curvilinear coordinates q, r and s are mapped into Cartesian coordinates within the physical space, where both systems have positive direction downward for the vertical coordinate. A boundary in the physical space presents a constant value of one of the curvilinear coordinates-be it a curve in two dimensions or a surface in three dimensions.

Boundary conforming grids may be of two fundamentally different types: structured and unstructured (or irregular) grids. A structured type grid (Figure 1) is characterized by having a fixed number of elements along each of the coordinate directions. The general element is a hexahedron in 3D, just as in the left panel of Figure 1. Neighboring elements in the physical space are also adjacent in the computational space, which is one of the great advantages of this type of grid. This property makes it relatively simple to implement in a computer. Structured grids are mainly used in finite difference and finite volume solvers. Here, we focus on structured boundary conforming grids. Several methods may be used to generate these grids, namely: Partial Differential Equation (PDE) methods, algebraic methods, co-normal mapping methods and variational methods. Here we use PDE methods (see Hvid, 1994; and Thompson et al., 1985 for details).

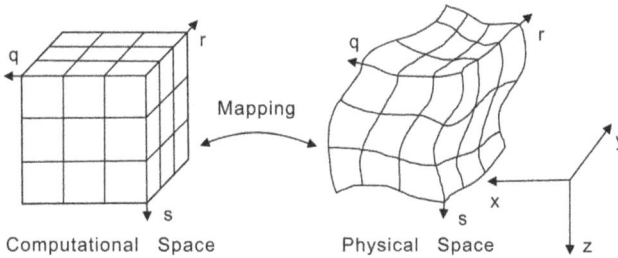

Figure 1. Mapping between computational and physical space in three dimensions (after Hvid, S., 1994).

After generating the boundary-conforming grid, the Cartesian coordinates of every grid point can be determined from the curvilinear coordinates through the equations:

$$x = x(q,r,s) \quad y = y(q,r,s) \quad z = z(q,r,s) \tag{1}$$

then, we can express the spatial derivatives in the Cartesian coordinate system (x, y, z) from the curvilinear coordinate system (q, r, s) following the chain rule:

$$\partial_x = q_x\partial_q + r_x\partial_r + s_x\partial_s \quad \partial_y = q_y\partial_q + r_y\partial_r + s_y\partial_s \quad \partial_z = q_z\partial_q + r_z\partial_r + s_z\partial_s \tag{2}$$

and similarly in other cases

$$\partial_q = x_q\partial_x + y_q\partial_y + z_q\partial_z \quad \partial_r = x_r\partial_x + y_r\partial_y + z_r\partial_z \quad \partial_s = x_s\partial_x + y_s\partial_y + z_s\partial_z \tag{3}$$

where q_x denotes $\partial q(x,y,z)/\partial x$ and the similar in other cases. These derivatives are called metric derivatives or simply the metric. We can also find the metric derivatives

$$q_x = \frac{1}{J}(y_r z_s - z_r y_s) \qquad q_y = \frac{1}{J}(z_r x_s - x_r z_s) \qquad q_z = \frac{1}{J}(x_r y_s - y_r x_s)$$

$$r_x = \frac{1}{J}(y_s z_q - z_s y_q) \qquad r_y = \frac{1}{J}(z_s x_q - x_s z_q) \qquad r_z = \frac{1}{J}(x_s y_q - y_s x_q) \qquad (4)$$

$$s_x = \frac{1}{J}(y_q z_r - z_q y_r) \qquad s_y = \frac{1}{J}(z_q x_r - x_q z_r) \qquad s_z = \frac{1}{J}(x_q y_r - y_q x_r)$$

where J is the Jacobian of the transformation that is written as

$$J = x_q y_r z_s - x_q y_s z_r - x_r y_q z_s + x_r y_s z_q + x_s y_q z_r - x_s y_r z_q$$

and whose detailed form can be found in Appendix A in Lan & Zhang (2011b).

It is worthful to note that even if the mapping equations (1) are given by an analytic function, the derivatives should still be calculated numerically to avoid spurious source terms due to the coefficients of the derivatives when the conservation form of the momentum equations are used (Thompson et al., 1985). In all examples presented in this paper the metric derivatives are computed numerically using second-order accurate finite difference approximations.

3. Elastic wave equations in Cartesian and curvilinear coordinate systems

In the following we consider a well-studied type of anisotropy in seismology, namely, a transversely isotropic medium. In the absence of external force, the elastic wave equations in the Cartesian coordinates are given by:

$$\rho\frac{\partial^2 u}{\partial t^2} = \frac{\partial}{\partial x}(c_{11}\frac{\partial u}{\partial x} + c_{12}\frac{\partial v}{\partial y} + c_{13}\frac{\partial w}{\partial z}) + \frac{\partial}{\partial y}(c_{66}\frac{\partial u}{\partial y} + c_{66}\frac{\partial v}{\partial x}) + \frac{\partial}{\partial z}(c_{44}\frac{\partial u}{\partial z} + c_{44}\frac{\partial w}{\partial x}) \quad (a)$$

$$\rho\frac{\partial^2 v}{\partial t^2} = \frac{\partial}{\partial x}(c_{66}\frac{\partial v}{\partial x} + c_{66}\frac{\partial u}{\partial y}) + \frac{\partial}{\partial y}(c_{11}\frac{\partial v}{\partial y} + c_{12}\frac{\partial u}{\partial x} + c_{13}\frac{\partial w}{\partial z}) + \frac{\partial}{\partial z}(c_{44}\frac{\partial v}{\partial z} + c_{44}\frac{\partial w}{\partial y}) \quad (b) \quad (5)$$

$$\rho\frac{\partial^2 w}{\partial t^2} = \frac{\partial}{\partial x}(c_{44}\frac{\partial w}{\partial x} + c_{44}\frac{\partial u}{\partial z}) + \frac{\partial}{\partial y}(c_{44}\frac{\partial w}{\partial y} + c_{44}\frac{\partial v}{\partial z}) + \frac{\partial}{\partial z}(c_{33}\frac{\partial w}{\partial z} + c_{13}\frac{\partial u}{\partial x} + c_{13}\frac{\partial v}{\partial y}) \quad (c)$$

where $c_{ij}(x, y, z)$ are elastic parameters and $c_{66} = 0.5(c_{11} - c_{12})$; u, v and w are the displacements in x-, y- and z-directions, respectively; $\rho(x, y, z)$ is density. Equations (5a)-(5c) are complemented by the initial data:

$$u(x,y,z,0) = u_0(x,y,z) \qquad v(x,y,z,0) = v_0(x,y,z) \qquad w(x,y,z,0) = w_0(x,y,z)$$

$$\frac{\partial u(x,y,z,0)}{\partial t} = u_1(x,y,z) \qquad \frac{\partial v(x,y,z,0)}{\partial t} = v_1(x,y,z) \qquad \frac{\partial w(x,y,z,0)}{\partial t} = w_1(x,y,z) \qquad (6)$$

Utilizing relationships (2), the wave equations (5a)-(5c) can be re-written in the curvilinear coordinate system in the following form (see Appendix B in Lan & Zhang, 2011b for details):

$$
\begin{aligned}
J\rho\frac{\partial^2 u}{\partial t^2} = {} & \frac{\partial}{\partial q}\{Jq_x[c_{11}(q_x\partial_q + r_x\partial_r + s_x\partial_s)u + c_{12}(q_y\partial_q + r_y\partial_r + s_y\partial_s)v + c_{13}(q_z\partial_q + r_z\partial_r + s_z\partial_s)w] \\
& + Jq_y[c_{66}(q_x\partial_q + r_x\partial_r + s_x\partial_s)v + c_{66}(q_y\partial_q + r_y\partial_r + s_y\partial_s)u] \\
& + Jq_z[c_{44}(q_z\partial_q + r_z\partial_r + s_z\partial_s)u + c_{44}(q_x\partial_q + r_x\partial_r + s_x\partial_s)w]\} \\
& + \frac{\partial}{\partial r}\{Jr_x[c_{11}(q_x\partial_q + r_x\partial_r + s_x\partial_s)u + c_{12}(q_y\partial_q + r_y\partial_r + s_y\partial_s)v + c_{13}(q_z\partial_q + r_z\partial_r + s_z\partial_s)w] \\
& + Jr_y[c_{66}(q_x\partial_q + r_x\partial_r + s_x\partial_s)v + c_{66}(q_y\partial_q + r_y\partial_r + s_y\partial_s)u] \\
& + Jr_z[c_{44}(q_z\partial_q + r_z\partial_r + s_z\partial_s)u + c_{44}(q_x\partial_q + r_x\partial_r + s_x\partial_s)w]\} \\
& + \frac{\partial}{\partial s}\{Js_x[c_{11}(q_x\partial_q + r_x\partial_r + s_x\partial_s)u + c_{12}(q_y\partial_q + r_y\partial_r + s_y\partial_s)v + c_{13}(q_z\partial_q + r_z\partial_r + s_z\partial_s)w] \\
& + Js_y[c_{66}(q_x\partial_q + r_x\partial_r + s_x\partial_s)v + c_{66}(q_y\partial_q + r_y\partial_r + s_y\partial_s)u] \\
& + Js_z[c_{44}(q_z\partial_q + r_z\partial_r + s_z\partial_s)u + c_{44}(q_x\partial_q + r_x\partial_r + s_x\partial_s)w]\}
\end{aligned}
\tag{7}
$$

$$
\begin{aligned}
J\rho\frac{\partial^2 v}{\partial t^2} = {} & \frac{\partial}{\partial q}\{Jq_x[c_{66}(q_x\partial_q + r_x\partial_r + s_x\partial_s)v + c_{66}(q_y\partial_q + r_y\partial_r + s_y\partial_s)u] \\
& + Jq_y[c_{11}(q_y\partial_q + r_y\partial_r + s_y\partial_s)v + c_{12}(q_x\partial_q + r_x\partial_r + s_x\partial_s)u + c_{13}(q_z\partial_q + r_z\partial_r + s_z\partial_s)w] \\
& + Jq_z[c_{44}(q_z\partial_q + r_z\partial_r + s_z\partial_s)v + c_{44}(q_y\partial_q + r_y\partial_r + s_y\partial_s)w]\} \\
& + \frac{\partial}{\partial r}\{Jr_x[c_{66}(q_x\partial_q + r_x\partial_r + s_x\partial_s)v + c_{66}(q_y\partial_q + r_y\partial_r + s_y\partial_s)u] \\
& + Jr_y[c_{11}(q_y\partial_q + r_y\partial_r + s_y\partial_s)v + c_{12}(q_x\partial_q + r_x\partial_r + s_x\partial_s)u + c_{13}(q_z\partial_q + r_z\partial_r + s_z\partial_s)w] \\
& + Jr_z[c_{44}(q_z\partial_q + r_z\partial_r + s_z\partial_s)v + c_{44}(q_y\partial_q + r_y\partial_r + s_y\partial_s)w]\} \\
& + \frac{\partial}{\partial s}\{Js_x[c_{66}(q_x\partial_q + r_x\partial_r + s_x\partial_s)v + c_{66}(q_y\partial_q + r_y\partial_r + s_y\partial_s)u] \\
& + Js_y[c_{11}(q_y\partial_q + r_y\partial_r + s_y\partial_s)v + c_{12}(q_x\partial_q + r_x\partial_r + s_x\partial_s)u + c_{13}(q_z\partial_q + r_z\partial_r + s_z\partial_s)w] \\
& + Js_z[c_{44}(q_z\partial_q + r_z\partial_r + s_z\partial_s)v + c_{44}(q_y\partial_q + r_y\partial_r + s_y\partial_s)w]\}
\end{aligned}
\tag{8}
$$

$$
\begin{aligned}
J\rho\frac{\partial^2 w}{\partial t^2} = {} & \frac{\partial}{\partial q}\{Jq_x[c_{44}(q_z\partial_q + r_z\partial_r + s_z\partial_s)u + c_{44}(q_x\partial_q + r_x\partial_r + s_x\partial_s)w] \\
& + Jq_y[c_{44}(q_z\partial_q + r_z\partial_r + s_z\partial_s)v + c_{44}(q_y\partial_q + r_y\partial_r + s_y\partial_s)w] \\
& + Jq_z[c_{33}(q_z\partial_q + r_z\partial_r + s_z\partial_s)w + c_{13}(q_x\partial_q + r_x\partial_r + s_x\partial_s)u + c_{13}(q_y\partial_q + r_y\partial_r + s_y\partial_s)v]\} \\
& + \frac{\partial}{\partial r}\{Jr_x[c_{44}(q_z\partial_q + r_z\partial_r + s_z\partial_s)u + c_{44}(q_x\partial_q + r_x\partial_r + s_x\partial_s)w] \\
& + Jr_y[c_{44}(q_z\partial_q + r_z\partial_r + s_z\partial_s)v + c_{44}(q_y\partial_q + r_y\partial_r + s_y\partial_s)w] \\
& + Jr_z[c_{33}(q_z\partial_q + r_z\partial_r + s_z\partial_s)w + c_{13}(q_x\partial_q + r_x\partial_r + s_x\partial_s)u + c_{13}(q_y\partial_q + r_y\partial_r + s_y\partial_s)v]\} \\
& + \frac{\partial}{\partial s}\{Js_x[c_{44}(q_z\partial_q + r_z\partial_r + s_z\partial_s)u + c_{44}(q_x\partial_q + r_x\partial_r + s_x\partial_s)w] \\
& + Js_y[c_{44}(q_z\partial_q + r_z\partial_r + s_z\partial_s)v + c_{44}(q_y\partial_q + r_y\partial_r + s_y\partial_s)w] \\
& + Js_z[c_{33}(q_z\partial_q + r_z\partial_r + s_z\partial_s)w + c_{13}(q_x\partial_q + r_x\partial_r + s_x\partial_s)u + c_{13}(q_y\partial_q + r_y\partial_r + s_y\partial_s)v]\}
\end{aligned}
\tag{9}
$$

4. Free boundary conditions in the Cartesian and curvilinear coordinate systems

At the free surface, the boundary conditions in the Cartesian coordinates are given by:

$$
\begin{bmatrix}
c_{11}\dfrac{\partial u}{\partial x}+c_{12}\dfrac{\partial v}{\partial y}+c_{13}\dfrac{\partial w}{\partial z} & c_{66}\dfrac{\partial u}{\partial y}+c_{66}\dfrac{\partial v}{\partial x} & c_{44}\dfrac{\partial u}{\partial z}+c_{44}\dfrac{\partial w}{\partial x} \\[2mm]
c_{66}\dfrac{\partial v}{\partial x}+c_{66}\dfrac{\partial u}{\partial y} & c_{11}\dfrac{\partial v}{\partial y}+c_{12}\dfrac{\partial u}{\partial x}+c_{13}\dfrac{\partial w}{\partial z} & c_{44}\dfrac{\partial v}{\partial z}+c_{44}\dfrac{\partial w}{\partial y} \\[2mm]
c_{44}\dfrac{\partial w}{\partial x}+c_{44}\dfrac{\partial u}{\partial z} & c_{44}\dfrac{\partial w}{\partial y}+c_{44}\dfrac{\partial v}{\partial z} & c_{33}\dfrac{\partial w}{\partial z}+c_{13}\dfrac{\partial u}{\partial x}+c_{13}\dfrac{\partial v}{\partial y}
\end{bmatrix}
\begin{bmatrix} n_x \\ n_y \\ n_z \end{bmatrix}=0 \quad (10)
$$

Here $\begin{bmatrix} n_x,n_y,n_z \end{bmatrix}^{T}$ is the inward normal of the free surface. Using relationships (2), the above boundary conditions in the curvilinear coordinates can be re-written as

$$
\begin{aligned}
&\bar{s}_x\Big[c_{11}(q_x u_q+r_x u_r+s_x u_s)+c_{12}(q_y v_q+r_y v_r+s_y v_s)+c_{13}(q_z w_q+r_z w_r+s_z w_s)\Big] \\
&+\bar{s}_y\Big[c_{66}(q_y u_q+r_y u_r+s_y u_s)+c_{66}(q_x v_q+r_x v_r+s_x v_s)\Big] \\
&+\bar{s}_z\Big[c_{44}(q_z u_q+r_z u_r+s_z u_s)+c_{44}(q_x w_q+r_x w_r+s_x w_s)\Big]=0
\end{aligned}
\quad (11)
$$

$$
\begin{aligned}
&\bar{s}_x\Big[c_{66}(q_x v_q+r_x v_r+s_x v_s)+c_{66}(q_y u_q+r_y u_r+s_y u_s)\Big] \\
&+\bar{s}_y\Big[c_{11}(q_y v_q+r_y v_r+s_y v_s)+c_{12}(q_x u_q+r_x u_r+s_x u_s)+c_{13}(q_z w_q+r_z w_r+s_z w_s)\Big] \\
&+\bar{s}_z\Big[c_{44}(q_z v_q+r_z v_r+s_z v_s)+c_{44}(q_y w_q+r_y w_r+s_y w_s)\Big]=0,
\end{aligned}
\quad (12)
$$

$$
\begin{aligned}
&\bar{s}_x\Big[c_{44}(q_x w_q+r_x w_r+s_x w_s)+c_{44}(q_z u_q+r_z u_r+s_z u_s)\Big] \\
&+\bar{s}_y\Big[c_{44}(q_y w_q+r_y w_r+s_y w_s)+c_{44}(q_z v_q+r_z v_r+s_z v_s)\Big] \\
&+\bar{s}_z\Big[c_{33}(q_z w_q+r_z w_r+s_z w_s)+c_{13}(q_x u_q+r_x u_r+s_x u_s)+c_{13}(q_y v_q+r_y v_r+s_y v_s)\Big]=0
\end{aligned}
\quad (13)
$$

Note that here the normal is represented by the normalized metric (evaluated along the free surface)

$$
\bar{s}_x=\frac{s_x}{\sqrt{s_x^2+s_y^2+s_z^2}},\quad \bar{s}_y=\frac{s_y}{\sqrt{s_x^2+s_y^2+s_z^2}},\quad \bar{s}_z=\frac{s_z}{\sqrt{s_x^2+s_y^2+s_z^2}}.
$$

5. A discretization scheme on curvilinear grid

To approximate (7)-(9) we discretize the rectangular solid (Figure 2)

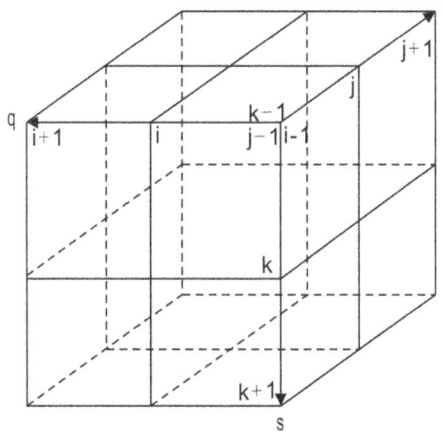

Figure 2. Grids distributions in curvilinear coordinate.

$$q_i = (i - 1)h_q \quad i = 1,...,N_q \quad h_q = l/(N_q - 1);$$
$$r_j = (j - 1)h_r \quad j = 1,...,N_r \quad h_r = w/(N_r - 1); \tag{14}$$
$$s_k = (k - 1)h_s \quad k = 1,...,N_s \quad h_s = h/(N_s - 1)$$

where l, w, h are the length of the rectangular solid in q-, r- and s-directions, respectively; $h_q, h_r, h_s > 0$ define the grid size in q-, r- and s-directions, respectively. The three components of the wave field are given by

$$[u_{i,j,k}(t), v_{i,j,k}(t), w_{i,j,k}(t)] = [u(q_i, r_j, s_k, t), v(q_i, r_j, s_k, t), w(q_i, r_j, s_k, t)]$$

and the derivation operators as

$$D_+^q u_{i,j,k} = \frac{u_{i+1,j,k} - u_{i,j,k}}{h_q}$$

$$D_-^q u_{i,j,k} = D_+^q u_{i-1,j,k} \qquad D_0^q u_{i,j,k} = \frac{1}{2}(D_+^q u_{i,j,k} + D_-^q u_{i,j,k})$$

$$D_+^r u_{i,j,k} = \frac{u_{i,j+1,k} - u_{i,j,k}}{h_r}$$

$$D_-^r u_{i,j,k} = D_+^r u_{i,j-1,k} \qquad D_0^r u_{i,j,k} = \frac{1}{2}(D_+^r u_{i,j,k} + D_-^r u_{i,j,k}) \tag{15}$$

$$D_-^s u_{i,j,k} = D_+^s u_{i,j,k-1} \qquad D_0^s u_{i,j,k} = \frac{1}{2}(D_+^s u_{i,j,k} + D_-^s u_{i,j,k})$$

$$D_+^s u_{i,j,k} = \frac{u_{i,j,k+1} - u_{i,j,k}}{h_s}$$

The right hand sides of eqs. (7)-(9) contain spatial derivatives of nine basic types, which are discretized according to the following equations

$$\frac{\partial}{\partial q}(a\omega_q) \approx D_-^q(E_{1/2}^q(a)D_+^q\omega) \quad \frac{\partial}{\partial q}(b\omega_r) \approx D_0^q(bD_0^r\omega) \quad \frac{\partial}{\partial q}(c\omega_s) \approx D_0^q(c\widetilde{D_0^s}\omega)$$

$$\frac{\partial}{\partial r}(d\omega_q) \approx D_0^r(dD_0^q\omega) \quad \frac{\partial}{\partial r}(e\omega_r) \approx D_-^r(E_{1/2}^r(e)D_+^r\omega) \quad \frac{\partial}{\partial r}(f\omega_s) \approx D_0^r(f\widetilde{D_0^s}\omega) \qquad (16)$$

$$\frac{\partial}{\partial s}(g\omega_q) \approx \widetilde{D_0^s}(gD_0^q\omega) \quad \frac{\partial}{\partial s}(m\omega_r) \approx \widetilde{D_0^s}(mD_0^r\omega) \quad \frac{\partial}{\partial s}(p\omega_s) \approx D_-^s(E_{1/2}^s(p)D_+^s\omega)$$

Here ω represents u, v or w; a, b, c, d, e, f, g, m and p are combinations of metric and material coefficients. We introduce the following averaging operators:

$$E_{1/2}^q(\gamma_{i,j,k}) = \gamma_{i+1/2,j,k} = \frac{\gamma_{i,j,k} + \gamma_{i+1,j,k}}{2}$$

$$E_{1/2}^r(\gamma_{i,j,k}) = \gamma_{i,j+1/2,k} = \frac{\gamma_{i,j,k} + \gamma_{i,j+1,k}}{2} \qquad (17)$$

$$E_{1/2}^s(\gamma_{i,j,k}) = \gamma_{i,j,k+1/2} = \frac{\gamma_{i,j,k} + \gamma_{i,j,k+1}}{2}$$

The cross terms which contain a normal derivative on the boundary are discretized one-sided in the direction normal to the boundary:

$$\widetilde{D_0^s}u_{i,j,k} = \begin{cases} D_+^s u_{i,j,k}, & k = 1, \\ D_0^s u_{i,j,k}, & k \geq 2. \end{cases} \qquad (18)$$

5.1. A discretization on curvilinear grid: Elastic wave equations

We approximate the spatial operators in eqs. (7)-(9) by (16). After suppressing grid indexes, this leads to

$$J\rho\frac{\partial^2 u}{\partial t^2} = D_-^q[E_{1/2}^q(M_1^{qq})D_+^q u + E_{1/2}^q(M_2^{qq})D_+^q v + E_{1/2}^q(M_3^{qq})D_+^q w]$$

$$+D_0^q[M_1^{qs}\widetilde{D_0^s}u + M_2^{qs}\widetilde{D_0^s}v + M_3^{qs}\widetilde{D_0^s}w]$$

$$+D_0^r[M_1^{rs}\widetilde{D_0^s}u + M_2^{rs}\widetilde{D_0^s}v + M_3^{rs}\widetilde{D_0^s}w]$$

$$+\widetilde{D_0^s}[M_1^{sq}D_0^q u + M_2^{sq}D_0^q v + M_3^{sq}D_0^q w]$$

$$+\widetilde{D_0^s}[M_1^{sr}D_0^r u + M_2^{sr}D_0^r v + M_3^{sr}D_0^r w] \qquad (19)$$

$$+D_0^q[M_1^{qr}D_0^r u + M_2^{qr}D_0^r v + M_3^{qr}D_0^r w]$$

$$+D_0^r[M_1^{rq}D_0^q u + M_2^{rq}D_0^q v + M_3^{rq}D_0^q w]$$

$$+D_-^r[E_{1/2}^r(M_1^{rr})D_+^r u + E_{1/2}^r(M_2^{rr})D_+^r v + E_{1/2}^r(M_3^{rr})D_+^r w]$$

$$+D_-^s[E_{1/2}^s(M_1^{ss})D_+^s u + E_{1/2}^s(M_2^{ss})D_+^s v + E_{1/2}^s(M_3^{ss})D_+^s w]$$

$$\equiv L^{(u)}(u, v, w)$$

$$J\rho\frac{\partial^2 v}{\partial t^2} = D_-^q[E_{1/2}^q(M_5^{qq})D_+^q v + E_{1/2}^q(M_2^{qq})D_+^q u + E_{1/2}^q(M_4^{qq})D_+^q w]$$

$$+ D_0^q[M_5^{qs}\widetilde{D_0^s}v + M_2^{sq}\widetilde{D_0^s}u + M_4^{qs}\widetilde{D_0^s}w]$$

$$+ D_0^r[M_5^{rs}\widetilde{D_0^s}v + M_2^{sr}\widetilde{D_0^s}u + M_4^{rs}\widetilde{D_0^s}w]$$

$$+ \widetilde{D_0^s}[M_5^{sq}D_0^q v + M_2^{qs}D_0^q u + M_4^{sq}D_0^q w]$$

$$+ \widetilde{D_0^s}[M_5^{sr}D_0^r v + M_2^{rs}D_0^r u + M_4^{sr}D_0^r w] \qquad (20)$$

$$+ D_0^q[M_5^{qr}D_0^r v + M_2^{rq}D_0^r u + M_4^{qr}D_0^r w]$$

$$+ D_0^r[M_5^{rq}D_0^q v + M_2^{qr}D_0^q u + M_4^{rq}D_0^q w]$$

$$+ D_-^r[E_{1/2}^r(M_5^{rr})D_+^r v + E_{1/2}^r(M_2^{rr})D_+^r u + E_{1/2}^r(M_4^{rr})D_+^r w]$$

$$+ D_-^s[E_{1/2}^s(M_5^{ss})D_+^s v + E_{1/2}^s(M_2^{ss})D_+^s u + E_{1/2}^s(M_4^{ss})D_+^s w]$$

$$\equiv L^{(v)}(u,v,w)$$

$$J\rho\frac{\partial^2 w}{\partial t^2} = D_-^q[E_{1/2}^q(M_3^{qq})D_+^q u + E_{1/2}^q(M_4^{qq})D_+^q v + E_{1/2}^q(M_6^{qq})D_+^q w]$$

$$+ D_0^q[M_3^{sq}\widetilde{D_0^s}u + M_4^{sq}\widetilde{D_0^s}v + M_6^{qs}\widetilde{D_0^s}w]$$

$$+ D_0^r[M_3^{sr}\widetilde{D_0^s}u + M_4^{sr}\widetilde{D_0^s}v + M_6^{rs}\widetilde{D_0^s}w]$$

$$+ \widetilde{D_0^s}[M_3^{qs}D_0^q u + M_4^{qs}D_0^q v + M_6^{sq}D_0^q w]$$

$$+ \widetilde{D_0^s}[M_3^{rs}D_0^r u + M_4^{rs}D_0^r v + M_6^{sr}D_0^r w] \qquad (21)$$

$$+ D_0^q[M_3^{rq}D_0^r u + M_4^{rq}D_0^r v + M_6^{qr}D_0^r w]$$

$$+ D_0^r[M_3^{qr}D_0^q u + M_4^{qr}D_0^q v + M_6^{rq}D_0^q w]$$

$$+ D_-^r[E_{1/2}^r(M_3^{rr})D_+^r u + E_{1/2}^r(M_4^{rr})D_+^r v + E_{1/2}^r(M_6^{rr})D_+^r w]$$

$$+ D_-^s[E_{1/2}^s(M_3^{ss})D_+^s u + E_{1/2}^s(M_4^{ss})D_+^s v + E_{1/2}^s(M_6^{ss})D_+^s w]$$

$$\equiv L^{(w)}(u,v,w)$$

in the grid points (q_i, r_j, s_k), $(i,j,k) \in [1, N_q] \times [1, N_r] \times [1, N_s]$. We have introduced the following notations for the material and metric terms in order to express the discretized equations in a more compact form:

$$M_1^{kl} = Jk_x l_x c_{11} + Jk_y l_y c_{66} + Jk_z l_z c_{44} \quad M_2^{kl} = Jk_x l_y c_{12} + Jk_y l_x c_{66}$$

$$M_3^{kl} = Jk_x l_z c_{13} + Jk_z l_x c_{44} \quad M_4^{kl} = Jk_y l_z c_{13} + Jk_z l_y c_{44} \qquad (22)$$

$$M_5^{kl} = Jk_x l_x c_{66} + Jk_y l_y c_{11} + Jk_z l_z c_{44} \quad M_6^{kl} = Jk_x l_x c_{44} + Jk_y l_y c_{44} + Jk_z l_z c_{33}$$

where k and l represent the metric coefficients q, r or s.

We discretize in time using second-order accurate centered differences. The full set of discretized equations is

$$\rho\left(\frac{u^{n+1} - 2u^n + u^{n-1}}{\delta_t^2}\right) = L^{(u)}(u^n, v^n, w^n)$$

$$\rho\left(\frac{v^{n+1} - 2v^n + v^{n-1}}{\delta_t^2}\right) = L^{(v)}(u^n, v^n, w^n) \tag{23}$$

$$\rho\left(\frac{w^{n+1} - 2w^n + w^{n-1}}{\delta_t^2}\right) = L^{(w)}(u^n, v^n, w^n)$$

where δ_t represents the time step.

5.2. A discretization on curvilinear grid: Free boundary conditions

The boundary conditions (11)-(13) are discretized by

$$\frac{1}{2}[(M_1^{ss})_{i,j,3/2}D_+^s u_{i,j,1} + (M_1^{ss})_{i,j,1/2}D_+^s u_{i,j,0}] + (M_1^{sq})_{i,j,1}D_0^q u_{i,j,1} + (M_2^{sq})_{i,j,1}D_0^q v_{i,j,1}$$

$$+ (M_3^{sq})_{i,j,1}D_0^q w_{i,j,1} + \frac{1}{2}[(M_2^{ss})_{i,j,3/2}D_+^s v_{i,j,1} + (M_2^{ss})_{i,j,1/2}D_+^s v_{i,j,0}] + (M_1^{sr})_{i,j,1}D_0^r u_{i,j,1} \tag{24}$$

$$+ (M_2^{sr})_{i,j,1}D_0^r v_{i,j,1} + (M_3^{sr})_{i,j,1}D_0^r w_{i,j,1} + \frac{1}{2}[(M_3^{ss})_{i,j,3/2}D_+^s w_{i,j,1} + (M_3^{ss})_{i,j,1/2}D_+^s w_{i,j,0}] = 0$$

$$\frac{1}{2}[(M_5^{ss})_{i,j,3/2}D_+^s v_{i,j,1} + (M_5^{ss})_{i,j,1/2}D_+^s v_{i,j,0}] + (M_5^{sq})_{i,j,1}D_0^q v_{i,j,1} + (M_2^{qs})_{i,j,1}D_0^q u_{i,j,1}$$

$$+ (M_4^{sq})_{i,j,1}D_0^q w_{i,j,1} + \frac{1}{2}[(M_2^{ss})_{i,j,3/2}D_+^s u_{i,j,1} + (M_2^{ss})_{i,j,1/2}D_+^s u_{i,j,0}] + (M_5^{sr})_{i,j,1}D_0^r v_{i,j,1} \tag{25}$$

$$+ (M_2^{rs})_{i,j,1}D_0^r u_{i,j,1} + (M_4^{sr})_{i,j,1}D_0^r w_{i,j,1} + \frac{1}{2}[(M_4^{ss})_{i,j,3/2}D_+^s w_{i,j,1} + (M_4^{ss})_{i,j,1/2}D_+^s w_{i,j,0}] = 0$$

$$\frac{1}{2}[(M_3^{ss})_{i,j,3/2}D_+^s u_{i,j,1} + (M_3^{ss})_{i,j,1/2}D_+^s u_{i,j,0}] + (M_3^{qs})_{i,j,1}D_0^q u_{i,j,1} + (M_4^{qs})_{i,j,1}D_0^q v_{i,j,1}$$

$$+ (M_6^{sq})_{i,j,1}D_0^q w_{i,j,1} + \frac{1}{2}[(M_4^{ss})_{i,j,3/2}D_+^s v_{i,j,1} + (M_4^{ss})_{i,j,1/2}D_+^s v_{i,j,0}] + (M_3^{sr})_{i,j,1}D_0^r u_{i,j,1} \tag{26}$$

$$+ (M_4^{rs})_{i,j,1}D_0^r v_{i,j,1} + (M_6^{sr})_{i,j,1}D_0^r w_{i,j,1} + \frac{1}{2}[(M_6^{ss})_{i,j,3/2}D_+^s w_{i,j,1} + (M_6^{ss})_{i,j,1/2}D_+^s w_{i,j,0}] = 0$$

$$i = 1, ..., N_q; j = 1, ..., N_r.$$

The key step in obtaining a stable explicit discretization is to use the operator $\widetilde{D_0^s}$ (which is one-sided on the boundary) for the approximation of the normal derivative in $\partial_q\partial_s, \partial_r\partial_s, \partial_s\partial_q$ and $\partial_s\partial_r$ cross derivatives. At first glance, it may appear that using a one-sided operator the accuracy of the method would be reduced to the first-order. However, as it was theoretically shown by Nilsson et al. (2007) (for a Cartesian discretization), a first-order error on the boundary in the differential equations (19)-(21) can be absorbed as a second-order perturbation of the boundary conditions (24)-(26).

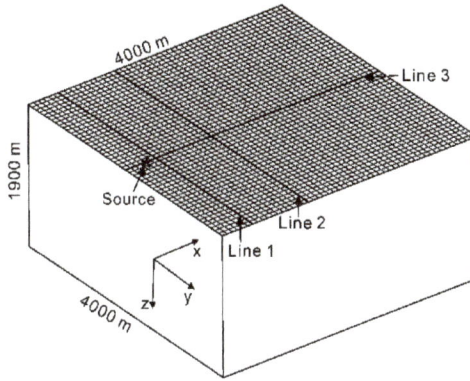

Figure 3. Model of a half-space with a planar free surface.

6. Accuracy and efficiency tests

6.1. Accuracy

The accuracy of the proposed method is examined by comparing numerical results with the analytical solution of the Lamb's problem, for a transversely isotropic medium with a vertical symmetry axis (VTI medium). The elastic parameters describing the VTI medium are given in Table 1. The analytical solution is obtained by convolving the free-surface Green-function with the source function (Payton, 1983). A vertical point source of the type

c_{11} (GPa)	c_{12} (GPa)	c_{13} (GPa)	c_{33} (GPa)	c_{44} (GPa)	ρ (g/cm^3)
25.5	2.0	14.0	18.4	5.6	2.4

Table 1. Medium parameters in the homogeneous half-space

$$f(t) = e^{-0.5f_0^2(t-t_0)^2} \cos\pi f_0(t - t_0) \tag{27}$$

with $t_0 = 0.5$ s and a high cut-off frequency $f_0 = 10$ Hz, is assumed to be located at (300 m, 2000 m) at the surface, which is marked as an asterisk in Figure 3. It should be mentioned that Carcione et al. (1992) and Carcione (2000) presented an analytical comparison of the point-source response in a 3-D VTI medium in the absence of the free surface. The comparisons are performed by first transforming the 3-D numerical results into a line-source response by carrying out an integration along the receiver line (Wapenaar et al., 1992) and then comparing the emerging results with the 2-D Lamb's analytical solutions. The numerical model contains 401 x 401 x 191 grid nodes in the x-, y- and z-direction, respectively. The grid spacings are 10 m in all directions. The solution is advanced using a time step of 1.25 ms for 3.5 s.

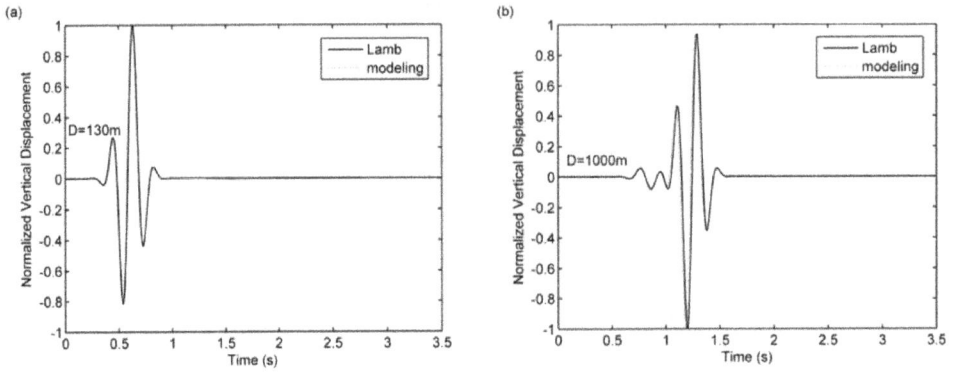

Figure 4. Comparison between numerical and analytical vertical components of the displacement for the VTI medium.

Three receiver lines are positioned on the free surface, two of which are parallel to the y-direction with respective normal distances of 130 (Line 1) and 1000 m (Line 2) away from the point source, the other crosses the source location and parallels to the x-direction (Line 3). The integrations are performed along the first two receiver lines, these represent 2-D results of 130 and 1000 m away from the source, respectively. Figure 4 shows the comparisons between the resulting numerical and 2-D analytical z-components of the displacement for the VTI medium. In spite of the errors resulting from the transformation of the point-source response into the line-source one, numerical and analytical results agree well in Figure 4. These comparisons demonstrate the accuracy of our corresponding algorithm.

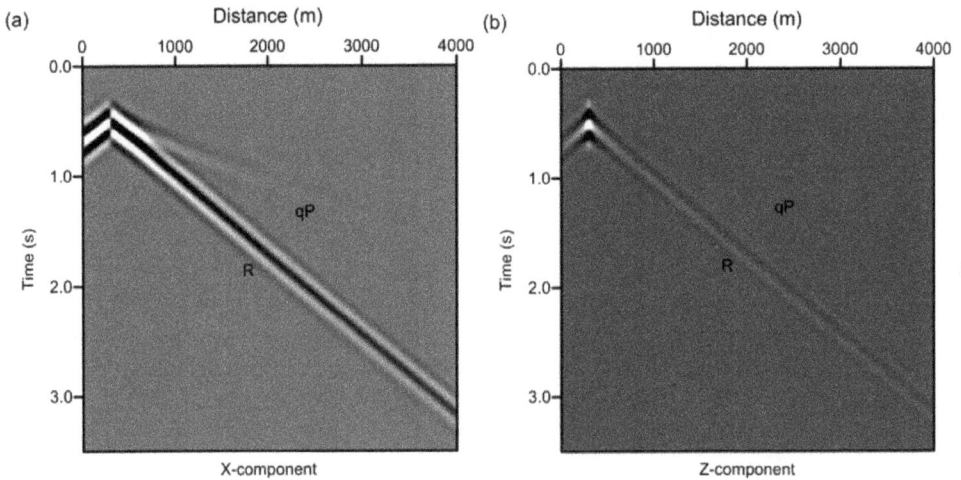

Figure 5. Seismogram sections at Line 3 for the planar surface model. Symbol qP indicates the qP wave and R indicates Rayleigh wave.

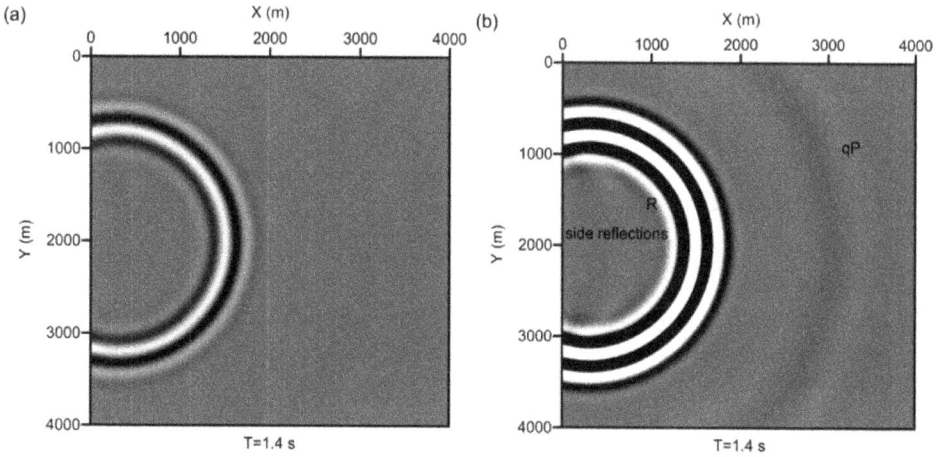

Figure 6. Snapshots of the vertical component of the wavefield at the surface (xy-plane) of the planar surface model.

Synthetic seismograms are computed at Line 3. The seismograms in Figure 5 show the direct quasi-P wave (qP) and a high-amplitude Rayleigh wave (R). Snapshots of the vertical component of the wavefield in horizontal (xy-) plane at the propagation time of 1.4 s are displayed in Figure 6. We define the incidence plane by the propagation direction and the z-axis, quasi-P wave and quasi-SV wave (qSV) motions lie in this plane, while SH motion is normal to the plane. Hence, the z-component does not contain SH motion. The xy-plane of a transversely isotropic medium is a plane of isotropy, where the velocity of the qP wave is about 3260 ms^{-1} and the velocity of the qSV wave is about 1528 ms^{-1}. The amplitude of the qP wave is so weak compared with that of the Rayleigh and qSV wave that one can hardly identify it in the snapshot (Figure 6a). In order to observe the qP wave, a gain has been given to the amplitudes of the wavefield. Owing to this, side reflections also appear in the photo, as shown in Figure 6b. As the velocity of the Rayleigh wave is very close to that of the qSV wave, the two waves are almost superimposed and it is difficult to distinguish between the two in synthetic seismograms and snapshots.

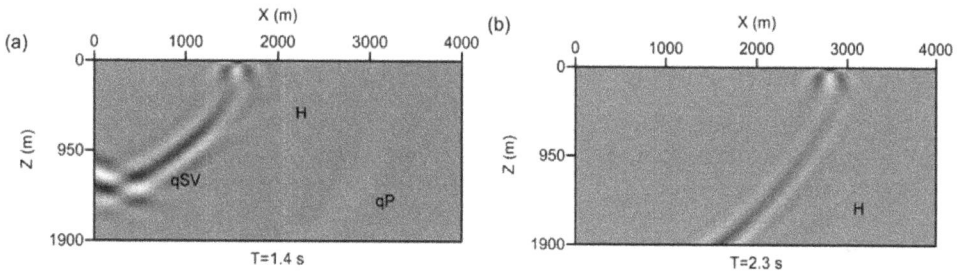

Figure 7. Snapshots of the x-component of the wavefield in the vertical (xz-) plane which contains the receiver line and the source at 1.4 (a) and 2.3 s (b) propagation times for the planar surface model.

Figure 7 shows the x-component of the wavefield in the vertical (xz-) plane at 1.4 and 2.3 s propagation times. The xz-plane contains the receiver line (Line 3) and the source location. Both snapshots show the wave front of the qP-wave and the qSV-wave. The former snapshot (1.4 s) shows the qSV-wave with the cusps. A headwave (H) can also be found in the photos, the headwave is a quasi-shear wave and is guided along the surface by the qP-wave.

6.2. Numerical simulations on an irregular (non-flat) free surface

Three numerical experiments with irregular free surfaces are now investigated. The first example is a test on smooth boundaries, while the second example consists of a hemispherical depression to test the ability of the method for non-smooth topography. For sake of simplicity both these examples are based on homogeneous half-spaces, i.e., the medium parameters are the same as in the case of flat surface (Table 1). The same source is located at the same place as in the planar surface model, the time step is 0.8 ms. The total propagation time is 3.5 s for the two models. Finally, we consider a two-layered model with a realistic topography.

6.2.1. Topography simulating a shaped Gaussian hill

The first model considered here is a half-space whose free surface is a hill-like feature (Figure 8). The shape of the hill resembles a Gaussian curved surface given by the function

$$z(x,y) = -150exp(-(\frac{x-2000}{150})^2 - (\frac{y-2000}{150})^2)m \quad (x,y) \in [0m, 4000m]^2 \qquad (28)$$

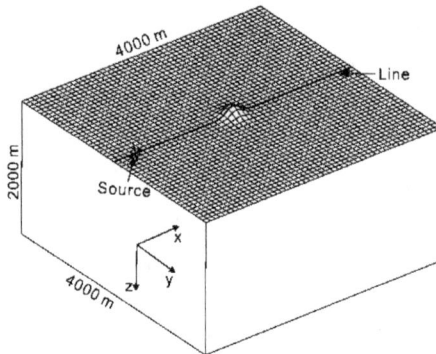

Figure 8. Model of a half-space with Gaussian shape hill topography.

The computational domain extends to depth z(x, y)=2000 m. The volume is discretized with equal grid nodes in each direction as in the planar surface one. The grid spacings are 10 m in the x, y-directions and about 10.5 m in the z-direction for average. The vertical spacing varies with depth, it is smaller toward the free surface and larger toward the bottom of the model. The minimum and maximum of the vertical spacings are 6 and 12 m, respectively.

The gridding scheme which shows the detailed cross-section of the grids along Line3 is shown in Figure 9. Synthetic seismograms are also computed at Line3 (Figure 10). As a result of the hill-shaped free surface (and compare with the synthetic seismograms in Figure 5), the amplitudes of the quasi-P wave and Rayleigh wave are reduced in the right-hand part of the sections. In addition, after the ordinary quasi-P wave a secondary quasi-P wave (RqPf) induced by the scattering of the direct Rayleigh wave can be observed. Similarly, a secondary Rayleigh wave (qPRf) which travels in front of the ordinary Rayleigh wave induced by the scattering of the direct quasi-P wave can also be distinguished. Some energy is scattered back to the left-hand side as a Rayleigh wave (qPRb, RR) and a quasi-P wave (RqPb).

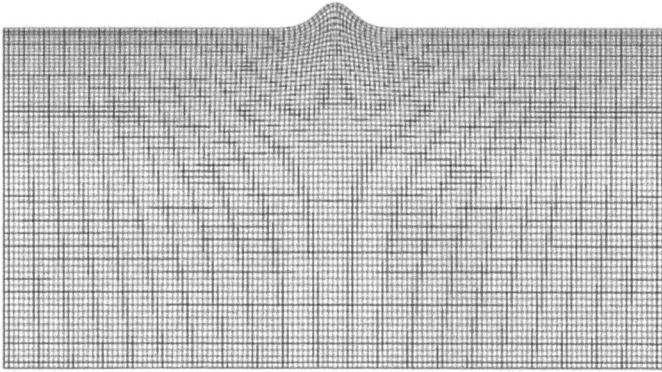

Figure 9. The gridding scheme which shows the detailed cross-section of the grids along Line3 in the Gaussian shape hill topography model. For clarity, the grids are displayed with a reducing density factor of 3.

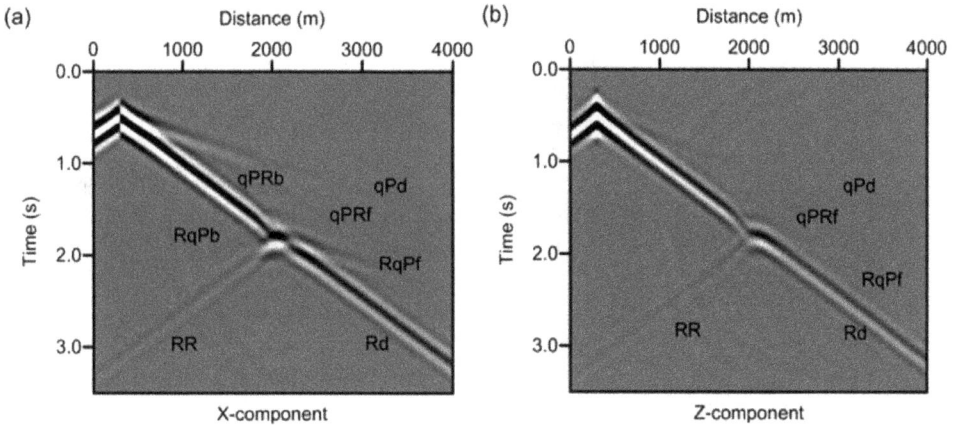

Figure 10. Seismograms along the receiver line for the Gaussian shape hill topography model: (a) x-component (horizontal) of the displacement; (b) z-component (vertical). Symbols mean the following: (qPd) qP wave diffracts to qP wave; (Rd) Rayleigh wave diffracts to Rayleigh wave; (qPRf) qP wave scatters to Rayleigh wave and propagates forward; (qPRb) qP wave scatters to Rayleigh wave and propagates backward; (RqPf) Rayleigh wave scatters to qP wave and propagates forward; (RqPb) Rayleigh wave scatters to qP wave and propagates backward; (RR) Rayleigh wave reflectes to Rayleigh wave.

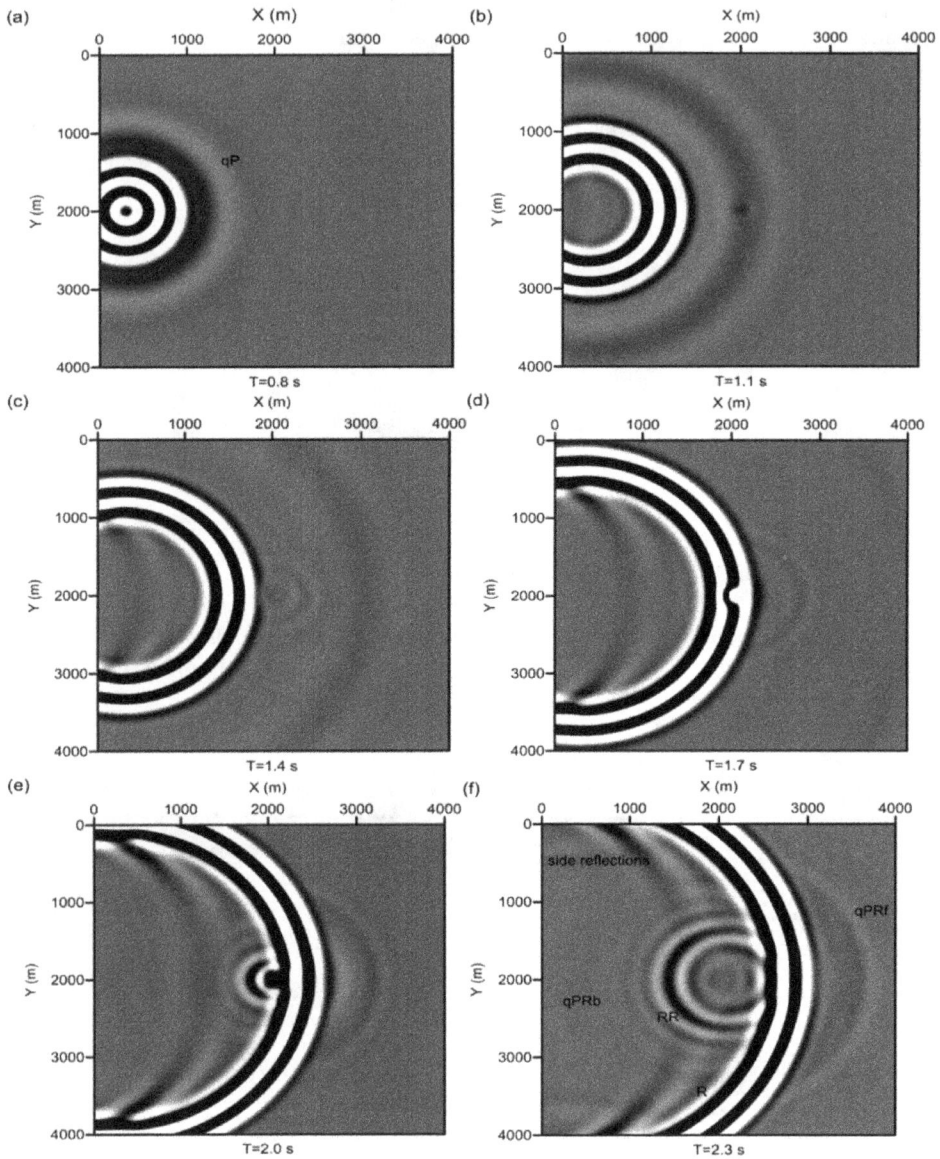

Figure 11. Snapshots of the vertical component of the wavefield at the surface (xy-plane) at different propagation times for the Gaussian shape hill topography model.

Snapshots of the wavefield in the horizontal (xy-) plane at different propagation times are displayed in Figures 11. The amplitudes are also gained. In the beginning the wavefield propagates undisturbed along the free surface. At 1.1 s the direct quasi-P wave hits the hill and generates a circular diffracted wave. This wave is a Rayleigh wave, which is marked as two parts, one travels forward (qPRf), and the other travels backward (qPRb). These can be seen clearly in the later snapshots (1.4 - 2.3 s). In addition, a reflected Rayleigh wave (RR) can be observed. The direct quasi-P wave (qP) and Rayleigh wave (R) are also marked in the figure. By the way, side reflections from the boundaries can also be noted in the plane. Figure 12 shows the x-component of the wavefield in the vertical (xz-) plane. The xz-plane contains the receiver line and source location. The snapshots show the diffracted quasi-P and quasi-SV waves clearly in the vertical plane.

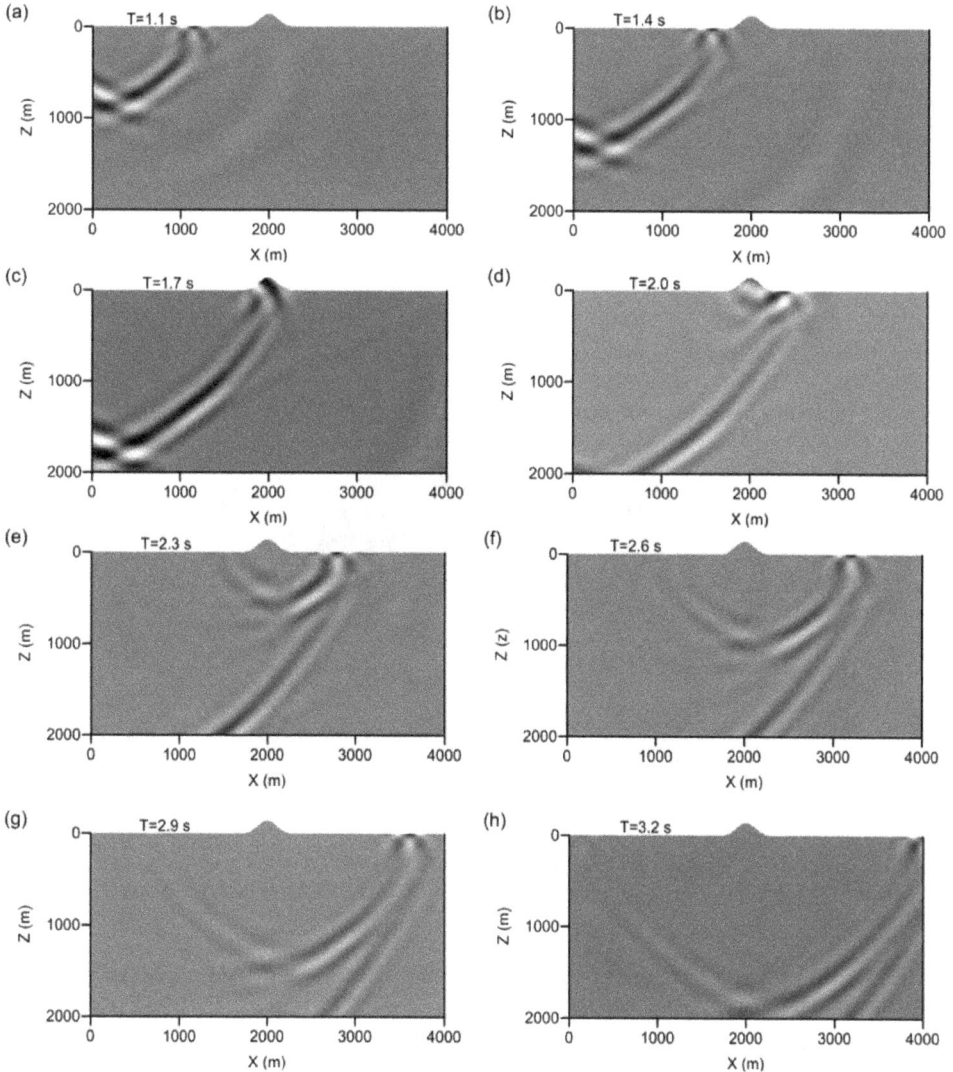

Figure 12. Snapshots of the x-component of the wavefield in the vertical (xz-) plane which contains the receiver line and the source at different propagation times for the Gaussian shape hill topography model.

6.2.2. Topography simulating a shaped hemispherical depression

In the second model, we consider a hemispherical depression model as illustrated in Figure 13. The first model that we have considered is of smooth topography, that is, with continuous and finite slopes everywhere. However, the shaped hemispherical depression here taken as reference is a case of extreme topography, such that the vertical-to-horizontal ratio of the depression is very large (1:2) and the slopes of the edges tend to infinity. The hemispherical depression is at the middle of the free surface and the radius is 150 m.

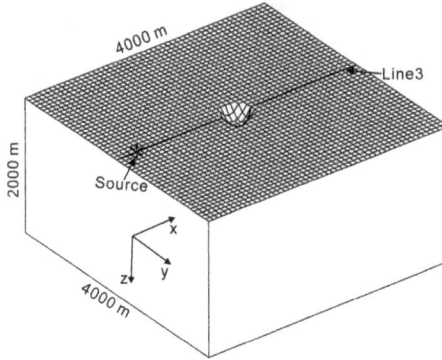

Figure 13. Model of a half-space with hemispherical shape depression topography.

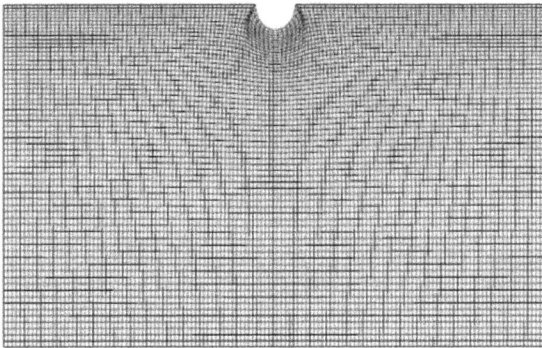

Figure 14. The gridding scheme which shows the detailed cross-section of the grids along Line3 in the hemispherical shape depression topography model. For clarity, the grids are displayed with a reducing density factor of 3.

The numerical model is discretized in the same way as in the hill topography model. The gridding scheme which shows the detailed cross-section of the grids along Line3 is shown in Figure 14. Owing to the existence of model edges with strong slopes at x=1850 and x=2150 m along the receiver line, both body and Rayleigh waves scattered by sharp changes in the topography can be clearly observed on the synthetic seismograms shown in Figure 15. Owing to its shorter wavelength, the scattering of Rayleigh wave is much stronger than that of the body wave when propagating through the hemispherical depression, this indicating that such sharp depression can affect the propagation of Rayleigh wave significantly.

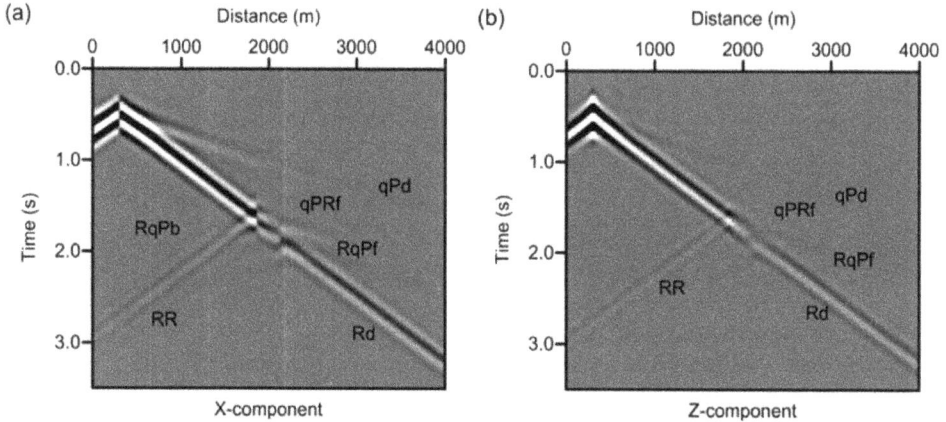

Figure 15. Seismograms at the receiver line for the hemispherical shape depression topography model: (a) x-component (horizontal) of the displacement; (b) z-component (vertical). Symbols mean the same as in Fig. 10.

The photos in Figure 16 show the vertical component of the wavefield in the horizontal (xy-) plane. Compared with the photos of the hill topography model, we can see the Rayleigh wave scattering at the edges of the hemispherical depression; it seems as if the reflected Rayleigh wave propagating faster in the hemispherical depression model than that in the hill topography model. What's more, the back scattered waves of Rayleigh wave in the hemispherical depression model are much stronger, this may also indicating that such sharp depression blocks the propagation of Rayleigh wave more significantly.

6.2.3. Real topography simulating

It is also interesting to study a realistic example. We consider a model in Tibet (Figure 17). The length and width of the model are 21.6 km, and the "average" height of the topography is roughly -3560 m (3560 m in the geodetic coordinate system). The computational domain is extended to depth $z(x, y)= 7200$ m. For simplicity we use a two-layered model with parameters given in the model sketch (Figure 17) instead of the "real" velocity structure under the realistic topography. It consists of 241×241×121 grid nodes in the x-, y-, and z-direction, respectively, with equal vertical grid nodes in each layer. A vertical point source like the used in previous models is loaded in the middle of the free surface (indicated by the asterisk in Fig. 17), where the high cut-off frequency has been changed to 2.7 Hz and the time-shift is 1.5 s. Two lines of receivers crossing the source location and paralleling to the x- and y-direction respectively are placed at the free surface. The time step is 5 ms, and the total propagation time is 8 s.

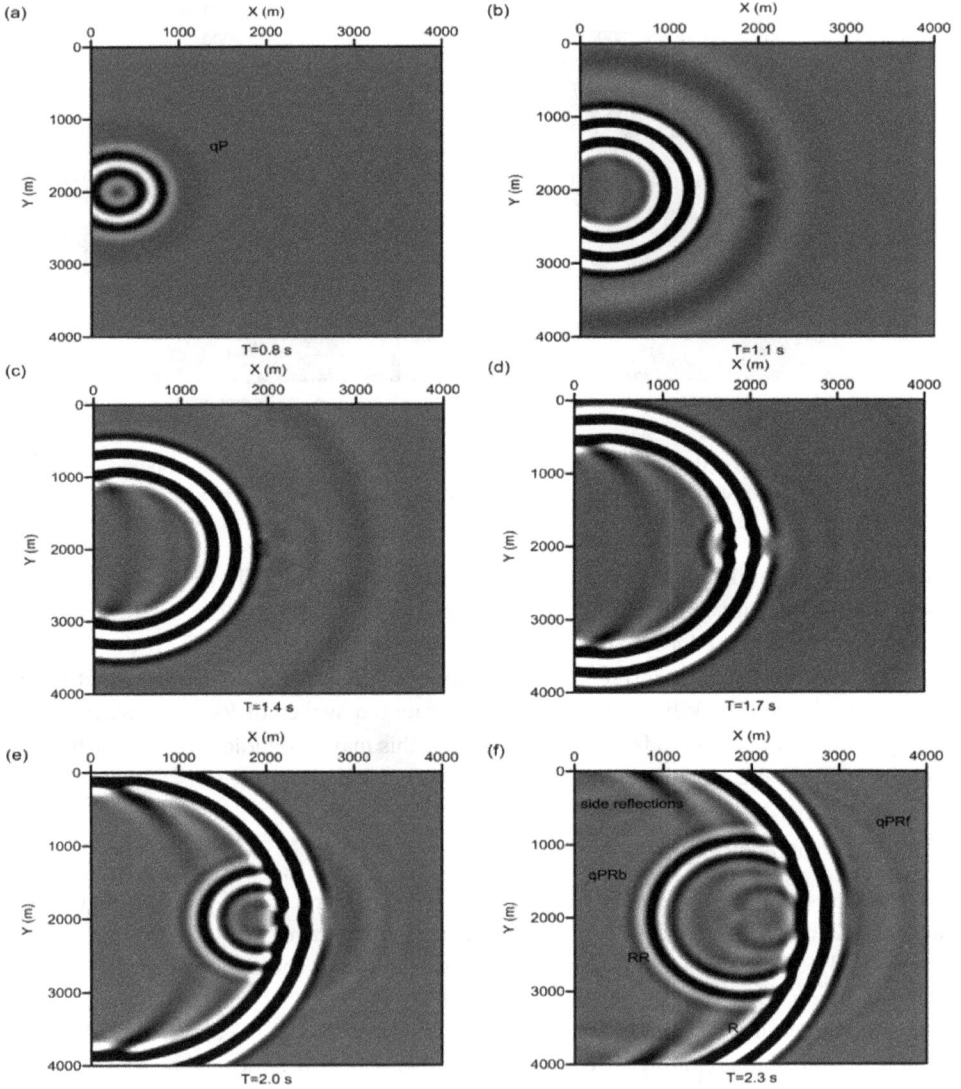

Figure 16. Snapshots of the vertical component of the wavefield at the surface (xy-plane) at different propagation times for the hemispherical shape depression topography model.

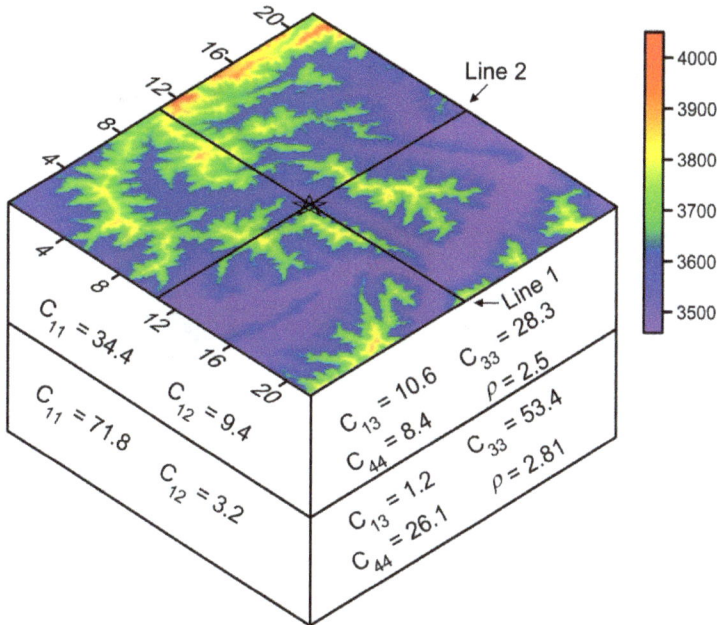

Figure 17. A two-layered model with a realistic topography. The medium parameters of each layer are also given in the figure. The units for the elasticity and density are GPa and g/cm^3, respectively.

Snapshots of the z-component of the wavefield in the vertical plane which contains receiver line Line1 and the source location are presented on Figure 18, and the seismograms of the z-component are also computed at the two receiver lines (Figure 19). We can see that the effect of the topography is very important, with strong scattered phases that are superimposed to the direct and reflected waves, which make it very difficult to identify effective reflections from subsurface interface. The scattering in the seismograms also reflect different features of the surface. The scattering in the seismograms at Line 1 (Figure 19a) is much stronger than that in the seismogram at Line 2 (Figure 19b), indicating that the surface along Line 1 is much rougher than that along Line 2, which also can be observed in Figure 17. What's more, the scattering in Figure 19a is approximately uniformly distributed while in Figure 19b it is mostly distributed in the vicinity of the shot. These may due to different distributions of the surface topography along these two lines.

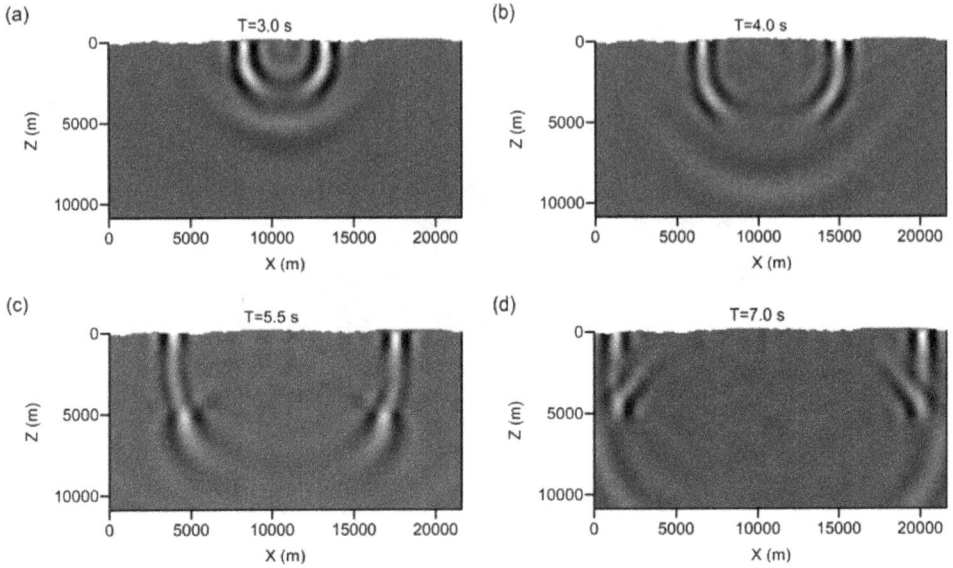

Figure 18. Snapshots of the vertical component of the wavefield in the vertical (xz-) plane along Line 1 at different propagation times for the two-layered model with a realistic topography.

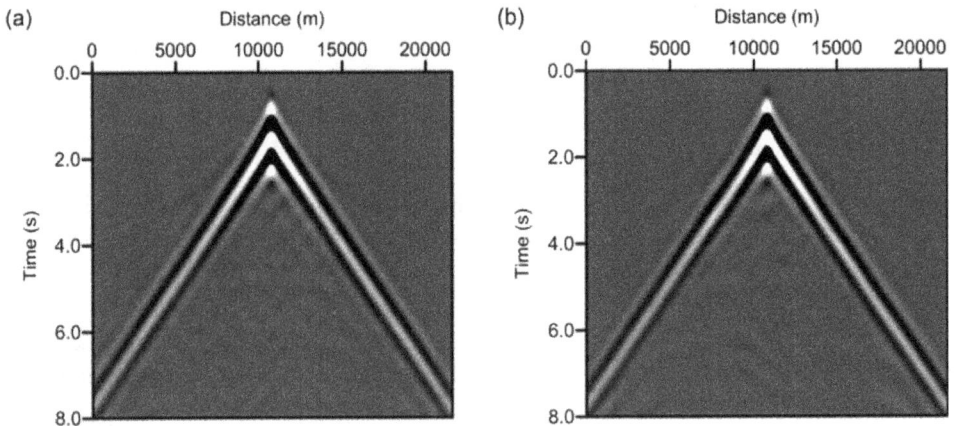

Figure 19. Vertical-component synthetic seismograms coming from the two-layered model with real topography represented in Figure 19: (a) at the receiver line (Line 1) that crosses the source location and is parallel to x-direction (Fig. 19); (b) at the receiver line (Line 2) that crosses the source location and is parallel to y-direction.

7. Conclusion

We propose a stable and explicit finite difference method to simulate with second-order accuracy the propagation of seismic waves in a 3D heterogeneous transversely isotropic medium with non-flat free surface. The method is an extension of the 2D method proposed by Appelo and Petersson (2009) to the 3D anisotropic case. The surface topography is introduced via mapping rectangular grids to curved grids. The accurate application of the free surface boundary conditions is done by using boundary-modified difference operators to discretize the mixed derivatives in the governing equations of the problem. Several numerical examples under different assumptions of free surface are given to highlight the complications of realistic seismic wave propagation in the vicinity of the earth surface. Synthetic seismograms and snapshots explain diffractions, scattering, multiple reflections, and converted waves provoked by the features of the free surface topography. The typical cuspidal triangles of the quasi-transverse (qS) mode also appear in the snapshots of the anisotropic medium.

The future directions of our research will include an extension of the schemes to the viscoelastic case. This will allow a realistic attenuation of the seismic waves due to the presence of a weathered layer to be included (2000).

Author details

Haiqiang Lan and Zhongjie Zhang
State Key Laboratory of Lithosphere Evolution, Institute of Geology and Geophysics,
Chinese Academy of Sciences, China P.R.

Acknowledgement

We are grateful to Jinlai Hao and Jinhai Zhang for their assistance and the facilities given in the course of this work. Fruitful discussions with Tao Xu, Kai Liang, Wei Zhang are also greatly appreciated. The Ministry of Science and Technology of China (SINOPROBE-02-02) supported this research.

8. References

Al-Shukri, H. J.; Pavlis, G. L. & Vernon F. L. (1995). Site effect observations from broadband arrays. *Bull. Seismol. Soc. Am.* 85, 1758-1769.

Appelo, D. & Petersson, N. A. (2009). A stable finite difference method for the elastic wave equation on complex geometries with free surfaces. *Commun. Comput. Phys.*, 5, 84–107.

Ashford, S. A.; Sitar, N.; Lysmer, J. & Deng N. (1997). Topographic effects on the seismic response of steep slopes. *Bull. Seismol. Soc. Am.*, 87, 701-709.

Backus, G. E. (1962). Long-wave elastic anisotropy produced by horizontal layering. *J. Geophys. Res.*, 67, 4427-4440.

Boore, D. M. (1972). A note on the effect of simple topography on seismic SH waves. *Bull. Seismol. Soc. Am.*, 62, 275-284.

Bouchon, M.; Campillo, M. & Gaffet, S. (1989). A boundary integral equation-discrete wavenumber representation method to study wave propagation in multilayered medium having irregular interfaces. *Geophysics*, 54, 1134-1140.

Bouchon, M.; Schultz, C. A. & Toksoz, M. N. (1995). A fast implementation of boundary integral equation methods to calculate the propagation of seismic waves in laterally varying layered medium. *Bull. Seismol. Soc. Am.*, 85, 1679-1687.

Campillo, M. & Bouchon, M. (1985). Synthetic SH seismograms in a laterally varying medium by the discrete wavenumber method. *Geophys. J. R. Astr. Soc.*, 83, 307-317.

Carcione, J. M.; Kosloff, D.; Behle, A. & Seriani G. (1992). A spectral scheme for wave propagation simulation in 3-D elastic-anisotropic medium. *Geophysics*, 57, 1593-1607.

Carcione, J. M. (2000). *Wave fields in real medium: Wave propagation in anisotropic, anelastic and porous medium*, Pergamon Amsterdam.

Crampin, S. (1981). A review of wave motion in anisotropic and cracked elastic-medium. *Wave motion*, 3, 343-391.

Dziewonski, A. M., and Anderson, D. L. (1981). Preliminary reference Earth model* 1, *Phys. Earth Planet. Inter.*, 25, 297-356.

Fornberg, B. (1988). The pseudo-spectral method: Accurate representation in elastic wave calculations. *Geophysics*, 53, 625-637.

Forsyth, D. W. (1975). The early structural evolution and anisotropy of the oceanic upper mantle. *Geophys. J. Int.*, 43, 103-162.

Frankel, A. & Vidale, J. (1992). A three-dimensional simulation of seismic waves in the Santa Clara Valley, California, from a Loma Prieta aftershock. *Bull. Seism. Soc. Am.*, 82, 2045-2074.

Gao, H. & Zhang, J. (2006). Parallel 3-D simulation of seismic wave propagation in heterogeneous anisotropic medium: a grid method approach. *Geophys. J. Int.*, 165, 875-888.

Galis, M.; Moczo, P. & Kristek J. (2008). A 3-D hybrid finite-difference—finite-element viscoelastic modelling of seismic wave motion. *Geophys. J. Int.*, 175, 153-184.

Helbig, K. (1984). Anisotropy and dispersion in periodically layered medium. *Geophysics*, 49, 364-373.

Hestholm, S. & Ruud, B. (1994). 2-D finite-difference elastic wave modeling including surface topography. *Geophys. Prosp.*, 42, 371-390.

Hestholm, S. & Ruud, B. (1998). 3-D finite-difference elastic wave modeling including surface topography. *Geophysics*, 63, 613-622.

Hudson, J. A. (1981). Wave speeds and attenuation of elastic waves in material containing cracks. *Geophys. J. Int.*, 64, 133-150.

Hvid, S. L. (1994). Three dimensional algebraic grid generation. *Ph.D. thesis*, Technical University of Denmark.

Jih, R. S.; McLaughlin, K. L. & Der, Z. A. (1988). Free-boundary conditions of arbitrary polygonal topography in a two-dimensional explicit elastic finite-difference scheme. *Geophysics*, 53, 1045.

Komatitsch, D. & Tromp, J. (1999). Introduction to the spectral element method for three-dimensional seismic wave propagation. *Geophys. J. Int.*, 139, 806-822.

Komatitsch, D. & Tromp, J. (2002). Spectral-element simulations of global seismic wave propagation-I. Validation. *Geophys. J. Int.*, 149, 390-412.

Lan, H. & Zhang, Z. (2011a). Comparative study of the free-surface boundary condition in two-dimensional finite-difference elastic wave field simulation. *Journal of Geophysics and Engineering*, 8, 275-286.

Lan, H. & Zhang, Z. (2011b). Three-Dimensional Wave-Field Simulation in Heterogeneous Transversely Isotropic Medium with Irregular Free Surface. *Bull. Seismol. Soc. Am.*, 101(3), 1354–1370.

Levander, A. R. (1990). Seismic scattering near the earth's surface. *Pure Appl. Geophys.*, 132, 21-47.

Liu, E.; Crampin, S.; Queen, J. H. & Rizer, W. D. (1993). Velocity and attenuation anisotropy caused by microcracks and microfractures in a multiazimuth reverse VSP. *Can. J. Explor. Geophys.*, 29, 177-188.

Lombard, B.; Piraux, J. ; Gélis, C. & Virieux, J. (2008). Free and smooth boundaries in 2-D finite-difference schemes for transient elastic waves. *Geophys. J. Int.*, 172, 252-261.

Moczo, P.; Bystricky, E.; Kristek, J.; Carcione, J. & Bouchon, M. (1997). Hybrid modelling of P-SV seismic motion at inhomogeneous viscoelastic topographic structures. *Bull. Seism. Soc. Am.*, 87, 1305-1323.

Nielsen, P.; If, F.; Berg, P. & Skovgaard, O. (1994). Using the pseudospectral technique on curved grids for 2D acoustic forward modelling. *Geophys. Prospect.*, 42, 321-342.

Nilsson, S., Petersson, N. A.; Sjogreen, B. & Kreiss, H. O. (2007). Stable difference approximations for the elastic wave equation in second order formulation. *SIAM J. Numer. Anal.*, 45, 1902-1936.

Payton, R. G. (1983). *Elastic wave propagation in transversely isotropic medium*. Martinus Nijhoff Publ.

Rial, J. A.; Saltzman, N. G. & Ling, H. (1992). Earthquake-induced resonance in sedimentary basins. *American Scientist*, 80, 566-578.

Robertsson, J. O. A. (1996). A numerical free-surface condition for elastic/viscoelastic finite-difference modeling in the presence of topography. *Geophysics*, 61, 1921.

Sanchez-Sesma, F. J. & Campillo, M. (1993). Topographic effects for incident P, SV and Rayleigh waves. *Tectonophysics*, 218, 113-125.

Sanchez-Sesma, F. J., Ramos-Martinez, J. & Campillo, M. (2006). An indirect boundary element method applied to simulate the seismic response of alluvial valleys for incident P, S and Rayleigh waves. *Earthquake Eng. Struct. Dynam.*, 22, 279-295.

Schoenberg, M. & Muir, F. (1989). A calculus for finely layered anisotropic medium. *Geophysics*, 54, 581-589.

Tessmer, E.; Kosloff, D. & Behle, A. (1992). Elastic wave propagation simulation in the presence of surface topography. *Geophys. J. Int.*, 108, 621-632.

Tessmer, E. & Kosloff, D. (1994). 3-D elastic modeling with surface topography by a Chebychev spectral method. *Geophysics*, 59, 464-473.

Thompson, J. F.; Warsi, Z. U. A. & Mastin, C. W. (1985). *Numerical grid generation: foundations and applications*. North-holland Amsterdam.

Thomsen, L. (1986). Weak elastic anisotropy. *Geophysics, 51,* 1954-1966.

Toshinawa, T. & Ohmachi, T. (1992). Love wave propagation in a three-dimensionai sedimentary basin. *Bull. Seism. Soc. Am.,* 82, 1661-1667.

Wapenaar, C.; Verschuur, D. & Herrmann, P. (1992). Amplitude preprocessing of single and multicomponent seismic data. *Geophysics,* 57, 1178-1188.

Zhang, W. & Chen, X. (2006). Traction image method for irregular free surface boundaries in finite difference seismic wave simulation. *Geophys. J. Int.,* 167, 337-353.

Zhang, Z.; Wang, G. & Harris, J. M. (1999). Multi-component wavefield simulation in viscous extensively dilatancy anisotropic medium. *Phys. Earth Planet. Inter.,* 114, 25-38.

Zhang, Z; Yuan, X.; Chen, Y.; Tian, X.; Kind, R.; Li, X.; Teng, J. (2010). Seismic signature of the collision between the east Tibetan escape flow and the Sichuan Basin. *Earth Planet Sci Lett,* 292, 254-264.

Chapter 5

Soil-Structure Interaction

Alexander Tyapin

Additional information is available at the end of the chapter

1. Introduction

First the very definition of the soil-structure interaction (SSI) effects is discussed, because it is somewhat peculiar: every seismic structural response is caused by soil-structure interaction forces, but only in certain situations they talk about soil-structure interaction (SSI) effects. Then a brief history of this research field is given covering the last 70 years. Basic superposition of wave fields is discussed as a common basis for different approaches – direct and impedance ones, first of all. Then both approaches are described and applied to a simple 1D SSI problem enabling the exact solution. Special attention is paid to the substitution of the boundary conditions in the direct approach often used in practice. Impedance behavior is discussed separately with principal differentiation of quasi-homogeneous sites and sites with bedrock. Locking and unlocking of layered sites is discussed as one of the main wave effects. Practical tools to deal with SSI are briefly described, namely LUSH, SASSI and CLASSI. Combined asymptotic method (CAM) is presented.

Nowadays SSI models are linear. Nonlinearity of the soil and soil-structure contact is treated in a quasi-linear way. Special approach used in SHAKE code is discussed and illustrated. Some non-mandatory additional assumptions (rigidity of the base mat, horizontal layering of the soil, vertical propagation of seismic waves) often used in SSI, are discussed. Finally, two of the SSI effects are shown on a real world example of the NPP building. The first effect is soil flexibility; the second effect is embedment of the base mat. Recommendations to engineers are summed in conclusions.

2. Peculiarities of the SSI definition

Soil-structure interaction (SSI) analysis is a special field of earthquake engineering. It is worth starting with definition. Common sense tells us that every seismic structural response is caused by soil-structure interaction forces impacting structure (by the definition of seismic excitation). However, engineering community used to talk about soil-structure interaction only when these interaction forces are able to change the basement motion as compared to the free-field ground motion (i.e. motion recorded on the free surface of the soil without structure). So, historically the conventional definition of SSI is different from simple occurrence of the interaction forces: these forces occur for every structure, but not always they are able to change the soil motion.

This simple fact leads to the important consequences. If a structure can be analyzed as based on rigid foundation with free-field motion at it, then they use to say that "no SSI effects occur" (though structure is in fact moved by the interaction forces, and the same forces impact the foundation). Looking at the variety of the real-world situations, we can conclude that only part of them satisfies the conventional definition of SSI.

The ability of the interaction forces to change the soil motion depends, of course, on two factors: value of the force and flexibility of the soil foundation. The value of the interaction force may be often estimated via the base mat acceleration and inertia of the structure. For given soil site and given free-field seismic excitation the heavier is the structure, the more likely SSI effects occur. Usually most of civil structures resting on hard or medium soils do not show the signs of considerable SSI effects.

From the inertial point of view, the heaviest structures we deal with are hydro-structures (like dams) and nuclear power plant (NPP) main structures – first of all, reactor buildings. So, the development of the SSI field in the earthquake engineering was historically linked to the development of these two fields of industry.

From the soil flexibility point of view, for the given structure and given free-field seismic excitation the softer is the soil, the more likely SSI effects occur. Soil shear module is a product of mass density and square of the shear wave velocity. Mass density of soil in practice varies around 2.0 t/m^3 in a comparatively narrow range, so the main characteristic of the soil stiffness is shear wave velocity V_s. Usually soil is considered "soft" when V_s is less than 300 m/s, and "hard" when V_s is greater than 800 m/s. If V_s is greater than 1100 m/s, they usually talk about "rigid" soil (no SSI effects – just rigid platform with a structure on it and excitation taken from the free field).

All these ranges are of course purely empirical. Obviously, one and the same soil can behave as rigid one towards very light and small structure (like a tent), and behave as a soft one towards heavy and rigid structure (like the NPP reactor building).

Sometimes to decide whether to account for SSI effects they compare the natural frequencies of the rigid structure on the flexible soil foundation with those of the flexible structure on a rigid foundation. If the lowest natural frequency of the first set is greater than the first dominant frequency of the second set two and more times, they do not consider SSI effects (e.g., see standards ASCE4-98 [1]). As SSI field is rather sophisticated, sometimes it is worth neglecting SSI, when allowed.

However, one should keep in mind another situation, when SSI effects occur. In soft soils seismic wave may have moderate wavelength comparable to the size of structure, so that the free-field motion over the soil-structure contact surface will have the so-called "space variability". In this case a comparatively stiff structure (even a weightless one) can impact the soil motion by structural rigidity (i.e., stiff contact soil-structure surface will not allow soil to move in the manner it used to move without structure). So, once more the presence of a structure changes the behavior of soil during seismic excitation, which is a SSI effect. This is a situation with considerably embedded structures; the same applies to the great base mats, "averaging" the travelling seismic waves [2]. But the most typical situation is a simple pile foundation – piles are used with soft soils, they are not rigid but usually stiff enough to change the foundation motion.

To conclude this part, let us talk a little about the so-called SSSI – structure-soil-structure interaction. When a group of structures is resting on a common soil foundation, the base mat motion of a structure may be changed (as compared to the free-filed motion) not because of this very structure only, but because of the neighboring structures. Of course, if all structures are comparatively light and no SSI effects occur for each of them standing alone, usually no SSSI effects occur for the group. But in case at least one of the structures standing alone can cause SSI effects, the additional waves in the soil (i.e. additional to the free-field wave picture) will spread from this structure, impacting neighbors. Sometimes (for the light/small neighbors) this situation can be analyzed without SSI but with special seismic excitation, accounting for the influence of the heavy neighbor. In other cases several structures should be analyzed together because of the mutual influence.

Thus, general goal of the SSI analysis is to calculate seismic response of structure based on seismic response of free field (sometimes free field may differ from the final soil). General format of the SSI analyses results is basements' motion obtained using the information about a) soil foundation (sometimes the initial one and the final one separately), b) structure, c) seismic excitation provided without structures. Another format of the SSI results is soil-structure interaction forces, necessary to estimate the soil foundation capability to withstand earthquake. Sometimes, the complete structural seismic response is obtained together with SSI in the format of response motion of different structural nodes and response structural internal seismic forces.

3. Brief history

As SSI field combines structures' and soil modeling, the level of such modeling is generally lower than in the classical soil mechanics and in the structural mechanics standing alone.

For several decades (up to 1960-s) only soil flexibility was considered without soil inertia – springs modeled soil. At that time mostly the machinery basements were analyzed for the dynamic interaction with soil foundation (the largest of them probably being turbines). In fact, it was a quasi-static approach – the well-known static solution for rigid stamps, beams and plates on elastic foundation was applied at every time step. The model was so simple, that nobody even used the special term "SSI" at that time. The key question of such a simple approach appeared to be damping. The material damping measured in the laboratories with soil samples proved to be considerably less than the damping measured in the dynamic field tests with rigid stamps resting on the soil surface.

The nature of this effect was discovered in 1930-40s (Reissner in Germany [3], Schechter in the USSR) and proved to be in inertial properties of the soil. Inertia plus flexibility always mean wave propagation. It turned out that in the field tests actual energy dissipation in the soil was composed of two parts: conventional "material" damping (the same as in laboratory tests) and so-called "wave damping". In the latter case the moving stamp caused certain waves in the soil, and those waves took away energy from the stamp, contributing to the overall "damping" in the soil-structure system. This energy was not transferred from mechanical form into heat (like in material damping case), but was taken to the infinity in the original mechanical form. In reality waves did not go to the infinity, gradually dissipating due to the material damping in the soil, but huge volumes of the soil were involved in this wave propagation. Even without any material damping in the soil this "wave damping" contributed a lot to the response of the stamp. In practice it turned out that the level of wave damping was usually greater than the level of material damping.

In parallel it turned out that when the base mat size is comparable to the wave length in the soil, not only damping, but stiffness also depends on the excitation frequency. This effect is invisible for comparatively small machine basements, but important for large and stiff structures.

To study these wave effects new soil-structure models with infinite inertial soil foundations should be considered. That was the moment (1960-s) when the very term "SSI" appeared [2,4,5]. It happened so, that NPPs were actively designed at that time, including seismic regions (e.g., California), and intensive research was funded in the US to study the SSI effects controlling the NPP seismic response. Earthquake Engineering Research Center (EERC) in the University of Berkley, California became a leader with such outstanding scientists as H.B.Seed and J.Lysmer on board [6].

The main result of these investigations was the development of new powerful tools to analyze more or less realistic models. The earlier SSI models considered homogeneous half-space with surface rigid stamp. They could be treated analytically or semi-analytically for simple stamp shapes (e.g., circle).

In practice soil is usually layered. Layering can lead to the appearance of the new wave types and change the whole wave picture. Important achievements of 1960-70s enabled to move from the homogeneous half-space to the horizontally–layered medium in soil modeling [7,8]. However, the infinite part of the foundation, excluding some limited soil volume around the basement, still remains a) linear, b) isotropic, c) horizontally layered. These limitations are due to the methods of the SSI analysis. The final masterpiece of Prof. Lysmer was SASSI code [9], further developed by F.Ostadan, M.Tabatabaie, D.Giocel et al. This code combined finite element modeling of the structure and limited volume of the soil with semi-analytical modeling of the infinite foundation (see below). Limitations on the embedment depth and on the shape of the underground part have gone.

Some other scientists have greatly contributed to this field approximately at the same time (J.Wolf [10,11], J.M..Roesset, E.Kausel [12] and J.Luco [4] should be mentioned).

After the Three Miles Island and Chernobyl accidents there was a long pause in the nuclear energy development in the West (e.g., in 2012 US NRC issued a first permission for a new NPP block in last 33 years) and in Russia (due to the economic problems of 1990s), though in Asia they continued to build new blocks. SSI investigations went forward in South Korea, China, India. Nowadays, in spite of the Fukushima accident, nuclear industry goes forward.

In parallel SSI was studied by hydro-engineers (for the dams design, first of all). However, this field separated from the "NPP field of SSI" about 40 years ago. The reason was that the SSI models usually applied in nuclear industry (horizontally layered soil, rigid or very stiff base mats) are often not applicable to the hydro-dams situations (rocky canyons, etc.). That is why both models and methods used in the SSI field usually are different in nuclear and hydro-industries.

In last decades, civil structures are gradually increasing in size and embedment. Effects like SSI and SSSI from time to time are considered during the design procedures of such structures. Naturally, they are closer to the NPP practice than to the hydro-energy practice.

The author has about 30 years of experience in nuclear industry, dealing with SSI problems. So, the subsequent text will be based on the "NPP" approach to the SSI problems.

4. Basic superposition

Today there exist three approaches to SSI problems, namely "direct", "impedance", and "combined" ones. To understand them all, let us start from common general approach based on the superposition of the wave fields.

Let us call the problem with seismic wave, soil and structure "problem A" and start with completely linear soil-structure model. Let Q be some surface surrounding the basement in the soil and dividing the soil-structure model into two parts: the "external" part V_{ext} and internal part V_{int}. Let $(-F)$ be additional external loads distributed over V_{int} and specially tuned so, to provide zero displacements in V_{int}. Then "problem A" can be split in the sum of two wave pictures: "problem A_1", including seismic excitation and loads $(-F)$, and "problem A_2" including only loads (F) without seismic wave – see Fig.1.

This simple superposition leads to a number of important conclusions.

1. As in "problem A_1" all displacements in the internal volume V_{int} are zero, the motion of V_{int} in "problem A_2" is the same as in "problem A". Hence, if we are interested in the motion of V_{int} only, we can substitute "problem A" with "problem A_2".
2. As in "problem A_1" all displacements in the internal volume V_{int} are zero, all the strains and internal forces in the internal volume V_{int} are zero, and the external loads $(-F)$ must be zero everywhere in V_{int}, except surface Q.
3. As in "problem A_1" all displacements, strains and internal forces in the internal volume V_{int} are zero, no forces are impacting Q from V_{int} (i.e., forces impacting Q from V_{ext} are

balanced by loads *(-F)*). Hence, V_{int} can be withdrawn or replaced by another medium (with zero displacements) without changing V_{ext}, seismic excitation, or loads *(-F)*. In particular, V_{int} can be replaced by initial soil without structure.

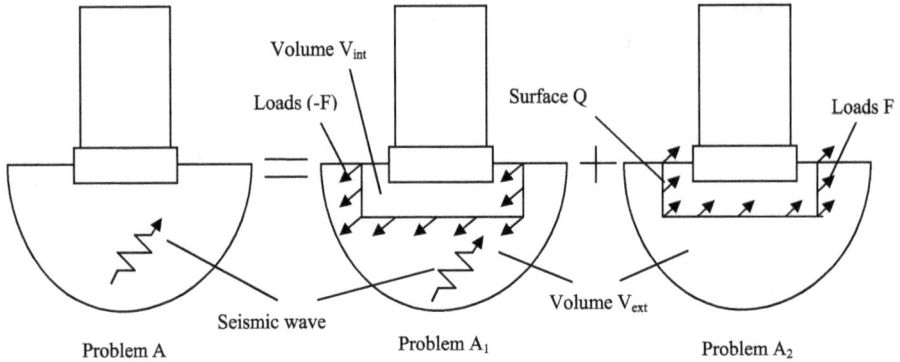

Figure 1. Basic superposition: split of the "problem A"

4.   We can withdraw structure from Fig.1 and call the problem with initial soil in V_{int} "problem B". This problem can be also split in "problem B_1" and "problem B_2" in the same manner as "problem A". Wave fields in V_{ext} and loads *(-F)* are equal in "problem A_1" and "problem B_1". However, in "problem A_2" and "problem B_2" wave fields are different in spite of similar loads F and similar V_{ext} in both problems. Generally, the motions of Q in "problem B_2" and "problem A_2" are different due to the waves, radiating from the structure in "problem A_2".

5.   Wave field in volume V_{int} in "problem B_2" is the same as in "problem B" (see conclusion 1 above). Very often this field is known apriori or easily calculated. This creates a powerful tool to verify models suggested for "problem A_2". Each of these models contains some description of the internal part V_{int}, external part V_{ext} and the loads F. It is useful to take the same V_{ext} and F and substitute the internal part V_{int} by the initial soil, thus coming from "problem A_2" to "problem B_2". The suggested V_{ext} and F must provide adequate solution for "problem B_2"; otherwise they cannot be applied to "problem A_2".

6.   Loads *(-F)* can be obtained from the wave field U_0 in "problem B" and dynamic stiffness operator G_0 for the initial unbounded soil as follows

$$-F(x) = G_0(x,y)[-U_0(y)] \tag{1}$$

Formula (1) uses operator G_0 in the time domain. This operator is applied to the displacement field in the volume V_{int} and provides the loads, distributed over the volume (this formula can be applied to the whole volume V_{int}, but for the internal nodes the result will be zero). For linear initial soil this operator in the frequency domain will turn into complex frequency-dependent dynamic stiffness function. Note that for the given surface Q the loads F in (1) can be split in two parts: loads F_{int} acting from V_{int} and loads F_{ext} acting from V_{ext}. Physical meaning is illustrated in Fig.2.

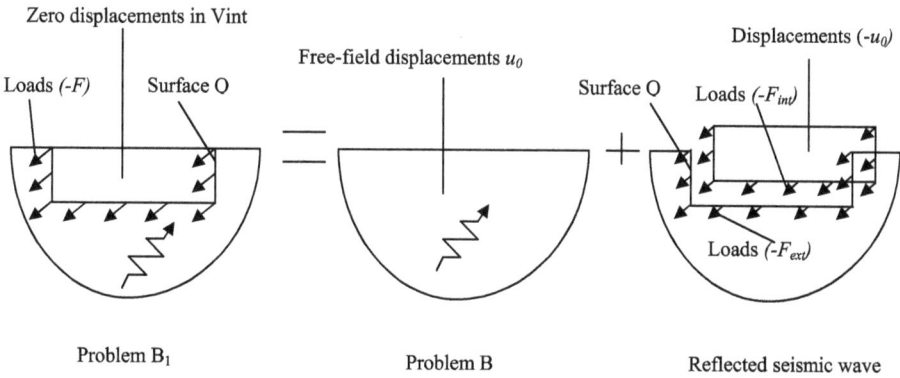

Figure 2. Split of loads F into F_{int} and F_{ext}

With given free-field wave field U_0 one can easily obtain F_{int} just as surface forces at Q corresponding to the internal stress field.

Theoretically we can change the content of the volume V_{int} in "problem B" (e.g., withdraw the medium completely, taking loads F_{int} to zero). This change will change operator G_0, change wave field U_0, but the left part of (1) will stay the same, as it is in fact fully determined by soil properties in V_{ext} and initial seismic wave.

Even if there exists some physical non-linearity in the model, and this non-linearity is localized inside V_{int}, initial "problem A" can still be substituted by "problem A_2" without changing the loads F, as compared to the linear case. This is a consequence of the logic of the previous point: the loads are fully determined by soil in V_{ext} and initial seismic wave.

5. Classification of methods: direct and impedance approaches

Current methods of SSI analysis can be classified according to the choice of surface Q. In "direct" method they try to put Q on some physical boundaries in the soil, usually apart from the basement. In "impedance" method they put Q right on the soil-structure contact surface, additionally presuming the rigidity of this surface. In "combined" method they put Q on the boundary of the "modified volume" (sometimes Q is a flexible soil-structure contact surface, but sometimes there exists some additional soil volume around the basement modified with the appearance of structure). Let us discuss some details of these methods one by one.

5.1. Direct approach

In direct method there always exists certain volume V_{int}. Usually the lower boundary (i.e., the bottom of V_{int}) is placed on the rock. In this case we presume that the additional waves radiating from the structure cannot change the motion of this boundary, as compared to the seismic field U_0. As a consequence, we can substitute boundary conditions in "problem A_2", fixing the motion of the bottom (and obtaining it from "problem B") instead of modeling V_{ext} and applying loads F at surface Q. This substitution of the boundary conditions is shown in Fig.3.

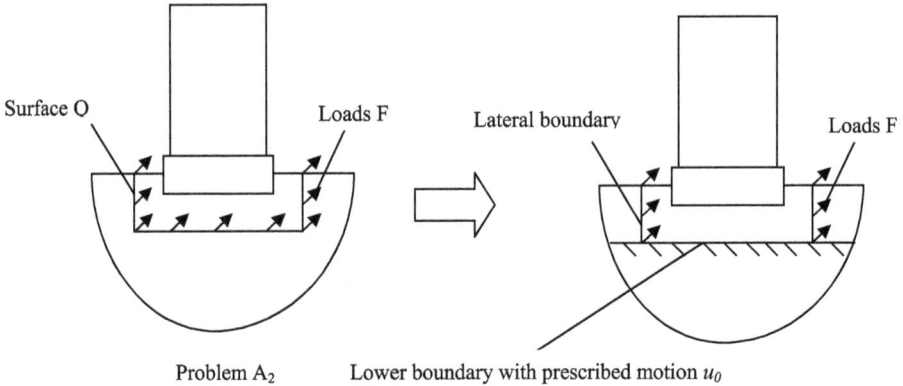

Problem A_2 Lower boundary with prescribed motion u_0

Figure 3. Substitution of the boundary conditions at the bottom of V_{int} in direct method

One should remember that such substitution is accurate for rigid rock only. Otherwise, an error occurs due to the reflection of the additional waves from such a boundary back into V_{int}. This error can be significant. The example will be shown below.

After the lower boundary is fixed, lateral boundaries (usually vertical) remain to be set. In 1970s, when direct method was popular, there was an evolution of these boundaries from "elementary" ones (fixed or free) towards more accurate ones. The first important step was so-called "acoustic boundaries" by Lysmer and Kuhlemeyer [13,14].

Physical base is as follows. 1D elastic (without material damping!) massive rod with shear or pressure waves can be cut in two, and one half of it can be accurately replaced by viscous dashpot. This is illustrated by Fig. 4.

Dashpot viscosity parameter is a product of mass density ρ and wave velocity c. So, Lysmer and Kuhlemeyer suggested to place distributed "shear" and "pressure" dashpots over vertical lateral boundary (three dashpots along three axes in each node). Practically they substituted the half-infinite layer in Fig.3 by number of horizontal 1D rods normal to the boundary.

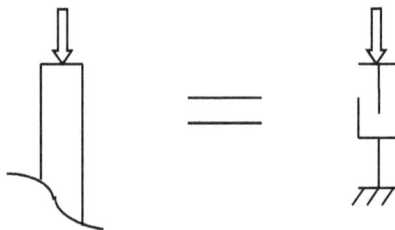

Figure 4. Analogue between half-infinite rod of unit cross-section area and viscous dashpot with viscosity parameter b=ρc

This boundary was far better than any elementary (fixed or free) boundary before. Horizontal body waves normal to the boundary did not have artificial reflection into V_{int} (that is why this boundary was called "non-reflecting boundary"). This boundary could work in the time domain. Besides, this boundary was "local": the response force acting to the node depended on the velocities of this node only, not neighbors.

This boundary was implemented in code LUSH [15], named after the authors (Lysmer, Udaka, Seed, Hwang). However, soon it turned out that the most important waves, radiating by structure in the soil underlain by rock are not horizontal body waves, but surface waves (see below). So, the artificial reflection at the boundary was not completely eliminated, and such a boundary could be placed only apart from the structure (3-4 plan sizes from the basements) to get more or less reasonable results.

One should keep in mind the important limitation: when wave process in V_{int} is modeled by ordinary finite elements, the element size should be about 1/8 (1/5 at most [1]) of the wavelength. It means that one cannot increase the finite element size going away from the structure (as they often do in statics). So, the increase in V_{int} leads to the considerable increase in the problem size and to the computational problems.

The next step was done by G.Waas [8,16,17]. Instead of replacing the infinite soil by number of horizontal rods he suggested to use "homogeneous solutions" – i.e. the waves in the horizontally-layered medium underlain by rigid rock. These waves can be evaluated in the frequency domain only. According to G.Waas, in 2D case each homogeneous displacement varies in the horizontal direction following complex exponential rule with a certain complex frequency-dependent "wave number"; in the vertical direction it varies according to the finite-element interpolation. Later it turned out that in 3D case cylindrical waves in V_{ext} along horizontal radius followed not exponents, but certain special Hankel functions with the same complex "wave numbers" as previously exponents in 2D case.

Any displacement of the lateral boundary (in the finite-element approximation) can be split into the sum of such waves in V_{ext}. Then stresses are calculated at the boundary and finally nodal forces impacting V_{int} from V_{ext} in response to boundary displacement can be evaluated. So, the final format of the Waas boundary was the dynamic stiffness matrix in the frequency domain, replacing V_{ext} in the model. Unlike previous boundaries, this boundary was not local: response forces from V_{ext} in the node depended not only on the motion of this very node, but on the motion of all the nodes. Thus, dynamic stiffness matrix became fully populated.

New boundaries (they were called "transmitting boundaries") proved to be far more accurate than previous ones. They could be placed right near the basement, decreasing V_{int} and accelerating the analysis. So, LUSH was converted into FLUSH (=Fast LUSH) [18] for 2D problems. Then appeared code ALUSH (=Axisymmetric LUSH) [19] to solve 3D problems with axisymmetric geometry. However, there remained two important limitations: hard rock at the bottom and axisymmetric geometry of structure in 3D case. The next step forward was again done by J.Lysmer in EERC. But this new approach will be discussed later (see "combined method").

5.2. Impedance approach

If we presume the rigidity of the soil-structure contact surface and place surface Q in Fig.1 right at this surface, we get six degrees of freedom for Q. Corresponding forces F in "problem A₂" are condensed to six–component integral forces, loading the immovable base mat during seismic excitation. In addition, when the mat moves, V_{ext} impacts the moving base mat by response forces set by "dynamic stiffness matrix" 6 x 6. This matrix is a linear operators' matrix in the time domain for linear soil (soil in V_{ext} was linear from the very beginning). In the frequency domain it becomes complex frequency dependent matrix called "impedance matrix". That is why the whole approach is called the "impedance" one. Impedances can be estimated in the field experiments with stamps excited by unbalanced rotors. They were the first parameters of the soil flexibility in the early times of SSI (when machinery base mats were analyzed).

Let $R_j(x)$ be real "rigid" vector displacements field of the contact surface Q, when one of six general coordinates of the surface Q (coordinate number j) gets a unit displacement. If some contact vector forces $F(x)$ act over Q, the condensed integral general force along coordinate j is given by

$$P_j = \int_Q R_j^T(x)\, F(x)\, dQ \tag{2}$$

Scalar product of two 3D vectors is used under the integral (2). If G_0 is the dynamic stiffness operator in the initial soil, and Q moves by unit along the general coordinate k, then the displacement field at Q is described by $R_k(x)$, and response forces impacting Q in the initial soil are given by $G_0 R_k$. However, for the embedded Q these total forces consist of the part $F_k(x)$ impacting Q from V_{ext} and another part impacting Q from V_{int} (see Fig.2). If G_{int} is the analogue of G_0 for the finite volume V_{int}, then $F_k = (G_0 - G_{int}) R_k$. In the frequency domain all linear operators become matrices. The impedance C_{jk}, meaning the condensed force acting from V_{ext} to Q along coordinate j in response to the unit rigid displacement of Q along coordinate k is then given by

$$C_{jk}(\omega) = \int_{Q_x}\int_{Q_y} R_j^T(x)\,[G_0(\omega) - G_{int}(\omega)]\, R_k(y)\, dQ_x dQ_y \tag{3}$$

Double integration in (3) reflects the fact that G provides the response forces in the node x due to the displacements in the node y.

The next step is to condense load F over rigid surface Q. If $U_k(x)$ is a transfer function from the control motion along coordinate k to the free-field wave in "problem B", then in the frequency domain the transfer function B_{jk} to the condensed load along coordinate j according to (1) is given by

$$B_{jk}(\omega) = \int_{V_x}\int_{V_y} R_j^T(x)\,[G_0(\omega)]\, U_k(y)\, dV_x dV_y \tag{4}$$

If U_0 is a control motion of a certain point in "problem B" (generally multi-component one, though usually only three-component one), then the condensed total forces (not just F!) acting from the soil to the basement are given by

$$P(\omega) = B(\omega) U_0(\omega) - C(\omega) U_b(\omega) \tag{5}$$

Here $U_b(\omega)$ is a six-component motion of the rigid base mat (i.e., soil-structure contact surface Q), $C(\omega)$ is 6 x 6 impedance matrix, $B(\omega)$ is usually 6 x 3 seismic loading matrix.

The important particular case is a surface basement – then V_{int} goes and G_{int} in right-hand part of (3) goes as well. If in addition we presume $U_k(y)=R_k(y)=1$, $k=1,2,3$ (it means that the whole future contact surface Q in "problem B" has no seismic rotations and has the same translations as the control point; this is the case for vertically propagating seismic body waves in the horizontally-layered media), then matrix B is composed of the first three columns of matrix C.

Let $K(\omega)$ and M be stiffness and mass matrices of the structure without soil foundation (K may be complex to account for the structural internal damping), and let $U_s(\omega)$ be absolute displacements of structure in the frequency domain. Then seismic motion in the soil-structure system is described by the equation

$$[K(\omega) - \omega^2 M] U_s(\omega) = P(\omega) \tag{6}$$

Matrix $[K-\omega^2 M]$ is huge in size. Loads $P(\omega)$ are non-zero only for the rigid basement's six degrees of freedom. As displacements U_b take part in (5), and they are at the same time part of U_s, they should better join the left-hand part of (6), making (6) look like

$$[K(\omega) - \omega^2 M + C(\omega)] U_s(\omega) = B(\omega) U_0(\omega) \tag{7}$$

In (7) $C(\omega)$ has to be huge matrix of the same size as K and M, but only 6 x 6 block of it referring to U_b is populated.

Now let us imagine the same rigid contact surface on the same soil with the same seismic excitation, with the only difference: structure with rigid basement is weightless. The right-hand part of (7) will stay in place; in the left-hand part only $C(\omega)$ represents dynamic stiffness, because weightless structure is moving rigidly together with the basement, and no forces occur due to K. Let $V_b(\omega)$ be the response of the rigid weightless basement (instead of $U_b(\omega)$ for massive structure:

$$C(\omega) V_b(\omega) = B(\omega) U_0(\omega) \tag{8}$$

Using (8) one can replace the right-hand part of (7) and come to the equation

$$[K(\omega) - \omega^2 M + C(\omega)] U_s(\omega) = C(\omega) V_b(\omega) \tag{9}$$

This equation is of great importance because it describes the so-called "platform" model, shown in Fig.5. Rigid platform is excited by a prescribed motion $V_b(\omega)$. Structural model is resting on a "soil support" with prescribed stiffness described by the impedance matrix $C(\omega)$.

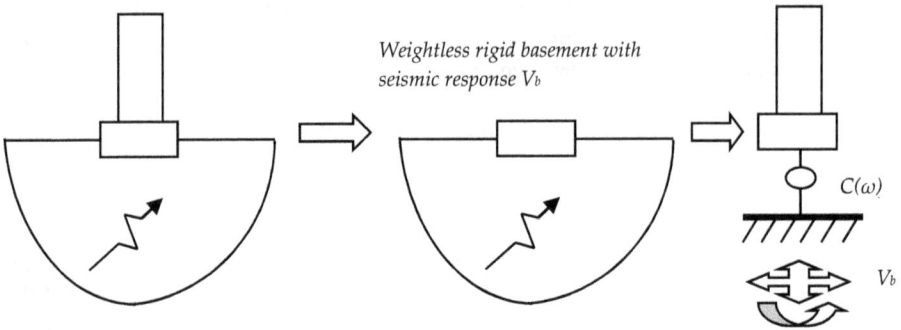

Figure 5. Scheme of platform model for SSI analysis

One should clearly understand that this platform is not physical: in the real world there is no rigid platform having such a motion. Often they try to imagine some hard rock somewhere – there is a mistake! This model is applicable, for example, for homogeneous half-space – no rock is present anywhere. In fact, platform model is just a mechanical analogue of the soil-structure model in terms of the structural response.

Let us work a little with this platform model. In the popular particular case with two assumptions mentioned above (i.e. surface rigid basement and vertical seismic waves in the horizontally layered medium) weightless rigid base mat will move exactly with "control" accelerations from free field. In other words $V_b=U_0$. This case is shown in Fig.6.

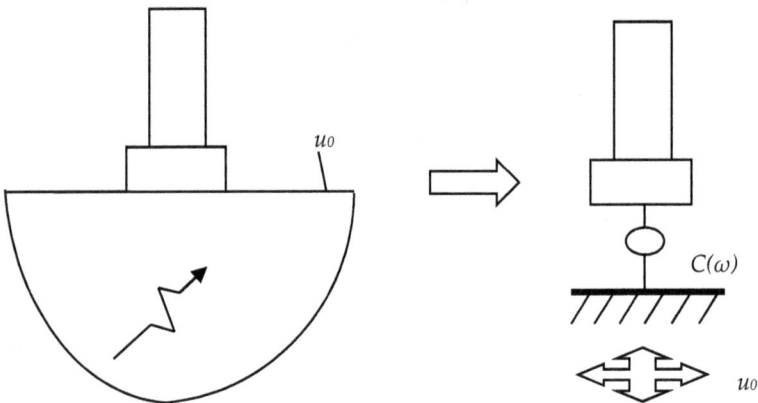

Figure 6. Scheme of platform model for SSI analysis with two assumptions: rigid surface basement and "rigid" free-field motion under it

If any of the two assumptions is not valid, one has to solve a special problem to obtain V_b – it is called "kinematic soil-structure interaction problem". For embedded basements it is allowed to neglect the embedment depth if it is less then 30% of the effective radius of the mat. In this case upper layers of the soil foundation are just withdrawn from the soil model. This may sometimes cause some changes in the free-field motion.

However, even for the surface basements the second assumption is not always correct. If the free-field motion U_0 of the soil surface is not "rigid" over Q, the rigid basement's response is somehow averaging it over the mat; in this case the translational response V_b may be less than control motion U_0 in a single node. On the other hand, the variability of horizontal U_0 under the mat may cause torsion in V_b, and the variability of vertical U_0 under the mat may cause rocking in V_b.

The second part of the problem (i.e. the dynamic analysis of the platform model from Fig.5) is called "inertial soil-structure interaction problem". Often they solve it putting the coordinate system on the platform (thus, coming to the fixed platform, but instead introducing inertial forces impacting all the masses of structure).

One can easily see that conventional structural seismic response without SSI is obtained from the platform model as well. The difference with Fig.5 is only in the soil support and in the platform excitation. If we make soil stiffer and stiffer, V_b comes closer and closer to U_0; "soil support" becomes stiff, and this difference also goes. So, we get a model without SSI as a limit case of the platform model with SSI with extremely hard soil.

Platform model can be extended to the case of several buildings having common soil foundation. The main limitations are the same – rigid basements and linear foundation/contact. Note that generally (with "kinematic interaction" on Fig.5) no requirements are made for soil layering, embedment depth, shape of the underground part, etc.

The alternative form of the platform model uses fixed platform, and dynamic loads are directly applied to the rigid basement. These loads correspond to the right-hand parts of (9), (8) or (7).

Platform model is convenient to work with in the frequency domain. Huge matrix $[K-\omega^2 M]$ can be condensed to the rigid base mat using natural modes and natural frequencies of the fixed-base structure [20]. As a result, we get $M(\omega)$ – the "dynamic inertia" complex frequency-dependent matrix 6 x 6 – similar in size to $C(\omega)$. Equation (7) turns to

$$[C(\omega) - \omega^2 M(\omega)] U_b(\omega) = B(\omega) U_0(\omega) \tag{10}$$

It is easily solved, because maximal size is 6. The response in the time domain is further obtained using Fast Fourier Transform (FFT). For multiple-base structure the dynamic stiffness is condensed in a somewhat different way [21], though the size of the resulting matrix is still 6 x (number of base mats).

However, usually engineers prefer the time domain for the dynamic problems. Platform model of Fig.5 and equation (9) can be transferred to the time domain. Structural part is transferred easily. Kinematic excitation V_b can be also transferred to the time domain using FFT. The only problem is to transfer to the time domain impedance matrix $C(\omega)$.

The most popular variant is just a set of six springs and six viscous dashpots (one spring and one dashpot along each of six coordinates). In the frequency domain an ordinary spring corresponds to the real frequency-independent impedance, and viscous dashpot corresponds to the frequency-linear purely imaginary impedance. So, matrix $C(\omega)$ corresponding to such a set of springs/dashpots will be diagonal complex matrix 6 x 6 with frequency-independent real parts and frequency-linear imaginary parts. Is it realistic for real-world soil foundations? This question deserves special discussion. But before we enter it, let us consider a very simple example, illustrating methodology of SSI problem as a whole and two basic different approaches to this problem (i.e., direct and impedance ones) described above.

6. Sample 1D SSI problem as an example of different approaches to SSI analysis

Let us consider 1D P-waves in a homogeneous massive rod, modeling soil. The only coordinate is x. Wave displacements $u(x,t)$ are described by the wave equation

$$c^2 \frac{\partial^2 u}{\partial x^2} = \frac{\partial^2 u}{\partial t^2} \tag{11}$$

Here c is wave velocity. In the frequency domain for certain frequency ω there exist two solutions of (11):

$$u_1 = U_1 \exp(-ix/\lambda); \quad u_2 = U_2 \exp(ix/\lambda); \quad \lambda = c/\omega \tag{12}$$

The first wave u_1 described by (12) goes up along Ox, the second wave u_2 goes down. Each wave has own amplitude U. Wave velocity c is determined by mass density ρ and constrained elasticity module E^0 as

$$c^2 = E^0 / \rho \tag{13}$$

Let us now apply basic principles discussed above to this simple model: solve "problem B", "problem A", and then solve "problem A2" by direct and impedance methods. All problems will be solved in the frequency domain, using (12).

6.1. The exact solution

Let us start with "problem B" – wave solution without structure. Let U^0 be displacement at the free surface $x=0$. As we are going to obtain coefficient U_1^0 of the upcoming wave and coefficient U_2^0 of the wave coming down, now we have the first of the two equations for them

$$u^0(0) = U_1^0 + U_2^0 = U^0 \tag{14}$$

The second equation comes from the description of the "free" condition of the surface; total stress in the soil must be zero:

$$E^0 \frac{\partial u^0}{\partial x}(0) = E^0 (i / \lambda)(-U_1^0 + U_2^0) = 0 \tag{15}$$

Two equations (14) and (15) give us the well-known "doubling rule": the upcoming wave reaching free surface reflects back. At the free surface displacements of the upcoming wave are doubled:

$$U_1^0 = U^0 / 2; \quad U_2^0 = U^0 / 2; \quad u^0(x) = U^0 [\exp(-ix / \lambda) + \exp(ix / \lambda)] / 2 = U^0 \cos(x / \lambda) \tag{16}$$

So, (16) gives the whole solution of the "problem B" linking wave field in any point to the "control motion" U_0.

Now let us move to the "problem A". Let structure be rigid and have mass m (this is a mass, related to the unit area of the cross-section of the rod). Schemes of "problem B" and "problem A" are shown on Fig.7.

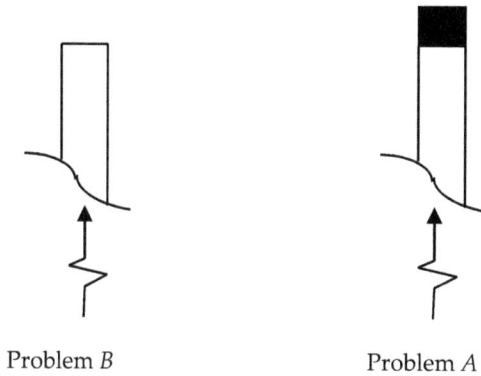

Problem B   Problem A

Figure 7. Layouts of "problem B" and "problem A"

General goal of the SSI analysis is to obtain the response motion of the basement from the given control motion of the free surface. In this simple case we can do it directly (it is not the "direct approach" so far, just a simple solution!). In "problem A" wave field in the soil is still described by (12), but with coefficients U^A_1 (upcoming wave) and U^A_2 (wave coming down):

$$u^A(x) = U_1^A \exp(-ix / \lambda) + U_2^A \exp(ix / \lambda) \tag{17}$$

Equation of motion for mass m in the frequency domain is

$$-\omega^2 m (U_1^A + U_2^A) = -E^0 [(-i / \lambda) U_1^A + (i / \lambda) U_2^A] \tag{18}$$

Having (18) one can express the reflected wave amplitude from the amplitude of the upcoming wave

$$U_2^A = U_1^A [iE^0 / \lambda + \omega^2 m] / [iE^0 / \lambda - \omega^2 m] = U_1^A [1 - i\frac{m}{\rho\lambda}] / [1 + i\frac{m}{\rho\lambda}] \tag{19}$$

"Problem A" and "problem B" have similar seismic excitation: in our case it means that the upcoming waves are similar:

$$U_1^A = U_1^0 = U^0 / 2 \qquad (20)$$

Hence we get the final expression linking the displacement of the mass in "problem A" to the control motion at the free surface in "problem B":

$$U^A = U^0\{1 + [1 - i\frac{m}{\rho\lambda}] / [1 + i\frac{m}{\rho\lambda}]\} / 2 = U^0 / [1 + i\frac{m\omega}{\rho c}] \qquad (21)$$

The transfer function in the frequency domain linking response displacements to the control displacements in (21) at the same time links response accelerations to the control accelerations. Having control accelerogram $a_0(t)$ one can get response accelerogram $a_A(t)$ easily using FFT technique.

6.2. Direct approach

Now we will show the direct approach for the same example, following p.4.1. Let surface Q be at depth H. There the free-field wave according to (16) will be

$$u^0(-H) = U^0 \cos(H / \lambda); \quad \frac{\partial u^0}{\partial x}(-H) = (U^0 / \lambda)\sin(H / \lambda) \qquad (22)$$

Loads (-F) may be obtained from "problem B_1" as shown on Fig.2. In the upper part the resulting sum must be zero; so, the part F_{int} of the load, balancing the reflected wave in the upper part of the soil, is just "mirror" of the free field:

$$F_{int} = -E^0 \frac{\partial u^0}{\partial x}(-H) = -(E^0 U^0 / \lambda)\sin(H / \lambda) = -\rho\omega c\, U^0 \sin(H / \lambda) \qquad (23)$$

The reflected wave u_1 in the lower part of the rod consists just of the single wave coming down. We know the displacement at Q; so, we can describe the additional wave in the lower part as

$$u^1(x) = -U^0 \cos(H / \lambda)\exp[i(x + H) / \lambda]; \quad \frac{\partial u^1}{\partial x}(x) = -(iU^0 / \lambda)\cos(H / \lambda)\exp[i(x + H) / \lambda] \qquad (24)$$

In this wave field at $x=-H$ we get the following force F_{ext}, impacting Q from the lower part of the soil

$$F_{ext} = E^0(iU^0 / \lambda)\cos(H / \lambda) = iU^0(\rho c^2 \omega / c) \cos(H / \lambda) = (i\omega\rho c)U^0 \cos(H / \lambda) \qquad (25)$$

Thus, the total load F in "problem B_1" is composed of two parts, given by (23) and (25):

$$F = F_{int} + F_{ext} = i\omega\rho c U^0[\cos(H / \lambda) + i\sin(H / \lambda)] = i\omega\rho c U^0 \exp(iH / \lambda) \qquad (26)$$

The same load F will impact Q in "problem B$_2$" and in "problem A$_2$". To complete the model we use the analogue between half-infinite rod and viscous dashpot, mentioned above (remember boundaries of Lysmer-Kuhlemeyer) and shown on Fig.4.

As a result, complete models for the direct approach look now like those shown on Fig.8

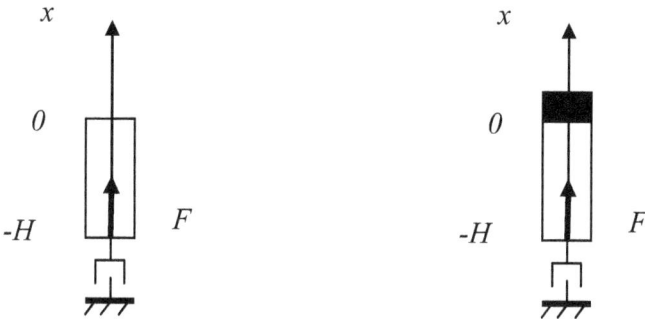

Figure 8. Direct approach models for "problem B$_2$" and "problem A$_2$"

As previously mentioned it is worth solving "problem B$_2$" first and compare the solution with free-field. Instead of that we can take "exact" displacement from exact wave field at Q, calculate total force impacting the upper part in the model on Fig.8 and check the equilibrium for Q. In our case the force impacting the upper part from dashpot is a product of exact velocity $[i\omega U_0(-H)]$ and viscous parameter of dashpot (ρc); we see that it annihilates with F_{ext} given by (25). The rest of the load F – the load F_{int} – annihilates with internal forces in the upper part due to the origin (23). So, "problem B$_2$" gives the exact solution.

Now let us check the equilibrium for Q in "problem A$_2$". The exact solution is given by (17,19,20):

$$U^A(x) = U^0\{\exp(-ix/\lambda) + [1 - i\frac{m}{\rho\lambda}]/[1 + i\frac{m}{\rho\lambda}]\exp(ix/\lambda)\}/2 \qquad (27)$$

So, Q is loaded by a) force from dashpot, b) internal force, acting from the upper part (not F_{int} any more, because in the upper part wave field has changed!) and c) the external load F given by (26):

$$-i\omega\rho c U^A(-H) + E^0\frac{\partial U^A}{\partial x}(-H) + F = 0 \qquad (28)$$

Let a reader obtain zero in (28) himself. General conclusion is that direct method provides exact results, if correctly applied.

Now let us perform the substitution of the boundary conditions in the direct approach, shown on Fig.3, i.e. let us fix the motion of Q as $U^0(-H)$. We see that "problem B$_2$" without structure will be solved exactly, providing $U^0(x)$ in the whole upper part. So, this check is OK. However, in "problem A2" exact displacement $U^A(-H)$ given by (27) is different from $U^0(-H)$ given by (22). As a result, both upcoming wave and the wave coming down in V_{int} will be different from the exact solution. If this new solution is marked with upper index "3", then the displacement of the mass is given by

$$U^3 = U_1^3 + U_2^3 = U^0 / [\cos(H / \lambda) - \frac{m}{\rho\lambda}\sin(H / \lambda)] \tag{29}$$

Note that this solution depends on H, while the exact solution does not (see (21)). As our boundary is artificial, such dependence cannot be physical.

We can calculate the "error coefficient" relating approximate solution (29) to the exact solution (21). With dimensionless frequency

$$\varpi = \frac{\omega m}{\rho c} = \frac{m}{\rho\lambda} \tag{30}$$

and dimensionless depth of Q

$$h = H\frac{\rho}{m} \tag{31}$$

this "error coefficient" makes

$$\mu = \frac{U^3}{U^A} = \frac{1 + i\varpi}{\cos(h\varpi) - \varpi\sin(h\varpi)} \tag{32}$$

Curves for different $\mu(\omega)$ for different h are shown in Fig.9.

We see that the solution in not satisfactory. The increase of the boundary depth h does not improve the situation. General conclusion is that one must be very careful in placing lower boundary in the direct approach when there is no rock seen in the depth.

6.3. Impedance approach

Now let us apply the impedance approach to the same system. As our structure rests on the surface, we can use Fig.6. If the displacement of the mass is U_b, then the equation of motion (10) turns to

$$[C(\omega) - \omega^2 m] U_b(\omega) = C(\omega) U^0(\omega) \tag{33}$$

Impedance is the same as in Fig.4: $C(\omega)=i\omega\rho c$. So, from (33) we at once get the ultimate result which turns to be similar to the exact one (21):

$$U_b = U^0 / [1 - \frac{\omega m}{i\rho c}] = U^0 / [1 + i\frac{m}{\rho\lambda}] = U^A \tag{34}$$

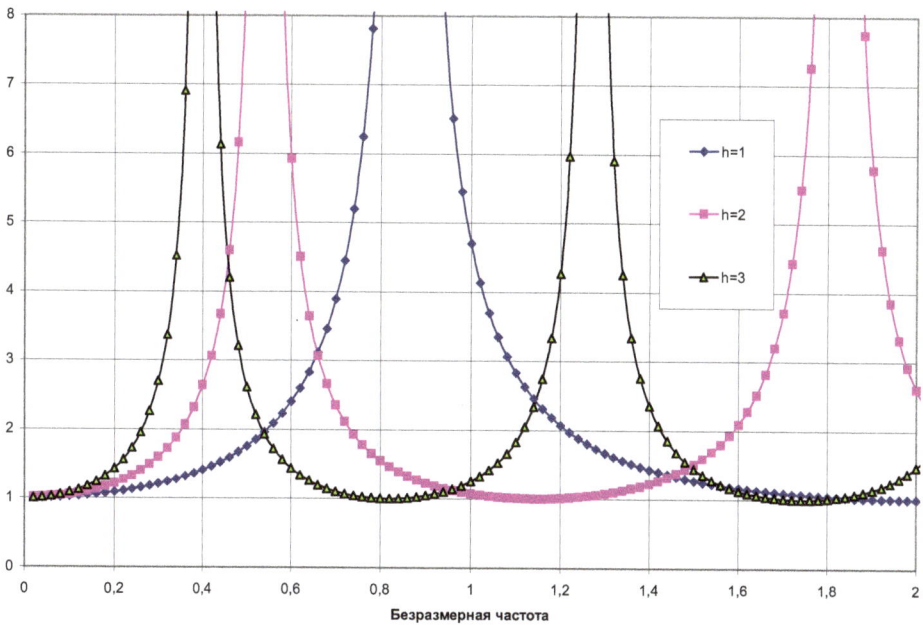

Figure 9. "Error coefficient" for direct approach after changing the lower boundary condition

General conclusion is that the correct application of the impedance method provides the exact solution.

The decisive factor in getting the exact solution both in direct and in impedance approaches for our 1D model was the ability to get exact model for substituting V_{ext} (dashpot in our case).

Unfortunately, this example is of methodology value only. While "problem B" (without structure) is often treated in practice with 1D models (though not homogeneous), "problem A" (with structure) is principally different. Great part of energy is taken from the structure by surface waves, which are not represented in 1D model.

7. Impedances in the frequency domain

Historically, the first problems for impedances considering soil inertia were solved semi-analytically for homogeneous half-space without the internal damping and for circular surface stamps. It turned out that horizontal impedances behaved more or less like pairs of springs and viscous dashpots (as mentioned above, in reality the dissipation of energy had completely different nature; mechanical energy was not converted in heat, as in dashpot, but taken away by elastic waves). One more "good news" was that the impedance matrix appeared to be almost diagonal. Though non-diagonal terms coupling horizontal translation with rocking in the same vertical plane were non-zero ones, their squares were considerably less in module than products of the corresponding diagonal terms of the impedance matrix.

These two facts created a base for using the above mentioned "soil springs and dashpots". There are several variants of stiffness and dashpot parameters. You can see below a table 1 from ASCE4-98 [1] for circular base mats.

Motion	Equivalent Spring Constant	Equivalent Damping Coefficient
Horizontal	$k_x = \dfrac{32(1-v)GR}{7-8v}$	$c_x = 0.576 k_x R\sqrt{\rho/G}$
Rocking	$k_\psi = \dfrac{8GR^3}{3(1-v)}$	$c_\psi = \dfrac{0.30}{1+B_\psi} k_\psi R\sqrt{\rho/G}$
Vertical	$k_z = \dfrac{4GR}{(1-v)}$	$c_z = 0.85 k_z R\sqrt{\rho/G}$
Torsion	$k_t = 16GR^3/3$	$c_t = \dfrac{\sqrt{k_t I_t}}{1+2I_t/\rho R^3}$

Notes: v = Poisson's ratio of foundation medium; G = shear modulus of foundation medium; R = radius of circular base mat; ρ = mass density of foundation medium; $B_\psi = 3(1-v)I_0/8\rho R^5$; I_0= total mass moment of inertia of structure and base mat about the rocking axis at the base; and I_t= polar mass moment of inertia of structure and base mat.

Table 1. Lumped Representation of Structure-Foundation Interaction at Surface for Circular Base

One more table of the same sort in the same standard [1] is given for rectangular base mats.

However, even for homogeneous half-space it turned out that vertical and angular impedances behaved differently from simple springs and dashpots. This can be seen in Fig.10, where the tabular impedances given by ASCE4-98 are compared to the wave solutions given by codes SASSI and CLASSI (number in the legend denotes number of finite elements along the side).

We understand now that the expressions given in the tables are just approximations for the real values. So, there cannot be "exact" or "true" expressions of this kind – different variants exist.

The expressions from the tables look particularly strange for angular damping, where parameters of the upper structure participate. It is real absurd from the physical point of view: impedances cannot depend on the upper structure; soil just "does not know" what is above rigid stamp. This is not an error, as some colleagues think. This is an attempt to take values from the frequency-dependent curves at certain frequencies. These frequencies estimate the first natural frequencies of rigid structure on flexible soil, so they depend on structural inertial parameters. Surely, such expressions are approximate.

The conclusion is that frequency dependence of the impedances exists even for the homogeneous half-space and spoils the spring/dashpot models in the impedance method.

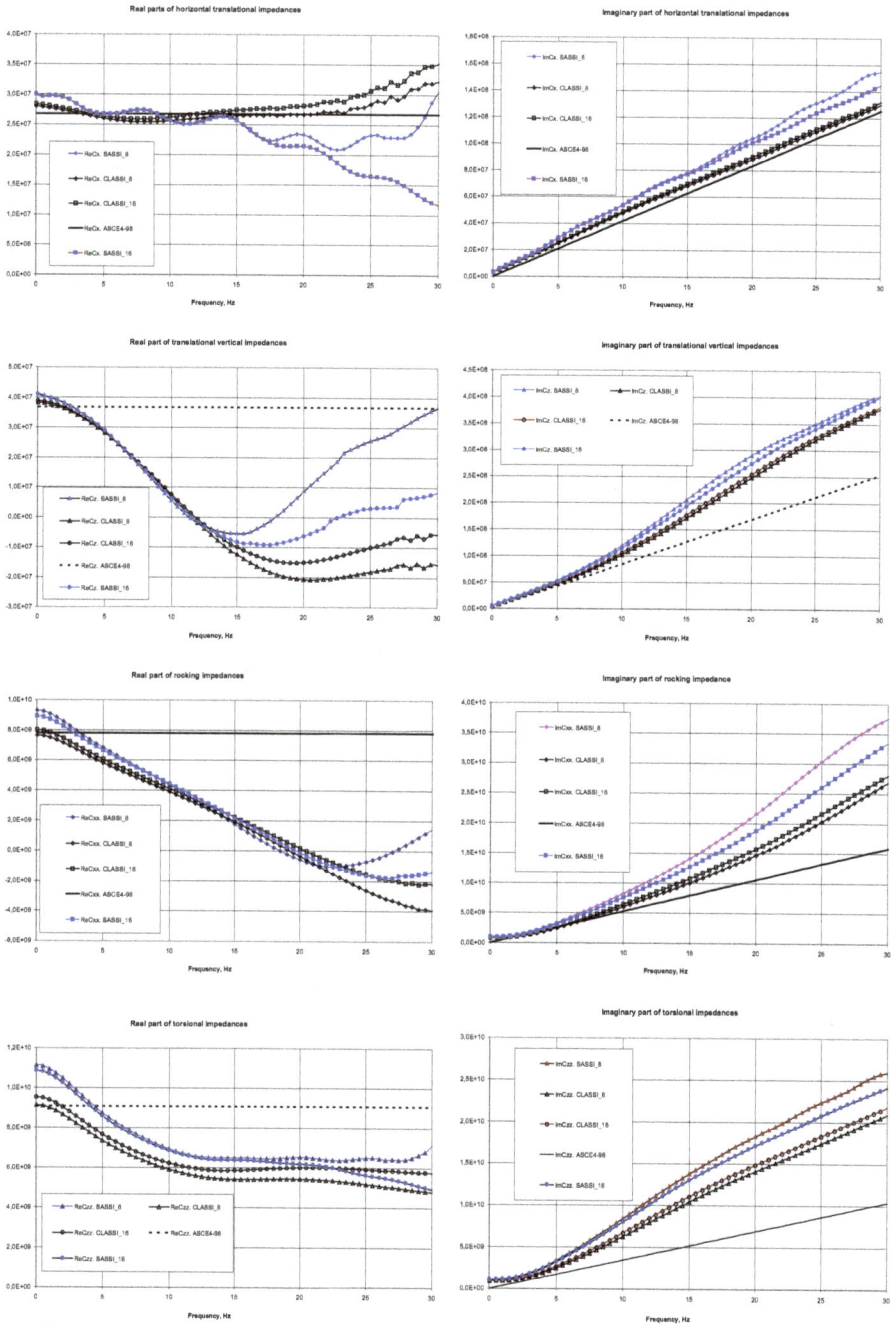

Figure 10. Impedances for a basement 30.6 x 30.6 meters on the surface of homogeneous half-space with mass density ϱ=2 t/m³, wave velocities V_s=400 m/s, V_p=1100 m/s, internal damping γ=5%.

One more comment should be added here. If a package of horizontal layers is underlain by rigid rock, surface waves behave in a completely different manner than for homogeneous half-space. Instead of two surface waves (Love and Rayleigh ones) there exist an infinite number of surface waves. Each of these waves for low frequencies cannot take energy from the basement – they are "geometrically dissipating" (even without internal soil damping), or "locked". However, when frequency goes up, each of these waves one by one transforms from a "locked" wave into a "running" wave capable to take energy to the infinity. This behavior depends on soil geometry only, not on structure.

It means that in the frequency domain there exists a certain low-frequency range, where the whole soil foundation is "locked". All the energy taken from the basement is only due to the internal damping. If there is no internal damping, complex impedances are completely real. In practice internal soil damping is several percent, so impedances are "almost real". After the first surface wave transforms into a running one, the soil foundation becomes "unlocked" – wave damping appears. Then one by one other surface waves turn into "running" ones, increasing the integral damping in the soil-structure system. The impedance functions for the same soil, but underlain by rigid rock at depth 26 m, are shown on Fig.11. One can see "locking phenomena" looking at the imaginary parts of the impedances.

The conclusion here is that the frequency dependence of the impedances may be rather sophisticated depending on soil layering. The attempts to "cover" the variety of soils by number of homogeneous half-spaces with different properties (usually this is an approach for "serial design" of structures) may lead to mistakes: there is always a possibility that real layered soil will not be "covered" by a set of half-spaces (e.g. no half-space can reproduce the "locking" effects described above).

The additional problem with springs and dashpots arises when the integral stiffness is distributed over the contact surface Q. Physically in every point of Q there are no distributed angular loads impacting basement from the soil. So, only translational springs and dashpots are usually distributed over Q, and angular impedances are the results of these distributed translational springs. For a surface basement vertical distributed springs are responsible for rocking impedances, and horizontal springs are responsible for torsional impedance. The problem here is that all attempts to find the distribution shape for vertical springs to represent rocking impedances simultaneously with vertical one have failed. If fact, integral rocking stiffness obtained from distributed vertical springs is always less than actual rocking stiffness; on the contrary, integral rocking damping obtained from distributed vertical dashpots is always greater than the actual one. Physical reason of this mismatch is the interaction between different points through soil. Spring/dashpot model is "local" in nature: the response is determined by motion of this very point, and not neighbors. This is not physically true.

The author found a way to treat both problems at once. The idea is to work in the time domain using a platform model of Fig.5 with conventional springs and dashpots (lumped or distributed). Of course we get some "platform" impedances $D(\omega)$, different from "wave" impedances $C(\omega)$, but the idea is to tune the platform excitation V_b so to account for the difference between "wave" impedances and "platform" impedances. Six components of the platform excitation may be tuned to reproduce six components of response – e.g., six components of the rigid base mat's accelerations. Such an approach combines the calculations in the frequency domain (platform seismic input) with calculations in the time domain (final dynamic analysis of the platform model), that is why this method is called "combined". Besides, this method is "exact" for rigid base mats only: the stiffer is a base mat, the more accurate are the results. That is why this method is also called "asymptotic" – full name is "combined asymptotic method" (CAM) [22].

The last item to discuss in this part is practical tools to obtain impedances and seismic loads (or weightless base mats' motions) in the frequency domain. At the moment the author uses one of two computer codes. For rigid surface basements on a horizontally-layered soil code CLASSI is the most appropriate. For the embedded basements with possible local breaks in horizontal layering code SASSI is used (SASSI can be used for surface base mats also, but is more sophisticated).

In both cases formula (3) is a basic equation for impedances, and the dynamic stiffness matrix G_0 linking set of nodes in the infinite soil is a key issue (the second matrix G_{int} is absent for surface basement in CLASSI and easily obtained by FEM for the embedded basement in SASSI). To get G_0, they first obtain a dynamic flexibility matrix, describing displacements due to the unit forces (this is a Green's function). Here is a difference between two codes.

Professor J.Luco managed [7] to develop Green's function analytically for the case of surface load and surface response node in horizontally-layered soil in the frequency domain. Then contact surface Q was covered with number of rectangular elements of different shapes. Loads were applied in the centers of each element one by one; response displacements were obtained in the centers of each element for each load. The additional convenience was that for the horizontally layered soil one can shift the loaded node and the response node horizontally, and the link between them stays the same. So, in fact one needs Green's functions only for a single loaded node and for the response nodes placed at a distance from the loaded node within maximal size of the mat. Zero (for vertical load) and first (for horizontal load) Fourier terms describe the displacement field along angular cylindrical coordinate, enabling to store Green's functions only along 1D radial line. Thus, it is not very time-consuming to obtain Green's function (it is done in a separate module "Green") and to use it in order to obtain the full flexibility matrix (in separate module "Claff"). Then this full flexibility matrix is turned into a full stiffness matrix G_0. Finally G_0 is condensed to the 6 x 6 impedance matrix C. The whole procedure is repeated for each frequency of the prescribed set. For a surface rigid basement and vertical waves there is no separate problem to find the load matrix B in equation (4): kinematical interaction does not change control motion, so B consists of the first three columns of C. This was a brief description of CLASSI ideology.

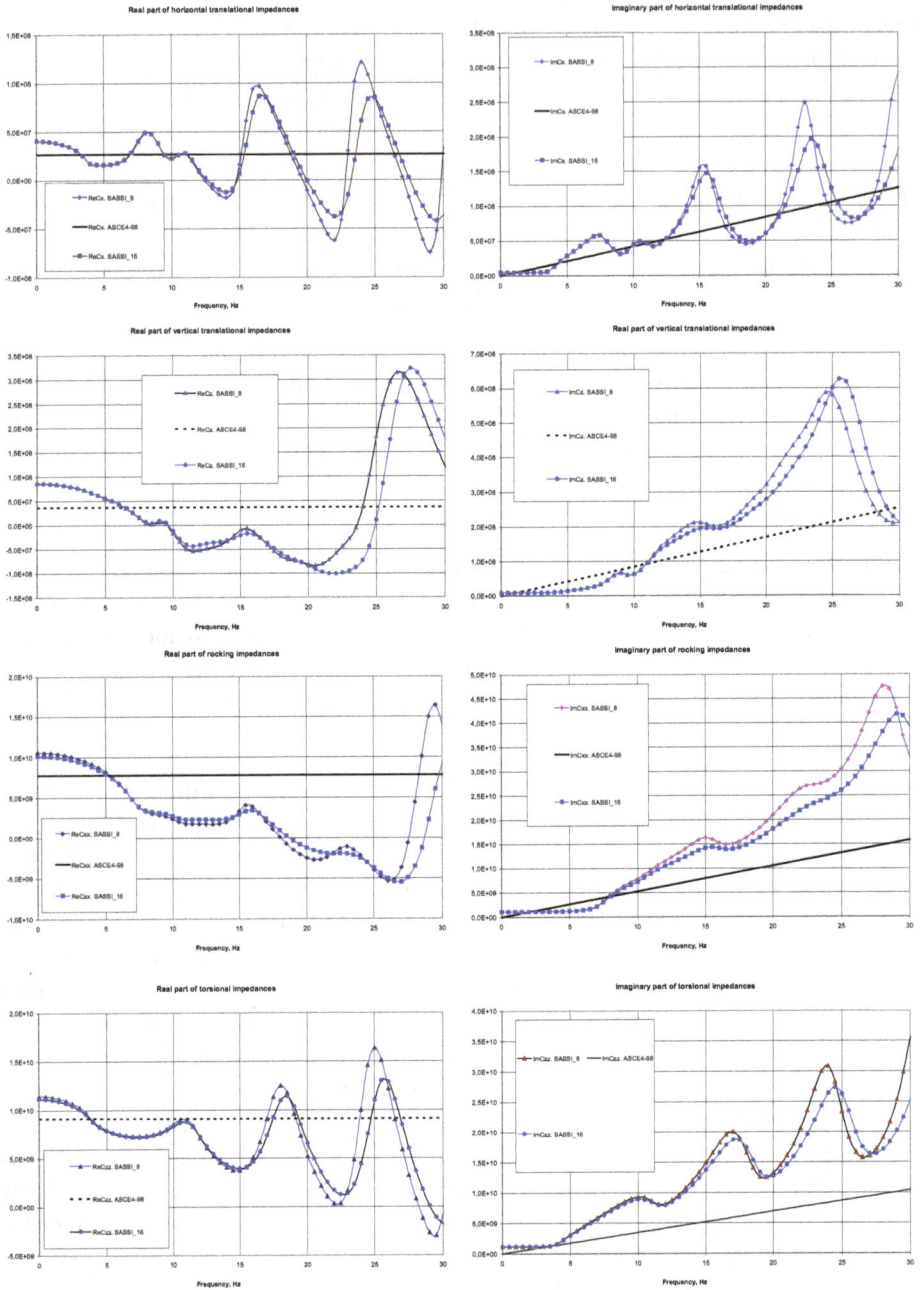

Figure 11. Impedances for a basement 30.6 x 30.6 meters on the surface of homogeneous layer with mass density ϱ=2 t/m³, wave velocities V_s=400 m/s, V_P=1100 m/s, internal damping γ=5%, thickness h=26 m, underlain by rigid rock

J.Lysmer for the same problem (but for the embedded loaded and response nodes) used direct approach described above. He made V_{int} a cylindrical column, with just a single element per radius. The bottom of this column was placed on a homogeneous half-space, so a dashpot was put there in each of three directions. Lateral boundaries were of Waas type. Loads were applied to the nodes at the vertical axis of cylinder. The response nodes were either at the same axis or out of the cylinder (for regular mesh the radius of cylinder was set equal to 0.9 of mesh size). In the first case, the response displacements were obtained from the FEM problem. In the second case the solution of the FEM problem gave the displacements of the nodes at the lateral boundary. Then the displacements in the whole infinite volume V_{ext} were obtained using "homogeneous wave fields" in cylindrical coordinates described above (with Hankel's functions along radius). Like in CLASSI, horizontal shift of the loaded and response nodes is allowed in the horizontally-layered soil. So only a single cylinder has to be studied. Like in CLASSI, it is done in a separate module (in SASSI it is called POINT). After a flexibility matrix for a set of "interacting nodes" covering V_{int} is obtained, it is turned into a stiffness matrix G_0. This matrix may be condensed into a part of the impedance matrix like in CLASSI (another part comes from G_{int}).

However, one can instead return to the "problem A2" without condensation (e.g., for flexible basement or locally modified soil around the basement). Bottom of surface Q is placed not on the rock, as it was in Fig.3, but where it is convenient (usually at the bottom of the modified soil volume, if any, or at the bottom of the basement). This approach may be called a "combined" one, because the direct approach is used to get Green's function only; further on, the impedance approach (extended for flexible underground volume V_{int}) is applied.

The problem of axial symmetry has gone both in CLASSI and in SASSI: though "basic problem" in Green or in POINT has an axisymmetric geometry and is solved in cylindrical coordinates, the "interaction nodes set" may be arbitrary in shape.

Another problem for SASSI is with flexible underlying half-space: Waas boundaries were developed for rigid half-space at the bottom only. Lysmer's team found a brilliant approximate solution. Energy for a cylinder in POINT is taken down in the half-space by vertical body waves (modeled by dashpots, see above) and by surface Rayleigh waves in the flexible half-space. These Rayleigh waves in the frequency domain have a certain depth where the wave displacements are close to zero. One can put "rigid" bottom at that very depth without spoiling the response in the upper part of the soil. The problem is that this depth is frequency-dependent: so, the model in POINT becomes non-physical. However, transmitting boundaries, as we remember, were developed by G.Waas in the frequency domain, so the problem is solved frequency by frequency. Hence, one can change the depth of "rigid boundary" at the bottom of the model according to the frequency change. In SASSI it is done automatically: soil model in POINT consists of the upper part with fixed layers, and of the lower part with frequency-dependent layers.

For surface basements one can compare the impedances given by SASSI and CLASSI. In the frequency range below 15 Hz the results are very close to each other. Examples were shown above (see Fig.10).

8. Soil non-linearity – SHAKE ideology – Contact non-linearity

Soil in reality is considerably non-linear. That is why H.B.Seed [23] suggested special method to describe seismic response of horizontally-layered soil to the vertically-propagating free-field seismic waves by equivalent linear model. Parameters of this model are obtained in iterative calculations, as modules and damping in each layer are strain-dependent. This method is used in the famous SHAKE code [24]. Usually 7-8 iterations are enough to get the error less than 1%.

SHAKE calculations may be started from the surface (i.e. surface motion is given; for each other layer the motion is to be calculated together with equivalent soil properties); this case is called "deconvolution". On the contrary, SHAKE calculations may be started from some depth where the motion is given; this case is called "convolution". Motion for one and the same point in depth may be given in one of two variants. The first variant is just a motion of the point inside soil composed of the upcoming wave and wave coming down (see the 1D example in p.5 above). The second variant is to set just the upcoming wave amplitude. Sometimes this motion is called "outcropped" motion, because they double the upcoming wave amplitude, like it is at the free surface (see the example in p. 6.1 above). However, such a "doubled" motion will be similar to the motion of the really outcropped surface only at the surface of the underlying homogeneous half-space, and not in the middle of the layered soil.

The typical case consists of two sequential SHAKE calculations. First, deconvolution is performed using the initial soil model and initial seismic motion of the free surface of the initial soil (the author usually calls it "foundation no.1"). Note, that the underlying half-space is not modified in the process of SHAKE iterations. That is why one should include into the modified part at least one upper layer of half-space. After the deconvolution is through, one should change the half-space manually making parameters equal to the lowest modified layer. The main result of deconvolution is an "outcropped" motion of the half-space surface. The second stage is a convolution. Half-space stays the same as in deconvolution (after manual modification), but the upper layers may change due to various reasons starting from the direct changes in the construction process (very often the upper several meters of the initial soil are withdrawn). Sometimes the weight of the future structure changes the properties of the soil. Anyhow, we generally have "foundation no.2" instead of the first soil foundation. In the convolution procedure the outcropped motion of the half-space surface is given, and the whole wave field with equivalent soil properties is obtained as a result together with motion of the new surface.

Typical strain-dependent characteristics of different soils are shown on Fig.12. Sometimes they call it "degradation curves".

One should remember that degradation curves for shear modulus G are given in relative terms G/G_0, where G_0 is a low-strains value. On the contrary, strain-dependent characteristics of the internal damping are given for dimensionless coefficients without using low-strain values.

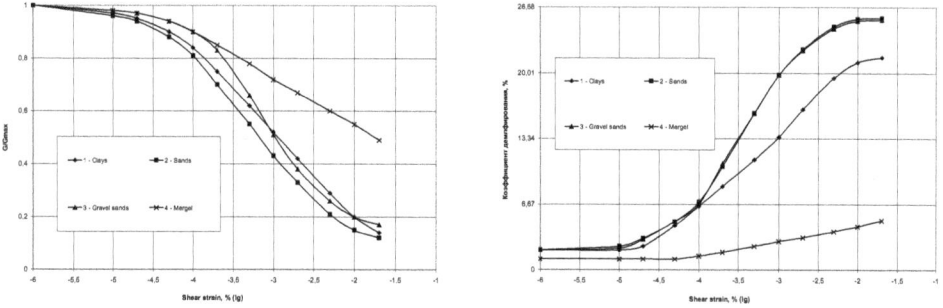

Figure 12. Typical strain-dependent characteristics of soils

Typical results of SHAKE calculations for three levels of seismic intensity are given on Fig.13. Abbreviations of the level names mean: OBE – operational basis earthquake; SSE – safe shutdown earthquake (operations and shutdown refer to nuclear reactors activity). The results are calculated separately in two vertical planes (OXZ and OYZ) and then averaged (usually in two planes they are close to each other for every layer).

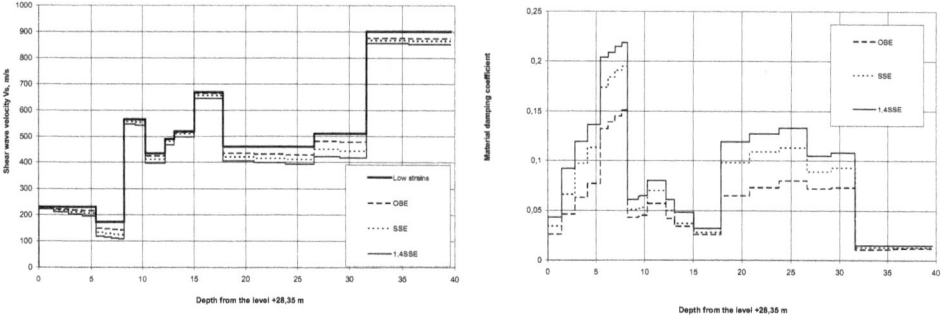

Figure 13. Typical results of the SHAKE calculations

We see that near the surface the equivalent soil characteristics are close to the low-strain values. This is because of zero strains at the free surface for every intensity level.

Usually the greatest relative shift in velocities and damping is at the bottom of soft soil layers. If the intensity is very high, the iterations during deconvolution may not converge. Physically it means that the given surface motion is no longer compatible with given soil properties; such motion cannot be transferred by given soil to the surface.

Both SHAKE results – profiles of wave velocities/damping and surface motion – are further on used in SSI analysis.

Theoretically a wide range of nonlinearities can be described by strain-dependent properties. However, these properties should be transient, i.e. they should vary from one time point to another during one seismic event. In the Seed's approach these properties in each soil layer are established once for the whole event duration (changing not from time to time, but from one linear run to another one). So, in fact soil properties depend not on the instant transient strains, but on some "effective" strains (in practice - on some portion of the maximal strain over the duration). In fact, Seed provided a tool to extrapolate the results of the lab test with harmonic excitation of the soil sample to another situation with non-harmonic wave in the same sample.

Nowadays seismologists are ready to provide more sophisticated approaches to the soil description. They can model accelerations in the soil more accurately with truly transient soil properties. But in the SSI problems (e.g., in SASSI or CLASSI) one will need linear soil with "effective" soil properties, so SHAKE still remains the best "pre-SSI" processor.

The result of SHAKE is obtained for horizontally-layered soil without structure. So, they say, that it contains "primary" non-linearity. The same equivalent strain-dependent properties can be further applied in the SSI problems to account for the "secondary" nonlinearity caused by structure, but the result will spoil horizontally-layered geometry of the soil. As we know, that will spoil CLASSI model and create additional problems in SASSI model.

That is why modern standards (e.g., ASCE4-98) require the consideration of the primary non-linearity, but do not require consideration of the secondary non-linearity of the soil.

Looking at soil-structure system, one can find nonlinearities not only in soil, but at the contact surface. Even if both soil and structure are modeled by linear systems, they may have various contact terms. Most often full contact is assumed – it does not spoil linearity. However, soil tension (unlike soil compression) is very limited, and that may cause a) uplift of the base mat from the soil, b) separation of the embedded basement walls from the soil.

Base mat uplift may be estimated if linear "full contact" vertical forces over contact surface Q are compared to the static vertical forces caused by structural weight. In practice the full uplift is seldom met, but rocking of the structure can cause dynamic tension near the edges of the base mat. This "partial" dynamic uplift usually occurs for stiff soils and sizable structures. Does it change the response motion considerably? Today they believe that if the area of partial dynamic uplift is less than 1/3 of the total contact area, one can neglect this uplift and still use SSI linear model with full contact.

Separation of vertical embedded walls is treated as follows. In the upper half of the embedment depth (but only up to 6 meters from the surface) they break soil-structure contact completely. Below this level they assume full contact.

9. Non-mandatory assumptions

So, basic assumptions currently used in the SSI analysis are a) linearity of the soil, of the structure and of the soil-structure contact; b) horizontal layering of the soil (except some limited volume near the structure).

There are two other assumptions - not mandatory, but usually used in the SSI analysis. The first assumption is the rigidity of the soil-structure contact surface. Usually base mats are not extremely rigid, but they are considerably enforced by rather dense and thick shear walls, so in fact their behavior is almost rigid. Standards ASCE4-98 allow the treatment of base mats of the NPP structures as rigid ones. However, SASSI can treat flexible base mats as well. Different parameters of structural seismic response show different sensitivity to the flexibility of the base mat. Some examples are presented in the author's reports in SMiRT-21 [25,26].

The second assumption is about seismic wave field in the soil without structure. Usually, one starts from the three-component acceleration recorded on the surface of the soil (in some "control point", as they say). As we saw, in SSI problem one needs to know the motion of a certain soil volume (at least the soil motion in the nodes of the future contact surface). So some additional assumptions are introduced. The most common assumption "is vertically propagating body seismic waves in horizontally-layered medium". This assumption means, that three components of the acceleration in the control point are produced by three separate vertically propagating waves: vertical acceleration is a result of the P-wave, two horizontal accelerations are the result of two S-waves in main coordinate vertical planes. Each wave can be analyzed separately by SHAKE, providing seismic motion of any point in depth. Another consequence of this assumption is that seismic motion depends on vertical coordinate, but not on the horizontal coordinates: every horizontal plane in the free field moves "rigidly".

Again, this assumption is not mandatory. In SASSI one can set up other assumptions linking the whole wave field to the control point motion. However, as SHAKE (usually used as a preprocessor for SASSI) implements that very assumption, most often this assumption is accepted for the whole SSI problem.

10. Some examples of the SSI effects in practice

Concluding the chapter, the author should like to give some practical examples. One of recently built NPPs was analyzed for different soil and excitation models. Let us look at the acceleration response spectra on the elevation +21.5 m.

The first comparison is for rigid soil and flexible soil (without embedment). As we remember, rigid soil means the absence of the SSI effects. Flexible soil in this case was of medium type. Structure was one of the NPP buildings. One and the same three-component seismic excitation (corresponding to the standard spectra described in RG1.60) was applied to the surface of the rigid and flexible soil. In the first case this very motion became the response motion of the base mat; in the second case, the response motion of the base mat was modified by the SSI effects. The comparison of the floor response spectra (enveloped over 8 corner nodes of the floor slab, smoothed and broadened 15% each side in the frequency range) at the structural level +21.5 m is shown in Fig. 14.

We see the considerable difference in the spectral shape: SSI effects form the main spectral peak, but for high frequency range (here - after 5 Hz) spectral accelerations with SSI are less than without SSI. As to the maximal accelerations (we see them in the right part of the spectral curves), SSI effects may decrease them (see Fig.14b) or not. It sometimes depends on the height of the floor considered.

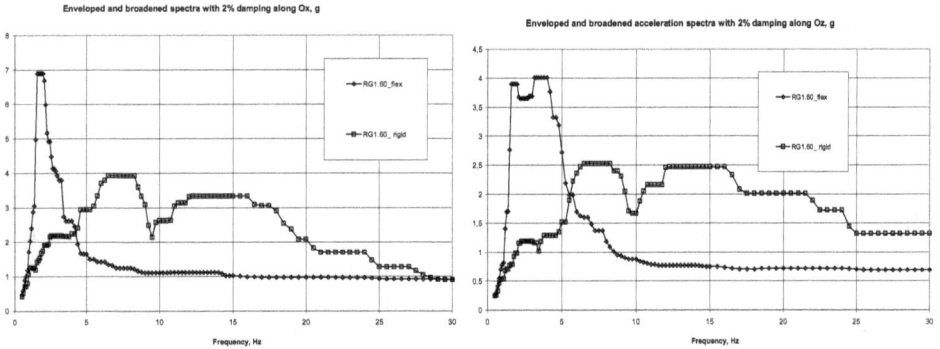

Figure 14. Comparison of enveloped broadened spectra with 2% damping at the level +21.5 m for excitation RG1.60 for rigid and flexible soil; spectral accelerations are shown along horizontal axis Ox (a) and vertical axis Oz (b)

The second comparison is for surface and embedded base mats. Site-specific three-component seismic excitation is "applied" at the free surface of two flexible soil foundations: first, real soil foundation; second, the same soil foundation without upper 10.4 meters of soil (corresponding to the embedment depth of the structure). The comparison of the response spectra is shown in Fig.15 in the same format as in Fig.14.

We see that the embedment considerably impacts the first spectral peaks, decreasing spectral accelerations. The physical reason is that the mass of the "outcropped soil" for the embedded basement in fact is subtracted from the mass of the basement when inertial loads are developed for the platform model of "inertial interaction" in the moving coordinate system placed on the platform.

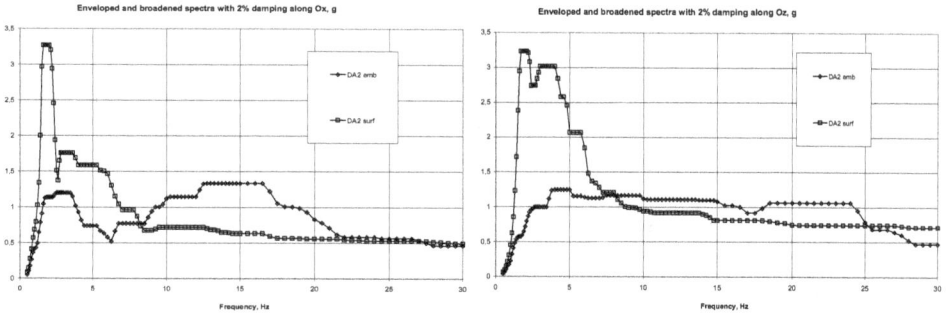

Figure 15. Comparison of enveloped broadened spectra with 2% damping at the level +21.5 m for site-specific excitation for surface and embedded structures; spectral accelerations are shown along horizontal axis Ox (a) and vertical axis Oz (b)

On the other hand, for higher frequencies there may appear spectral accelerations for the embedded structure greater than those for the surface structure (see Fig.14a). Still the overall effect of the small embedment is considered conservative; that is why standards ASCE4-98 [1] allow neglecting the embedment, if the embedment depth is less than 30% of the equivalent radius of the basement. Interesting to note that in the sample shown in Fig.15, this limit is almost met (10.4 m of embedment versus 34.7 m of equivalent radius). Nevertheless, the difference in the spectral shape in Fig.15 is dramatic.

11. Conclusions

Concluding the chapter about soil-structure interaction (SSI), the author should like to give several recommendations to engineers.

1. At the very beginning one should estimate the importance of SSI and decide whether it should be considered at all. The answer depends on the soil data (wave velocities in the soil, first of all), base mat size/embedment and inertia of the structure. For civil structures most often SSI can be omitted (i.e. structure can be analyzed using a platform model with platform kinematical excitation given by seismologists and without soil springs).

2. If SSI is to be considered, one should examine whether some simple assumptions can be applied. Main assumptions: homogeneous half-space or a layer underlain by rigid rock as a soil model (depends on real-world geotechnical data), surface base mat, rigid base mat. General recommendation is as follows. One should start with the simplest model allowed by standards. Only if the results seem overconservative, one should try to go to more sophisticated models, accounting to various specific SSI effects. Usually there is a trade between the complexity of the model and the conservatism of the results.

3. SSI effects are frequency-dependent. Most of effects are valid in a certain frequency range. Out of this range they may lead to the opposite changes.
4. SSI analysis requires special tools. General-purpose codes, including FEM soft, cannot treat SSI properly because of the infinite geometry of the initial problem. Special-purpose tools like CLASSI, SASSI, etc. should be used.
5. If direct approach is used, special attention should be paid to the boundaries. Preliminary analysis of test examples (e.g., initial soil without structure with the same boundaries and excitation) is strongly recommended.
6. Wave nature of SSI effects requires special attention when FEM is used: element size for the soil and time step must be compared with frequency ranges of interest. Otherwise, the most significant effects may be missed.
7. Non-linearity of different kinds is to be treated properly. Primary non-linearity of the soil is handled by SHAKE. Contact non-linearity is treated approximately as described above. If a structure itself is considerably non-linear, usually one has to omit wave SSI at all.

Nowadays, the research goes forward. The current goal is to combine non-linearity inside V_{int} (including contact non-linearity and structural nonlinearity) with linearity of infinite soil in V_{ext}.

Author details

Alexander Tyapin
"Atomenergoproject", Moscow, Russia

12. References

[1] Seismic Analysis of Safety-Related Nuclear Structures and Commentary. ASCE4-98. Reston, Virginia, USA. 1999.

[2] A Methodology for Assessment of Nuclear Power Plant Seismic Margin (Revision 1). EPRI NP-6041-SL. Revision 1. Project 2722-23. August 1991. California, USA.

[3] Reissner E (1936) Stationare, axialsymmetriche durch eine Shuttelnde Masse erregte Schwingungen eines homogenen elastischen Halbraumes. Ingenieur-Archiv. 7/6: 381-396.

[4] Luco J.E (1982) Linear Soil-Structure Interaction: A Review. Earthquake Ground Motions and Its Effects on Structures. Applied Mechanics Division, ASME 53: 41-57.

[5] Gulkan P, Clough R /Editors (1993) Developments in dynamic soil-structure interaction/Ed. NATO Advanced Institutes Series. Series C: Mathematical and Physical Sciences. Vol.390. Dordrecht/Boston/London: Kluwer Academic Publishers. 439 p.

[6] Seed H, Lysmer J (1977) Soil-Structure Interaction Analysis by Finite Element Method. State of the Art. Transactions of the International Conference on Structural Mechanics in Reactor Technology (SMiRT-4). San Francisco. Vol.K. K2/1.

[7]   Luco J (1976) Vibrations of a Rigid Disc on a Layered Viscoelastic Medium. Nuclear Engineering and Design. 36: 325-340.

[8]   Kausel E, Roesset J, Waas G (1975) Dynamic Analysis of Footings on Layered Media. J. of Engineering Mechanics Div., ASCE. 101. EM5: 679-693.

[9]   Lysmer J, Tabatabaie R, Tajirian F, Vahdani S, Ostadan F (1981) SASSI - A System for Analysis of Soil-Structure Interaction. Research Report GT 81-02. University of California, Berkeley.

[10] Wolf J.P (1985) Dynamic Soil-Structure Interaction. Prentice-Hall, Englewood Cliffs, NJ.

[11] Wolf J.P (1988) Soil-Structure Interaction Analysis in Time Domain. Prentice-Hall, Englewood Cliffs, NJ.

[12] Kausel E, Roesset J (1977) Semianalytic Hyperelement for Layered Strata. J. of Engineering Mechanics Div., ASCE. 103. EM4: 569-588.

[13] Kuhlemeyer R, Lysmer J (1973) Finite Element Method Accuracy for Wave Propagation Problems. J. of Soil Mechanics and Foundations Div., ASCE. 99. SM5: 421-427.

[14] Lysmer J, Kuhlemeyer R (1969) Finite Dynamic Model for Infinite Media. J. of Engineering Mechanics Div., ASCE. 95. EM4: 859-877.

[15] Lysmer J, Udaka T, Seed H, Hwang R (1974) LUSH - a Computer Program for Complex Response Analysis of Soil-Structure Systems Rep. EERC 74-4. Berkeley, California.

[16] Lysmer J, Waas G (1972) Shear Waves in Plane Infinite Structures. J. of Engineering Mechanics Div., ASCE. 98. EM1: 85-105.

[17] Waas G (1972) Linear Two-Dimensional Analysis of Soil Dynamics Problems in Semi-infinite Layered Media. Ph.D. Dissertation. University of California, Berkeley.

[18] Lysmer J, Udaka T, Tsai C, Seed H (1975) FLUSH - A Computer Program for Approximate 3-D Analysis of Soil-Structure Interaction Problems. Report No.75-30. University of California, Berkeley.

[19] Berger E, Lysmer J, Seed H (1975) ALUSH - A Computer Program for Seismic Response Analysis of Axisymmetrical Soil-Structure Systems. Report No.75-31. University of California, Berkeley.

[20] Tyapin A (2007) The frequency-dependent elements in the code SASSI: a bridge between civil engineers and the soil-structure interaction specialists. Nuclear Engineering and Design. 237: 1300-1306.

[21] Tyapin A (2011) Modal approach to the condensed dynamic stiffness evaluation // Journal of Mechanics Engineering and Automation. 1. No.1: 1-9.

[22] Tyapin A (2010) Combined Asymptotic Method for Soil-Structure Interaction Analysis. J. of Disaster Research. 5. No.4: 340-350.

[23] Schnabel P, Seed H, Lysmer J (1972) Modifications of Seismograph Records for Effects of Local Soil Conditions. Bulletin of Seismological Society of America. 62. N 6: 1649-1664.

[24] Schnabel P, Lysmer J, Seed H (1972) SHAKE - a Computer Program for Earthquake Response Analysis of Horizontally Layered Sites. Rep. EERC 72-12. Berkeley, California.

[25] Tyapin A (2011) The effects of the base mat's flexibility on the structure's seismic response. Part I: wave solution. SMiRT21. New Delhi. #85.

[26] Tyapin A (2011) The effects of the base mat's flexibility on the structure's seismic response. Part II: platform solutions. SMiRT21. New Delhi. #266.

Chapter 6

Full-Wave Ground Motion Forecast for Southern California

En-Jui Lee and Po Chen

Additional information is available at the end of the chapter

1. Introduction

The damages and loses caused by earthquakes are increased as urbanization increased in past decades. However, scientists currently cannot predict the time, location and magnitude of an earthquake accurately. Currently, the earthquake predictions are not yet reliable, so long-term probabilistic earthquake hazard analysis and rapid post-earthquake early warning are two alternative solutions to reduce potential earthquake damages [1-3].

For long-term ground motion forecasts, seismic hazard maps are widely used for estimating probabilities of ground motion exceeding certain amount in different areas in 50 years. In addition, the ground motion estimations in seismic hazard maps are useful for different applications, including, for instance, building codes, insurance rates, land-use policies and education of earthquake response. In United States, the U.S. Geological Survey (USGS) periodically updates the National Seismic Hazard maps that provide a 50 years ground motion estimations for United States[4, 3]. To consider effects form different aspects, the National Seismic Hazard maps incorporate both geological and geophysical information in ground motion estimations[3]. Recent advances in computational seismology allow us to simulate wave propagation in complex velocity structure models and then open the probability of physics-based long-term ground motion estimations. The Southern California Earthquake Center (SCEC) has developed a methodology which considers both source and structure effects in ground motion simulations for long-term ground motion estimations in Los Angles region[5].

The earthquake early warning systems are designed for short-term ground motion forecasts. The idea of earthquake early warning systems is based on the transmission speed of electromagnetic signal is much faster than the propagation speed of seismic shear-waves and surface waves that usually generate strong ground motion [2]. The Earthquake Alarms Systems, ElarmS, is the earthquake early warning systems designed for California region [1,

2]. The ElarmS uses the signal of the first arrived primary waves to estimate magnitudes, locations and then estimate peak ground motions of earthquakes [1, 2]. Here we propose a rapid full-wave Centroid Moment Tensor (CMT) inversion method for earthquakes in Southern California [6]. The algorithm has potential for (near) real time CMT inversion and then use optimal CMT solutions to generate peak ground motion maps for earthquake early warning purposes. In addition, the full-wave ground motion forecasts, which include basin amplification effects, source effects and wave propagation effects in the 3D structure model, will provide more accurate and detailed estimations.

2. USGS National Seismic Hazard Maps

In the United States, the USGS incorporates different geophysics and geological information to continually update the National Seismic Hazard Maps for log-term ground motion forecasts[4, 3]. In USGS hazard maps, source models, including seismicity models and faults source models, and attenuation relations are two main components[3]. The Southern California is included in the western U.S. hazard maps, so here we take western U.S. as an example for explaining the procedures of hazard maps of USGS.

To estimate potential seismicity, we need to consider earthquake recurrence in or near the locations of past earthquakes occurred and the possibility of earthquake occurrences in areas never have earthquakes. First, the gridded-seismicity models are based on earthquake catalogs and historical earthquakes. The seismicity rates in each grid ($0.1°$ longitude by $0.1°$ latitude) are based on the number of earthquake in it [3]. To smooth the seismicity rates, a 2D Gaussian function is applied to the model[3]. In most of areas the correlation distance is 50 kilometers, but in high seismicity regions the correlation distances parallel to the seismicity trends is 75 kilometers and normal to the seismicity trend is 10 kilometers to avoid effecting the seismicity estimations near the fault zones[3]. The uniform background seismicity models are used to estimate the possibilities of random earthquakes in aseismic regions [3]. The western U.S. region is separated into few sub-regions and the uniform background seismicity rate in each sub-region is based on the annual seismicity rates of earthquakes with Mw ≥ 4 since 1963 [3]. Now, there are two seismicity rate estimations in each grid cell. If the uniform background seismicity rate is larger than the gridded-seismicity rate in a grid, the final seismicity rate is the sum of 67% gridded-seismicity rate and 33% of uniform background seismicity rate in that grid; otherwise, the final seismicity rate just equals to the gridded-seismicity rate in the grid [3].

Existing fault zones have relative high possibilities of occurring destructive earthquakes. The fault source models are based on geological fault studies, geodesy and seismological date to estimate geometries, maximum magnitudes and recurrence periods for fault zones [3, 7]. To obtain fault geometries, the geological surveys and earthquake location distributions are used for estimating fault areas. The maximum magnitudes in fault zones could be inferred from relationships between fault areas and magnitudes or historical magnitudes [3, 7]. The Gutenberg-Richter magnitude-frequency distribution and the characteristic rate on a fault, ratio of the slip rate to the slip of the characteristic earthquake of the fault, are used in earthquake recurrence estimations [3]. In California region, USGS gives 67% on the characteristic rate and 33% on the Gutenberg-Richter [3]. The Uniform California Earthquake Rupture Forecast, Version 2 (UCERF 2) [7] presented in 2007 Working Group on California Earthquake Probabilities (WGCEP) is used as the fault source model in California region. In seismic hazard maps, fault sources only consider type-A faults that have information on fault geometries, slip rates and earthquake data and type-B faults that only have information on fault geometries and slip rates[3].

In California region, the gridded-seismicity model is derived form earthquake catalog and estimates probabilities of earthquakes between Mw 5 to 7.0 [3]. In addition, the fault models also estimate the possibilities of earthquakes with Mw larger than 6.5 to consider the possibilities of destructive earthquakes in fault zones [3]. When the two types of source models are put into seismic hazard maps the probabilities of earthquakes between Mw 6.5 to 7.0 may over estimated. For more accurate estimations, the seismicity rates of Mw ≥ 6.5 in gridded-seismicity model reduced by one-thirds in fault zones [3].

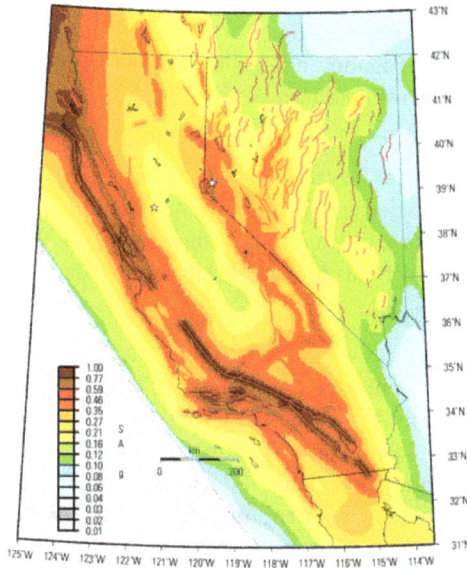

Figure 1. The California seismic hazard map of 1 Hz spectral acceleration (SA) for 2% exceedance probability in 50 years. Adopted from [3].

The Next Generation Attenuation (NGA) database developed by Pacific Earthquake Engineering Research Center (PEER) is used in USGS hazard maps as attenuation relations for ground motion predictions[8, 9, 3]. The NGA database is not only an empirical ground motion model derived from selected recordings but also includes 1D ground motion simulations, 1D site response, and 3D basin response results from other studies [8]. So, the database includes many essential effects, including, for example, basin response, site response, earthquake rupture properties and style of faulting.

Hazard curves, exceedance probability as a function of ground motion, are derived from source models and attenuation relations of grids. The final seismic hazard maps are made by interpolating annual exceedance probabilities form hazard curves in the model. On the California 1 Hz spectral acceleration (SA) hazard map [Figure 1], high hazard level regions are controlled by the major faults in California.

3. A Physics-based seismic hazard model: CyberShake

The CyberShake, one of the Southern California Earthquake Center's (SCEC) projects, is a seismic hazard model that uses full-wave method to simulate ground motions in Southern California. Here the term "full-wave" means using numerical solutions to compute the exact wave equation, rather than approximations. Recent advances in computational technology and numerical methods allow us to accurately simulate wave propagations in 3D strongly heterogeneous media [10, 11], and opened up the possibility of simulation-based seismic hazard models [5],extracting more information from waveform recordings for seismic imaging [12-14] and earthquake source inversions [15, 6]. For seismic hazard model, these physics-based simulations consider factors that affect ground motion results, for example, source rupture and wave propagation effects in a 3D velocity structure and then provide more accurate ground motion estimations.

The Los Angeles region is one of the most populous cities in the United States. The city is in a basin region and near active fault systems, so a reliable seismic hazard model is important for the city. The CyberShake selected 250 sites and simulated potential earthquake ruptures in Los Angeles region to build a seismic hazard model [5]. The SCEC Community Velocity Model, Version 4 (CVM4) which has detailed basins and other structures is used as the 3D velocity model in simulations [16].

The potential earthquake ruptures within 200km and Mw larger than 6.0 in the Los Angeles region are selected from the Uniform California Earthquake Rupture Forecast, Version 2 (UCERF 2) for ground motion simulations in Cybershake [5]. The earthquake ruptures in UCERF2 only provide possible magnitudes in faults, without information of rupture process. To consider the earthquake rupture effects, each earthquake rupture selected from UCERF2 could convert to a kinematic rupture description for numerical simulations [5] based on Somerville et al.'s method [17].

In CyberShake, ground motion predictions are based on physics-based simulations rather than empirical attenuation relations. The qualified rupture sources are more than 10,000 in the Los Angeles region [5]. However, when the uncertainties of earthquake ruptures are considered, the number of earthquake rupture increases to more than 415,000. It will take a lot of computational resources and time to simulate all rupture models [5]. An efficient method is storing receiver Green's tensors (RGTs) of selected sites in the model and applying reciprocity to generate synthetic seismograms of rupture models [18, 5]. The RGTs called strain Green's tensors (SGTs) in CyberShake project [5]. Following Zhao et al. [18], the displacement field from a point source located at $\mathbf{r}'$ with moment tensor M_{ij} can be expressed as [19]

$$u_k(\mathbf{r},t;\mathbf{r}') = M_{ij}\partial_j{}'G_{ki}(\mathbf{r},t;\mathbf{r}'), \tag{1}$$

where $\partial_j{}'$ denotes the source-coordinate gradient with respect to $\mathbf{r}'$ and the Green's tensor $G_{ki}(\mathbf{r},t;\mathbf{r}')$ relates a unit impulsive force acting at location $\mathbf{r}'$ in direction $\hat{e}_i$ to the displacement response at location $\mathbf{r}$ in direction $\hat{e}_k$. Taking into account the symmetry of the moment tensor, we also have

$$u_k(\mathbf{r},t;\mathbf{r}') = \frac{1}{2}\Big[\partial_j{}'G_{ki}(\mathbf{r},t;\mathbf{r}') + \partial_i{}'G_{kj}(\mathbf{r},t;\mathbf{r}')\Big]M_{ij}. \tag{2}$$

Applying reciprocity of the Green's tensor

$$G_{ki}(\mathbf{r},t;\mathbf{r}') = G_{ik}(\mathbf{r}',t;\mathbf{r}), \tag{3}$$

equation (2) can be written as

$$u_k(\mathbf{r},t;\mathbf{r}') = \frac{1}{2}\Big[\partial_j{}'G_{ik}(\mathbf{r}',t;\mathbf{r}) + \partial_i{}'G_{jk}(\mathbf{r}',t;\mathbf{r})\Big]M_{ij}. \tag{4}$$

For a given receiver location $\mathbf{r} = \mathbf{r}_R$, the receiver Green tensor (RGT or SGT) is a 3rd-order tensor defined as the spatial-temporal strain field

$$H_{jik}(\mathbf{r}',t;\mathbf{r}_R) = \frac{1}{2}\Big[\partial_j{}'G_{ik}(\mathbf{r}',t;\mathbf{r}_R) + \partial_i{}'G_{jk}(\mathbf{r}',t;\mathbf{r}_R)\Big]. \tag{5}$$

Using this definition, the displacement recorded at receiver location $\mathbf{r}_R$ due to a source at $\mathbf{r}_S$ with moment tensor $\mathbf{M}$ can be expressed as

$$u_k(\mathbf{r}_R,t;\mathbf{r}_S) = M_{ij}H_{jik}(\mathbf{r}_S,t;\mathbf{r}_R) \text{ or } \mathbf{u}(\mathbf{r}_R,t;\mathbf{r}_S) = \mathbf{M}:\mathbf{H}(\mathbf{r}_S,t;\mathbf{r}_R), \tag{6}$$

and the synthetic seismogram due to a source at $\mathbf{r}_S$ with the basis moment tensor $\mathbf{M}_m$ can be expressed as

$$\mathbf{g}_m(\mathbf{r}_R,t;\mathbf{r}_S) = \mathbf{M}_m:\mathbf{H}(\mathbf{r}_S,t;\mathbf{r}_R). \tag{7}$$

In CyberShake, the SGTs can therefore be computed through wave-propagation simulations of two orthogonal horizontal components with a unit impulsive force acting at the receiver location $\mathbf{r}_R$ and pointing in the direction $\hat{e}_k$ in each simulation and store the strain fields at all spatial grid points $\mathbf{r}'$ and all time sample t. The synthetic seismogram at the receiver due to any point source located within the modeling domain can be obtained by retrieving the strain Green's tensor at the source location from the SGT volume and then applying equation (6).

In CyberShake project, one of objectives is improving the Ground Motion Prediction Equations (GMPEs), which are widely used in seismic hazard analysis, by replacing empirical ground motion database with physics-based simulated ground motions. Some advantages in physics-based simulation results could be found by comparing hazard curves among different methods. The hazard curves derived from Boore and Atkinson's [20] method and Campbell and Bozorgnia's [8] method that consider basin effects in GMPEs are selected for comparisons. However, the earthquake rupture directivity effects are not considered in these methods.

Here, three hazard curves which show exceedance probability for spectral acceleration (SA) at 3 seconds period are used to discuss differences among results [Figure 2]. At the PAS site [Figure 2], a rock site, the hazard curves among the three methods are similar. At the STNI site [Figure 2], a basin site, the hazard curves of CyberShake and Campbell and Bozorgnia's [Figure 2] method which consider basin amplification effects are similar, but the hazard curve of Boore and Atkinson's [20] method is significantly lower than the other two curves. However, at WNGC site, the hazard curve of CyberShake has higher hazard level than the other two. The WNGC site is at the region that channeling energy from earthquake ruptures in the southern San Andreas fault into Los Angeles basin, and the factors are included in physics-based simulations. The channeling phenomenon also can be found from other studies [21, 22]. The CyberShake seismic hazard map [Figure 3] is derived from the 250 sites used in simulations [5]. In the physics-based hazard map, some effects don't include in attenuation relations, including, for example, earthquake rupture effects, basin amplification effects, and wave propagation phenomena in 3D complex structures.

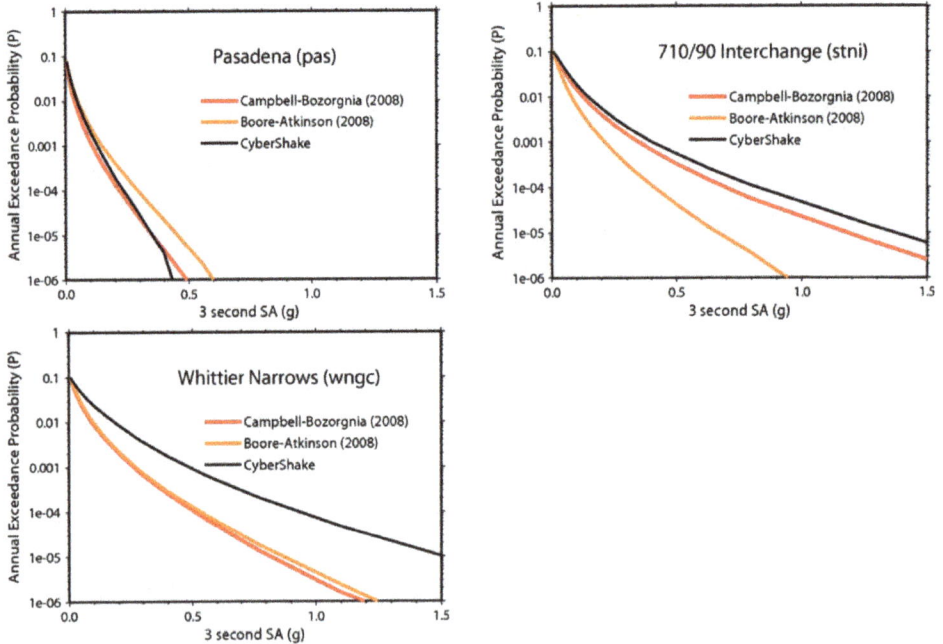

Figure 2. Hazard curves derived form three different methods at three sites, PAS, STNI and WNGC, in Los Angeles region. The red lines represent the results of using Campbell and Bozorgnia's [8] method; the orange lines represent the results of Boore and Atkinson's [20] method; the black lines represent the results of CyberShake. Adopted from [5].

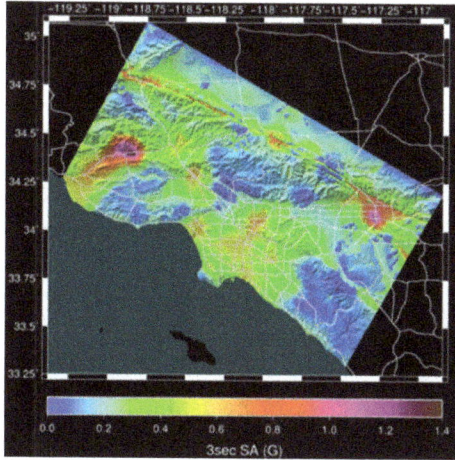

Figure 3. The CyberShake hazard map for Los Angeles region of 3 seconds period spectral acceleration (SA) for 2% exceedance probability in 50 years. Adopted from [5].

4. Comparisons between USGS and CyberShake hazard maps

There are many differences between the hazard maps of USGS and CyberShake, including, procedures of making hazard maps, required computational resources and results [3, 5]. The USGS National Seismic Hazard maps in California region are derived form source models based on seismological data, geological surveys and earthquake rupture models, and the Next Generation Attenuation (NGA) database [8, 9]. The CyberShake hazard map is constructed by physics-based simulations in the 3D velocity model for all potential earthquake ruptures with Mw ≥ 6.0 near Los Angeles region [5]. The computational resources requirements for generating USGS hazard maps do not mention in the 2008 report of seismic hazard maps update, but the hazard maps should be able to done without a super computer. To generate the CyberShake hazard map, lots of wave propagation simulations are required to build a database for generating synthetic seismograms of potential earthquake ruptures [5]. The computational resource of physics-based seismic hazard maps is much higher than the computational requirement of USGS hazard maps. However, the advances in computer sciences make the computational requirements affordable for CyberShake, also accurate estimations of ground motions are important for a city with a large population. The seismic hazard levels are quit different in the Los Angeles region between two hazard maps. In the USGS hazard map [Figure 1], the high hazard level regions are almost along the fault zones and hazard values decrease as the distance between a site and fault zones increases. In the Los Angeles basin region, the hazard level is about the same in the USGS hazard map [Figure 1]. In the CyberShake hazard map, the hazard values along the San Andreas fault are high, but the width of high hazard zones is narrower. In addition, the CyberShake hazard map in the Los Angeles basin has more details [5]. This probably reflects the source and structure effects in ground motion predictions.

5. ElarmS: Earthquake alarms systems for California

In Southern California, an earthquake-prone area, many cities are under earthquake risks, hence earthquake early warning systems are becoming an important role in earthquake disaster mitigation [2]. Allen [2] developed earthquake early warning systems called Earthquake Alarms Systems (ElarmS) for California. In the ElarmS methodology, three steps are designed for rapid estimations of earthquake source parameters and prediction of peak ground motions [1, 2]. First, using the time information of first-arrived signal to locate earthquakes and estimate the warning time. Second, using frequency information of first four seconds of P-wave to estimate magnitudes of earthquakes. Third, using attenuation relations and the earthquake source information, an estimated location and magnitude, to generate ground shaking maps.

In the ElarmS, the arrival times of P-waves are used to rapidly locate earthquake locations. The possible areas of an earthquake location could be inferred by using the information of the first two or three stations trigged by an earthquake. To locate a more accurate earthquake location, including, longitude, latitude, depth and origin time, the first arrival time form four stations are required. A grid search algorithm is used to find an optimal earthquake location that has minimum arrival time misfits. The warning time, the remaining time before the peak ground motion arrived, can be estimated by using the predicted S-wave arrival times of sites. Peak ground motions are usually caused by S-wave or surface wave, so use predicted S-wave arrival times as peak ground motion times may provide additional warning time for some sites.

The magnitude, which represents the released energy of an earthquake, is an important parameter in earthquake early warning systems. The rapid magnitude estimation method of an earthquake by using the frequency information of the first four seconds of P-wave is adopted in the ElarmS [1, 2]. Basically, the magnitude estimations take two procedures. The first step is finding the maximum predominant period within the first 4 seconds of the vertical component P-wave waveforms, and then use linear relations to scale the maximum predominant period value to an estimated earthquake magnitude [1, 2]. As the number of the maximum predominant period value from different receivers increases, the average magnitude errors will decrease [2].

When the location and magnitude of an earthquake is available, attenuation relations can be used to estimate ground motions of sites and then generate a ground motion prediction map for whole California. In the ElarmS, the attenuation relations are based on the recordings of earthquakes with magnitude larger than 3.0 in California [2]. However, the empirical attenuation relations used in the ElarmS do not account effects of wave propagation in 3D structures, for example, basin amplification effects.

6. Rapid full-wave CMT inversion

In Southern California, preliminary 3D earth structure models are already available, and efficient numerical methods have been developed for 3D anelastic wave-propagation simulations. We develop an algorithm to utilize these capabilities for rapid full-wave centroid moment tensor (CMT) inversions. The procedure relies on the use of receiver Green tensors (RGTs), the spatial-temporal displacements produced by the three orthogonal unit impulsive point forces acting at the receivers. Once we have source parameters of earthquakes, a near-real time full-wave ground motion map, that considers both source and wave-propagation effects in a 3D structure model, may also available for earthquake early warning purposes.

In our CMT inversion algorithm, the RGTs are computed in our updated 3D seismic structure model for Southern California using the full-wave method that allows us to account for 3D path effects in our source inversion. The efficiency of forward synthetic calculations could be improved by storing RGTs and using reciprocity between stations and any spatial grid point in our model. In our current model, we will use three component broadband waveforms below 0.2 Hz to invert source parameters. Based on Kikuchi and Kanamori's [23]source inversion method, any moment tensor can be expressed as linear combination of 6 elementary moment tensors. In our current coordinate (x=east, y=north, z=up), the moment tensor can be expressed as below:

$$\mathbf{M1} = \begin{bmatrix} 0 & 1 & 0 \\ 1 & 0 & 0 \\ 0 & 0 & 0 \end{bmatrix}; \quad \mathbf{M2} = \begin{bmatrix} -1 & 0 & 0 \\ 0 & 1 & 0 \\ 0 & 0 & 0 \end{bmatrix}; \quad \mathbf{M3} = \begin{bmatrix} 0 & 0 & -1 \\ 0 & 0 & 0 \\ -1 & 0 & 0 \end{bmatrix}.$$

$$\mathbf{M4} = \begin{bmatrix} 0 & 0 & 0 \\ 0 & 0 & -1 \\ 0 & -1 & 0 \end{bmatrix}; \quad \mathbf{M5} = \begin{bmatrix} 0 & 0 & 0 \\ 0 & -1 & 0 \\ 0 & 0 & 1 \end{bmatrix}; \quad \mathbf{M6} = \begin{bmatrix} 1 & 0 & 0 \\ 0 & 1 & 0 \\ 0 & 0 & 1 \end{bmatrix}$$

(8)

There are two main advantages of using this method. First, different subsets of 6 elementary moment tensors could represent different source parameter assumptions such as M1~M6 could recover general moment tensors and M1~M5 could represent pure-deviatoric moment tensors [23]. From efficiency point of view, we only need to generate synthetic waveforms of 6 elementary moment tensors at grid points close to initial sources locations for receivers to invert an optimal CMT solution.

For centroid location x_1 and centroid time t_1, the synthetic seismograms of 6 elementary moment tensors could be defined as:

$$S_{mi}^r(t; x_1, t_1) , \quad m:1\text{~}6$$

(9)

where r is receiver, m is index of 6 moment tensor, i is component index. The synthetic seismogram can be expressed as:

$$u_i^r(t) = \sum_{m=1}^{6} a_m S_{mi}^r(t; x_1, t_1)$$

(10)

From inversion point of view, the phases have less structure heterogeneous effects can reduce the nonlinear effects caused by complex 3D structure such as body wave phases that propagate through relative simple deep structure and surface wave phases that propagate along free surface and average out the heterogeneity. We apply our seismic waveform segmentation algorithm that is based on continuous wavelet transforms and a topological watershed method to observed seismograms and then select the first (potential body wave) and biggest (potential surface wave) time-localized waveforms to invert source parameters.

In source inversion, we applied a multi-scale grid-searching algorithm based on Bayesian inference to find an optimal solution [Figure 4]. We consider a random vector H composed of 6 source parameters: the longitude, latitude and depth of the centroid location r_s, and the strike, dip and rake of the focal mechanism. We assume a uniform prior probability $P_0(H)$ over a sample space Ω_0, which is defined as a sub-grid in our modeling volume centered around the initial hypocenter location provided by the seismic network with grid spacing in three orthogonal directions given by a vector θ_0 and a focal mechanism space with the ranges given by $0° \le$ strike $\le 360°$, $0° \le$ dip $\le 90°$ and $-90° \le$ rake $\le 90°$ and with angular intervals in strike, dip and rake specified by a vector θ_0.

We apply Bayesian inference in three steps sequentially. In the first step, the likelihood function is defined in terms of waveform similarity between synthetic and observed seismograms. We quantify waveform similarity using a normalized correlation coefficient (NCC) defined as

$$NCC_n = \max_{\Delta t}\left[\int_{t_n^0}^{t_n^1} \overline{s}_n(t)s_n(t-\Delta t)dt \middle/ \sqrt{\int_{t_n^0}^{t_n^1} \overline{s}_n^2(t)dt \int_{t_n^0}^{t_n^1} s_n^2(t-\Delta t)dt} \right]. \tag{11}$$

where n is the observation index, $\overline{s}_n(t)$ and $s_n(t)$ are the filtered observed seismogram and the corresponding synthetic seismogram, $\left[t_n^0, t_n^1\right]$ is the time window for selecting a certain phase on the seismograms for cross-correlation (Figure 4b). We allow a certain time-shift Δt between the observed and synthetic waveforms. To prevent possible cycle-skipping errors, we restrict $|\Delta t|$ to be less than half of the shortest period. We assume a truncated exponential distribution for the conditional probability

$$P(NCC_n \mid H_q) = \frac{\lambda_n \exp\left[-\lambda_n(1-NCC_n)\right]}{1-\exp(-2\lambda_n)}, \ -1 < NCC_n \le 1, \ H_q \in \Omega_0, \tag{12}$$

where λ_n is the decay rate. Assuming the NCC observations are independent, the likelihood function can be expressed as

$$L_0\left(H \mid \bigcap_{n=1}^{N} NCC_n\right) = \exp\left[-\sum_{n=1}^{N}\lambda_n(1-NCC_n)\right]\prod_{n=1}^{N}\left\{\lambda_n\left[1-\exp(-2\lambda_n)\right]^{-1}\right\} \tag{13}$$

where N is the total number of NCC observations. The posterior probability for the first step can then be expressed as

$$P_0\left(H \mid \bigcap_{n=1}^{N} NCC_n\right) = \frac{P_0(H)\exp\left[-\sum_{n=1}^{N}\lambda_n(1-NCC_n)\right]\prod_{n=1}^{N}\left\{\lambda_n\left[1-\exp(-2\lambda_n)\right]^{-1}\right\}}{P_0\left(\bigcap_{n=1}^{N} NCC_n\right)} \quad (14)$$

where

$$P_0\left(\bigcap_{n=1}^{N} NCC_n\right) = \sum_q P\left(\bigcap_{n=1}^{N} NCC_n \mid H_q\right)P_0(H_q). \quad (15)$$

We note that the λ_n in front of the $(1-NCC_n)$ in equation (14) can be used as a weighting factor for various purposes, such as to account for different signal-to-noise ratios in observed seismograms and to avoid the solution to be dominated by a cluster of closely spaced seismic stations.

The probability for individual measurements

$$P_0(NCC_n) \propto \sum_q P\left(NCC_n \mid H_q\right)P_0(H_q). \quad (16)$$

can be used for rejecting problematic observations. In practice, we only accept observations with

$$P_0(NCC_n) \geq Q_0 \quad (17)$$

A very low $P_0(NCC_n)$ indicates that the nth observed waveform cannot be fit well by any solutions in our sample space. This may be due to instrumentation problems or unusually high noise levels in the observed waveform data.

In the second step, we apply the same algorithm on another measurements, time-shifts between the observed and synthetic waveforms when the NCC is the maximum in allowed time-shift range. The last step is applying the same processes to the amplitude ratio measurements. By using the Bayesian approach, we can obtain the probability density functions of source parameters that contain uncertainties information rather than a single best solution. Our optimal source parameter solution is the one with highest probability. In Figure 4, examples of the marginal probabilities for some of the source parameters are shown for the 3 September 2002 Yorba Linda earthquake.

For earthquake early warning purposes, there are few approaches to make our CMT inversion method toward (near) real time and then use optimal CMTs for generating full-wave peak ground motion maps. To save some time in generating synthetic seismograms, we can store synthetic seismograms of 6 elementary moment tensors rather than extract them from RGTs. Destructive or larger earthquakes tend to occur in existing fault zones or regions where earthquakes occurred. Based on the assumption above, rather than store synthetic seismograms of all grid points in our model, we can store the synthetic seismograms of grid points near fault zones or high seismicity regions. Another possibility is to save time of inversion by using other efficient inversion algorithms. Full-wave ground motion prediction maps could be generated based on the synthetic seismograms of optimal CMT solutions.

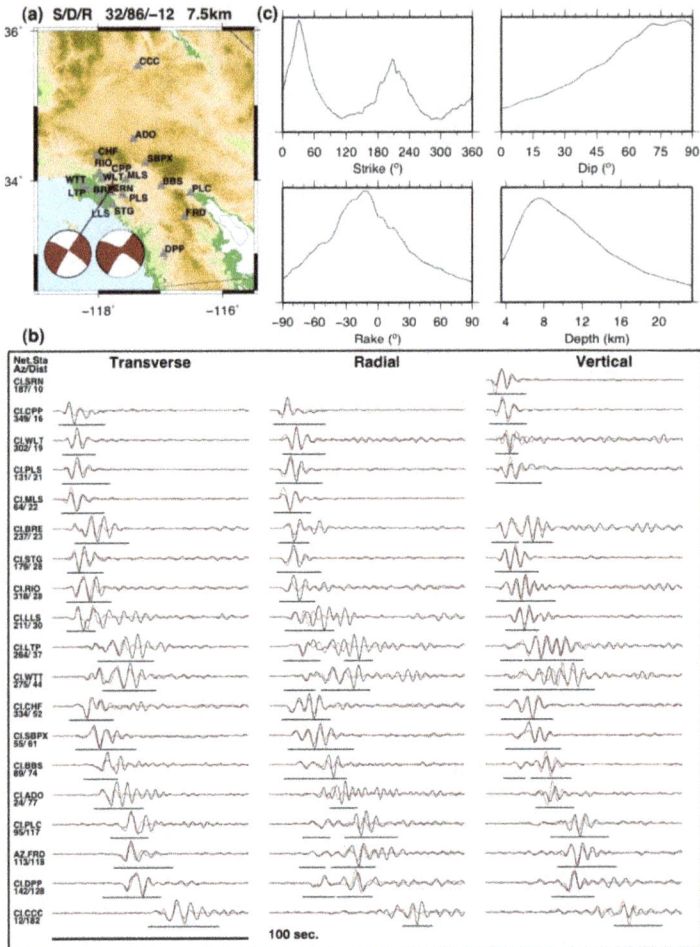

Figure 4. An example of our CMT inversion procedure. (a) The map shows epicenter of the 3 September 2002 M_w 4.3 Yorba Linda earthquake (the star), the best-fit double-couple solution (the red beachball) and stations (gray triangles) selected for this inversion. (b) Examples of the waveforms selected for in the CMT inversion. The black lines are observed seismograms and the red lines are synthetic seismograms. The black bars indicate the selected waveform segments for CMT inversion. (c) The marginal probability densities for strike, dip, rake and depth obtained after our grid-search step.

The speed of earthquake source parameters estimation and accurate ground motion predictions are both play essential role in earthquake early warning systems. In the ElarmS, earthquake locations, origin time and magnitudes could be inverted in very short time. Peak ground motion maps can be generated shortly by using empirical attenuation relations. The CMT inversion method we proposed has potential for (near) real time inversion and then solutions could be used for (near) real time full-wave peak ground motion maps. In addition, the Bayesian approach used in our CMT inversion has uncertainty of solutions and could be projected into ground motion estimations.

7. Conclusion

In this chapter, we compare full-wave based and non-full-wave based methods of ground motion forecast for (southern) California. There are advantages and disadvantages to different methods. Since the full-wave methods involve numerical simulations of wave propagation in 3D velocity models, the computational resource requirements are much higher than non-full-wave methods. However, numerical simulations are usually affordable in most of super computers. In general, ground motion estimations of non-full-wave methods are usually based on empirical attenuation relations. In full-wave methods, the ground motion estimations are based on numerical simulations that considered source effects, basin amplification effects and wave propagation effects in a 3D complex velocity [5, 6]. Those effects may play very important roles in ground motion estimations. For example, if a large earthquake occurs on southern San Andreas fault, the released energy will channel into Los Angeles region, one of the most populous cities in the United States, and basin effects will amplify the ground motion [22, 5]. Full-wave based ground motion forecast should able to provide more accurate and detailed ground motions and this will benefit cities under earthquake risks, such as Los Angeles city.

Author details

En-Jui Lee and Po Chen

Department of Geology and Geophysics, University of Wyoming, USA

Acknowledgement

The full-wave CMT inversion research used resources of the Argonne Leadership Computing Facility at Argonne National Laboratory, which is supported by the Office of Science of the U.S. Department of Energy under contract DE-AC02-06CH11357. En-Jui Lee is supported by the Southern California Earthquake Center. Po Chen is supported jointly by the School of Energy Resources and the Department of Geology and Geophysics at the University of Wyoming. Comments from the book editor improved our manuscript.

8. References

[1] Allen RM, Kanamori H (2003) The potential for earthquake early warning in southern California. Science 300:786–789

[2] Gasparini P, Manfredi G, Zschau J (eds) (2007) Earthquake early warning systems. Springer, Berlin

[3] Petersen MD, Frankel AD, Harmsen SC, et al (2008) Documentation for the 2008 update of the United States national seismic hazard maps. U.S. Geological Survey Open-File Report

[4] Frankel AD, Field EH, Petersen MD, et al (2002) Documentation for the 2002 update of the national seismic hazard maps. U.S. Geological Survey Open-File Report

[5] Graves R, Jordan T, Callaghan S, Deelman E (2010) CyberShake: A Physics-Based Seismic Hazard Model for Southern California. Pure appl geophys 168:367–381

[6] Lee E, Chen P, Jordan T, Wang L (2011) Rapid full-wave centroid moment tensor (CMT) inversion in a three-dimensional earth structure model for earthquakes in Southern California. Geophysical Journal International 186:311–330

[7] Field EH, Dawson TE, Felzer KR, et al (2009) Uniform California Earthquake Rupture Forecast, Version 2 (UCERF 2). Bulletin of the Seismological Society of America 99(4):2053–2107

[8] Campbell KW, Bozorgnia Y (2008) NGA ground motion model for the geometric mean horizontal component of PGA, PGV, PGD and 5% damped linear elastic response spectra for periods ranging from .... Earthquake Spectra 24:139–171

[9] Chiou BS-J, Youngs RR (2008) An NGA model for the average horizontal component of peak ground motion and response spectra. Earthquake Spectra 24:173–215

[10] Olsen K (1994) Simulation of three-dimensional wave propagation in the Salt Lake Basin. University of Utah

[11] Tromp J, Komatitsch D, Liu Q (2008) Spectral-element and adjoint methods in seismology. Commun Comput Phys 3(1):1–32

[12] Chen P, Zhao L, Jordan TH (2007) Full 3D tomography for the crustal structure of the Los Angeles region. Bulletin of the Seismological Society of America 97(4):1094–1120

[13] Tape C, Liu Q, Maggi A, Tromp J (2009) Adjoint tomography of the southern California crust. Science 325:988–992

[14] Fichtner A, Kennett B, Igel H, Bunge H (2009) Full seismic waveform tomography for upper-mantle structure in the Australasian region using adjoint methods. Geophysical Journal International 179:1703–1725

[15] Liu Q, Polet J, Komatitsch D, Tromp J (2004) Spectral-Element Moment Tensor Inversions for Earthquakes in Southern California. Bulletin of the Seismological Society of America 94(5):1748–1761

[16] Kohler MD, Magistrale H, Clayton RW (2003) Mantle Heterogeneities and the SCEC Reference Three-Dimensional Seismic Velocity Model Version 3. Bulletin of the Seismological Society of America 93(2):757–774

[17] Somerville P, Irikura K, Graves R, Sawada S, Wald D, Abrahamson N, Iwasaki Y, Kagawa T, Smith N, Kowada A (1999) Characterizing crustal earthquake slip models for the prediction of strong ground motion. Seismological Research Letters 70:199–222

[18] Zhao L, Chen P, Jordan T (2006) Strain Green's tensors, reciprocity, and their applications to seismic source and structure studies. Bulletin of the Seismological Society of America 96(5):1753–1763

[19] Aki K, Richards PG (2002) Quantitative Seismology, 2nd ed. University Science Books

[20] Boore DM, Atkinson GM (2008) Ground-Motion Prediction Equations for the Average Horizontal Component of PGA, PGV, and 5%-Damped PSA at Spectral Periods between 0.01 s and 10.0 s. Earthquake Spectra 24(1):99

[21] Olsen KB, Day SM, Minster JB, Cui Y, Chourasia A, Faerman M, Moore R, Maechling P, Jordan T (2006) Strong shaking in Los Angeles expected from southern San Andreas earthquake. Geophysical Resaerch Letters 33:

[22] Graves RW, Aagaard BT, Hudnut KW, Star LM, Stewart JP, Jordan TH (2008) Broadband simulations for Mw 7.8 southern San Andreas earthquakes: Ground motion sensitivity to rupture speed. Geophys. Res. Lett 35:

[23] Kikuchi M, Kanamori H (1991) Inversion of complex body waves-III. Bulletin of the Seismological Society of America 81(6):2335–2350

Seismic Performance and Simulation of Behavior of Structures

Chapter 7

Seismic Performance of Historical and Monumental Structures

Halil Sezen and Adem Dogangun

Additional information is available at the end of the chapter

1. Introduction

Historical, religious and monumental structures and their susceptibility to damage in recent earthqaukes in Turkey are presented and discussed in this chapter. Turkey has a very large number of historical structures and is located in one of the most seismically active regions of the world. Some of these historical and monumental masonry and reinforced concrete structures suffered substantial damage or collapsed during two major earthquakes in 1999. The Kocaeli (M_w7.4) and Düzce (M_w7.2) earthquakes occurred on August 17 and November 12, 1999, and ruptured approximately 110 km and 40 km of the 1550-km-long North Anatolian fault, respectively. This chapter describes briefly the construction materials and techniques for historical religious and monumental structures and state-of-practice in Turkey, and presents dynamic analyses of a masonry minaret example. The seismic performance of the mosques and minarets (tall slender towers) during the 1999 earthquakes is presented.

2. Seismic design and construction practice in Turkey

Prior to the 1999 earthquakes, two codes governed the design and construction of reinforced concrete and masonry buildings in Turkey: Turkish Earthquake Code (1998) and the Turkish Building Code, TS-500 (1985). The earthquake code included procedures for calculating earthquake loads on buildings. The ductility requirements and details described in the earthquake code were rarely observed in religious or monumental structures inspected by the authors after the 1999 earthquakes. Details of the seismic design and building construction practice prior to the 1999 earthquakes are provided in Sezen et al. (2003).

The descriptions of ground motion characteristics, structural damage, and performance of structures during these earthquakes are provided in Sezen et al. (2003). Response spectra for selected acceleration histories for 5% damping are presented in Figure 1. The ground motions included in Figure 1 were recorded at: SKR station in Adapazari (Peak Ground

Acceleration, PGA 0.41g, stiff soil); YPT station in Yarimca (PGA 0.23g, soft soil); and DZC station in Düzce (PGAs: EW-Aug.17 0.36g, NS-Aug.17 0.31g, EW-Nov.12 0.54g, and NS-Nov.12 0.35g, soft soil). Figure 1 also provides a comparison between the linear elastic acceleration response spectra calculated for rock and that calculated for soft soil sites using the provisions of the Uniform Building Code (UBC, 1997) and those of the Turkish seismic code (1998) with 5 percent damping for the highest seismicity in the United States and in Turkey, respectively. From Figure 1, it can be concluded that, for a structure with given periods of vibration, the difference between the base shear calculated using the 1998 Turkish seismic code and the base shear demand obtained from the recorded acceleration response spectra do not differ significantly. If the static lateral load distribution over the height is suggested by the code is assumed to be credible, structures designed and detailed according to the Turkish code should not have collapsed or suffered severe damage during the 1999 earthquakes.

Specifically, according to Figure 1, a structure with a fundamental period of 0.5 seconds would be subjected to seismic forces larger than those specified in the code. The period of older masonry structures with thick and shorter walls tends to be relatively small, probably 0.5 seconds or less. On the other hand, the minarets or very slender towers tend to be very flexible with relatively large fundamental periods. Figure 1 shows that the recorded spectral accelerations are significantly large at large periods. The authors' field observations after the 1999 earthquakes, especially the November 12 Duzce earthquake (DZC), showed more damage in slender structures like minarets. In addition, among other factors leading to inadequate performance, the extent of damage observed in most reinforced concrete structures including minarets after the 1999 earthquakes is probably related to poor engineering and lack of conformance to the relatively new Turkish seismic code.

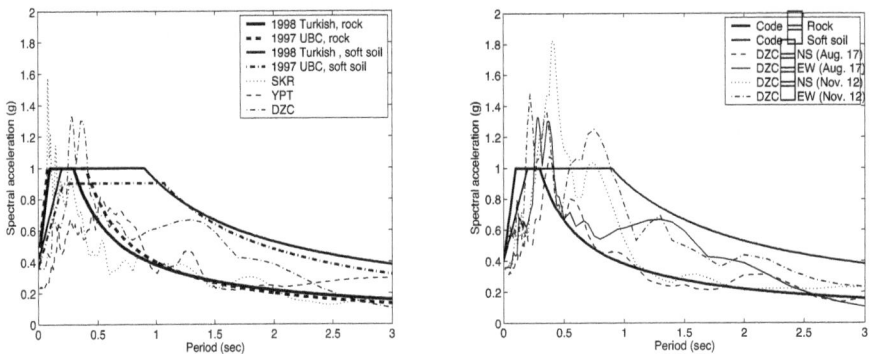

Figure 1. Absolute Acceleration Response Spectra

2.1. Construction materials and deterioration of historical structures

In the region affected by the earthquakes, in general, construction materials include stone or brick masonry and wood for older mosques and minarets, steel and reinforced concrete in more recently constructed structures. Typically locally available stone blocks were used. A special mortar called Horasan mortar was used in historical masonry structures. The Horasan mortar is made by mixing lime with baked clay powder that could be obtained by grinding clay tiles or bricks. The Horasan mortar was commonly used in the Middle East, where use of earthenware, pottery and brick construction were widespread (Çamlıbel 1988).

Hollow clay tiles or lightweight concrete blocks are used as infill wall material in recently built reinforced concrete (RC) moment frame mosques. Solid clay bricks were used in infill walls of RC mosques built before early 1970s. Almost all historical religious structures in Turkey were constructed using cut stone, masonry blocks or combination of these two materials. Timber construction is rare and reinforced concrete is relatively recent. The structural and geometrical properties of each masonry structure depend on many factors including the structural knowledge and applications at the time of construction, experience of the architect or engineer, seismicity of the region, and availability of construction materials in that area. Recent earthquakes in Turkey have shown that most masonry monumental structures as well as buildings (Ural et al. 2012) in high seismic regions are vulnerable to structural damage and collapse.

Historical structures deteriorate over time mainly due to environmental effects, and therefore may experience failure or collapse under gravity loads or seismic loads lower than those predicted by the design codes. As discussed in Dogangun and Sezen (2012) and Sezen and Dogangun (2009), in many cases, the following factors trigger or exacerbate the damage: (1) Surface or rain water runoff. If the structure and drainage system are not maintained properly, grass or fungus may grow and weaken the structural materials. Water accumulated on or penetrated into structural members may cause cracks due to freezing and thawing. (2) Soil settlement and relative movement of foundation. (3) Insufficient material strength. Layers of clay or other impure materials inside stone blocks may eventually lead to wearing, spalling or cracking. In stone masonry structures, the properties of the mortar significantly influence the strength of the entire structural component such as a load bearing wall. Deterioration of mortar binding the stone blocks, especially poor quality mortar including mud or low quality lime, can reduce the strength and stiffness of the wall considerably. (4) Other problems. Historical structures can be subjected to various environmental and loading conditions depending on their use and geographic location. For example, timber is more susceptible to humidity and temperature variations. Loading from continuous traffic and heavy trucks can lead to vibrations and excessive loads on foundations because the streets and other structures in historical cities are not designed for modern day traffic. Similarly, the use or occupancy of the structure may change and create larger unexpected loads. Parts of older structures are sometimes used as storage, in which the magnitudes of loads are usually much higher. Other local and environmental effects, such as acid rains, may adversely affect construction materials.

Most of the factors presented here result in gradual deterioration of materials or the load carrying structural system, which can be prevented as the damage progresses and becomes visible in many cases. On the other hand, structural damage, failure or total structural collapse occurs suddenly during moderate or strong earthquakes. Thus, it is essential to evaluate the capacity of existing historical structures and to retrofit them before an expected earthquake strikes.

2.2. Construction practice and techniques

Overall plan dimension of a typical mosque generally varies between 12 and 25 meters. Depending on the plan dimensions, height ranges from 7 m to 15 m not including the dome. Height of the dome, a half sphere, is typically half of the plan dimension (Figure 2). Lateral seismic loads are typically resisted by relatively thick unreinforced stone masonry walls in historical mosques. Addition to load bearing masonry walls, most mosques include a few columns typically carrying the gravity loads.

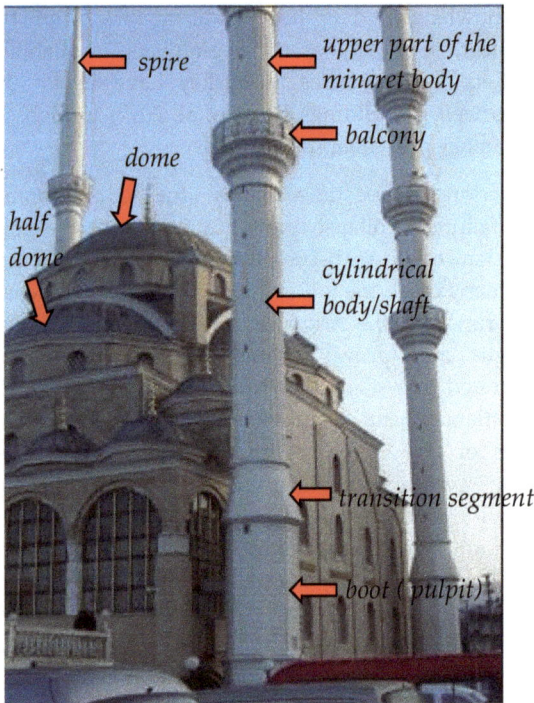

Figure 2. Typical mosque and its minarets in Turkey

Minarets can be separate or contiguous and integral with the mosque structure, and are typically built using stone, brick, wood or reinfroced concrete. They typically include cylindrical or polygonal body/shafts, one or two balconies, and a conical roof or spire (Figure 2). In masonry minarets or slender tower structures, rather small tensile strength of mortar placed between the masonry blocks presents a major problem in regions of high seismicity. The brick or stone blocks have fairly large compressive strength, however unreinforced masonry lacks tensile strength required to resist bending moments imposed by the lateral earthquake loads. Older masonry minarets were typically constructed using stone blocks or solid clay bricks or a combination of two, whereas unreinforced lightweight stone blocks are preferred in new construction.

After a major earthquake in 1509, Ottoman architects tackled the problem of constructing tall earthquake resistant minarets (Oğuzmert, 2002). They started to use a special technique for linking adjacent stone blocks with iron bars and clamps in the vertical and horizontal directions as shown in Figure 3 (Doğangün et al. 2007). Use of iron clamps in the two perpendicular directions (transverse and vertical) has improved the lateral load carrying capacity of slender masonry minarets significantly under earthquake loads. The clamps and vertical bars were placed inside anchorage holes in the stone blocks, and melted lead was poured inside the hole to provide bond between the stone and iron clamp or vertical bars. Depending on the properties and dimensions of the stone units, different clamps were developed. For example, as shown in Figure 4a, curved clamps were used within the circular stone wall on the minaret perimeter. Sharper and thinner clamps in thin stone blocks (Figure 4b) and shorter clamps were used if low tensile stresses are expected (Figure 4c). Approximately 2000 kilograms of this heavy metal, lead was used for the construction of a typical masonry minaret. Lead performs as intended for a very long time because it does not corrode or is hardly ever influenced by the adverse environmental conditions.

3. Post-earthquake surveys

The data presented here are based on observations from two surveys conducted after the August 17 and November 12, 1999 earthquakes. Damage to historical and recently constructed mosques and minarets was documented to investigate the seismic performance of these structures. The main objectives of these surveys were to provide detailed information about the characteristics of the observed damage, and to study the relative vulnerability of these historical and modern structures to strong ground motions. The parameters considered in the survey included: type of construction material, minaret height, location and description of damage, and location (coordinates, if available). Vast majority of the surveyed minarets and mosques were located in the cities of Düzce and Bolu. The peak ground accelerations recorded in Düzce during both earthquakes were larger than 0.30g (Figure 1), whereas that recorded in Bolu was reported as 0.82g after the November 12, 1999 earthquake.

Structural performance levels and corresponding representative sample damage descriptions for mosques and minarets are shown in Table 1. It should be noted that non-structural damage is irrelevant for minarets. In general, damage to the non-structural components of the mosques was insignificant as compared to the total structural damage.

A total of 59 sites were visited after the October 12, 1999 Duzce earthquake (second earthquake). The name, location, construction date, and observed damage levels are provided in Tables 2 and 3. The first 22 mosques listed in Table 2 were located in the city of Duzce and neighboring town of Kaynasli. Mosques numbered 23 through 44 were located in the city of Bolu.

Preparation of stone blocks in at least two rows without anchorage

iron clamb

vertical bars

Iron reinforcement elements with thicknesses around 20-25 mm

≈40   ≈60 mm

Carving out anchorage holes, inserting vertical iron bars, and pouring lead into the holes. The top face will be the bottom of the units in the structure

Stone block to be used as a step in stairs

Turn the units upside down

Horasan mortar

Inserting iron clamps and pouring lead into the holes over the top face (after the vertical barsare inserted and units are turned upside down)

After the clamps are inserted and mortar is placed between the blocks, one ring or layer of minaret is complete.

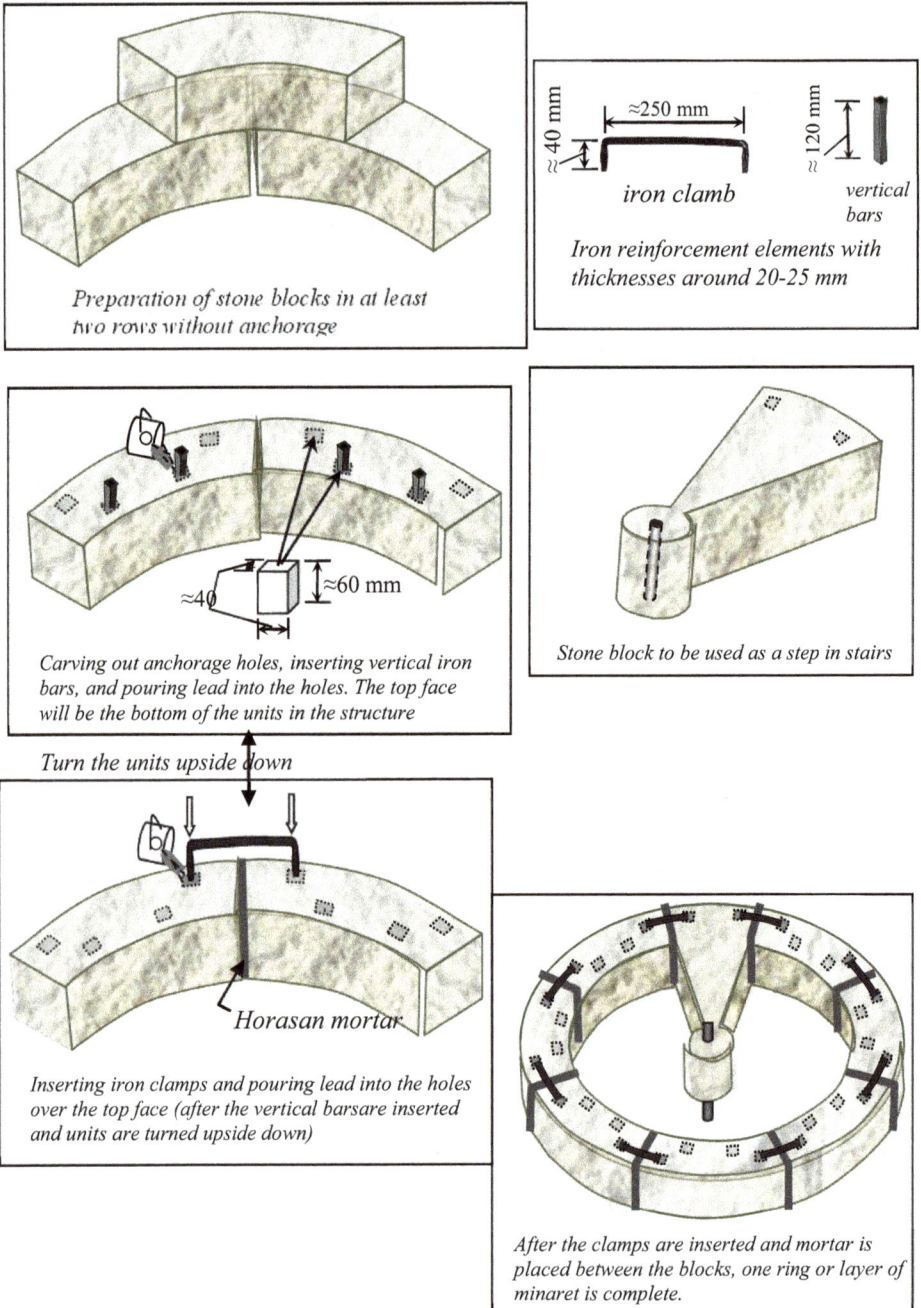

Figure 3. Construction of traditional Turkish minarets using stone blocks reinforced and anchored with iron bars and clamps (Dogangun et al. 2007)

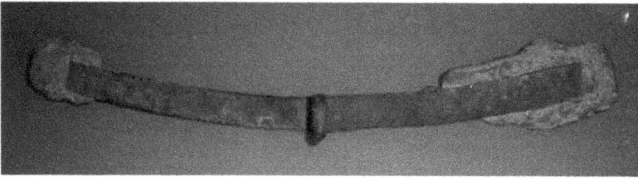

(a)

(b)

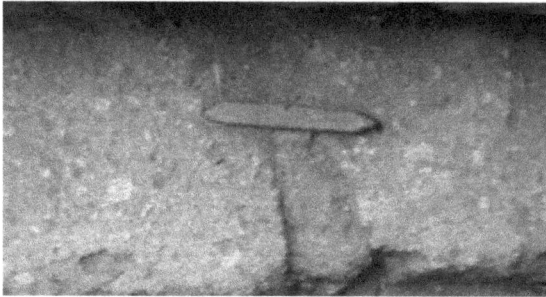

(c)

Figure 4. Variations in iron clamps used in historical stone masonry walls

Performance level	Damage classification	Sample damage description (mosque)	Sample damage description (minaret)
I	None	Negligible	Negligible
II	Light	Minor cracks in primary structural components	Minor cracks in masonry or RC minarets
III	Moderate	Significant cracks in RC members or masonry walls	Significant cracks especially around the minaret base
IV	Major	Hinge formation and wide cracks in primary RC members Infill wall collapse	Permanent visible drift Wide cracks and concrete spalling
V	Collapse	Partial or total collapse	Collapse

Table 1. Structural damage description and classification for mosques and minarets

| Name | Location (coordinates) | | Construction date | Damage level | |
	North	East		Mosque	Minaret
1. Sirali Koyu	40° 49.246′	31° 11.544′	1987	Light	Collapse
2. Kocyazi Koyu	40° 50.569′	31° 10.249′	1977-79	Light	Collapse
3. Karaca	40° 50.733′	31° 09.950′	1970s	Collapse	Collapse
4. Hamidiye	40° 50.838′	31° 09.518′	1980s	Light	Light
5. Rumelipalas	40° 50.877′	31° 08.444′	1972	None	None
6. Uzun Mustafa	40° 50.755′	31° 08.736′	1989	None	Collapse
7. Kultur Mahallesi	40° 50.623′	31° 09.257′	-	None	Moderate
8. Otopark	40° 50.446′	31° 09.119′	1971-73	None	Moderate
9. Aydinpinar	40° 49.796′	31° 09.190′	1994	Light	Collapse
10. Yesil	40° 49.966′	31° 09.480′	1990	None	Collapse
11. Asar	40° 50.012′	31° 09.650′	1977	None	Moderate
12. AzmimilliYeni	40° 49.897′	31° 10.087′	1988	None	Collapse
13. Mimar Sinan Nur	40° 50.006′	31° 10.236′	1988-91	None	Collapse
14. Maresal F. Cakmak	40° 49.928′	31° 10.525′	1990	Light	Collapse
15. Huzur	40° 49.680′	31° 11.324′	1986	None	None
16. Topalakli	40° 49.606′	31° 11.539′	1951	Light	None
17. Kirazli Koyu	40° 48.918′	31° 12.844′	1965	None	None
18. Doganli Koyu	40° 48.182′	31° 14.190′	1981-94	None	None
19. Uckopru Merkez	40° 47.662′	31° 15.018′	1985	None	Light
20. Yesiltepe	40° 46.690′	31° 17.467′	1986-90	Collapse	Collapse
21. Karacaali	40° 46.496′	31° 18.204′	1992-94	None	Moderate
22. Kaynasli merkez	40° 46.406′	31° 19.138′	-	Collapse	Collapse
23. Sanayi	40° 44.280′	31° 37.562′	1988	None	None
24. Kultur	40° 44.507′	31° 36.340′	1990	None	Light
25. Oksuztekke	40° 44.488′	31° 35.851′	1993	Major	Collapse
26. Ozayan	40° 44.638′	31° 35.405′	1996	None	Moderate
27. Beskonaklar Yeni	40° 44.263′	31° 35.599′	1997	Moderate	-
28. Pasakoy Berberler	40° 43.942′	31° 34.096′	1987	None	None
29. Pasakoy Eniste	40° 43.803′	31° 34.689′	1997	None	Light
30. Sumer	40° 43.575′	31° 35.587′	1992-95	None	Light
31. Sumer Mah. Yeni	40° 43.468′	31° 35.297′	1983	None	None
32. Karacayir Siteler	40° 43.577′	31° 36.033′	1998-99	None	-
33. Semsi AhmetPasa	40° 43.852′	31° 36.635′	14th cent	Major	Collapse
34. Sarachane	40° 43.935′	31° 36.513′	17th cent	Light	None
35. Yildirim Bayezid	40° 44.040′	31° 36.576′	1804	Major	None
36. Kadi	40° 43.901′	31° 36.459′	1499	Major	Collapse
37. Aslahaddin	40° 43.981′	31° 36.732′	1978	None	None
38. Balci	40° 43.885′	31° 37.111′	1990	None	None
39. Aktas	40° 43.783′	31° 36.650′	1900s	Major	None

| Name | Location (coordinates) | | Construction date | Damage level | |
	North	East		Mosque	Minaret
40. Karacayir	40° 43.825′	31° 36.515′	1946	None	None
41. Kabaklar	40° 44.859′	31° 36.111′	1981	None	None
42. Camli	40° 44.243′	31° 35.841′	1980s	Major	None
43. Sultanzade	40° 44.249′	31° 35.817′	1930s	Major	Major
44. Yesil	40° 44.170′	31° 36.170′	1966	Major	None

Table 2. Damage to the minarets and mosques surveyed in Duzce and Bolu

No.	Name	Location	Mosque	Minaret
45	Cumhuriyet Mah.	Duzce, downtown	None	Collapse
46	Merkez	Duzce, downtown	Major	Collapse
47	Cedidiye Merkez	Duzce, downtown	Light	Collapse
48	Yuvacik	Yuvacik village, near Golcuk	Major	-
49	Asagi Yuvacik	Yuvacik village, near Golcuk	None	None
50	Yeni	Adapazari	None	Collapse
51	Yalova	Yalova, downtown	Light	None
52	Izmit (1)	Izmit, next to highway E5	Light	Collapse
53	Izmit (2)	Izmit, next to highway TEM	None	Light
54	Izmit (3)	Izmit, downtown	Major	Light
55	Golyaka	Golyaka, downtown	Major	Collapse
56	Suleymanbey	5 km east of Golyaka	Collapse	-
57	Golcuk (1)	4 km west of downtown	Major	-
58	Golcuk (2)	Near Ford plant, west of city	Collapse	Moderate
59	Dariyeri Hasanbeyi	East of Duzce (a village)	Major	Light

Table 3. Damage to the other mosques and minarets visited (coordinates not available)

Before mid-1960s, major construction materials were wood and stone or brick masonry. Most of the recently constructed mosques and minarets were reinforced concrete. All of the reinforced concrete mosques were built after 1965. In older mosques, solid bricks were used in infill walls. Hollow clay tiles were used as infill material in the mosques built after late 1970s. As shown in Figure 5, 84 percent of the minarets surveyed in Duzce and Kaynasli were reinforced concrete, whereas only 46 percent of the minarets were reinforced concrete in Bolu. Unreinforced stone masonry was commonly used in old minarets as well as in the minarets constructed in recent years.

4. Observed earthquake damage

The Figure 6 shows the damage distribution for the mosques and minarets surveyed. Damage distribution for RC mosques and minarets are presented in Figures 6c and 6d separately.

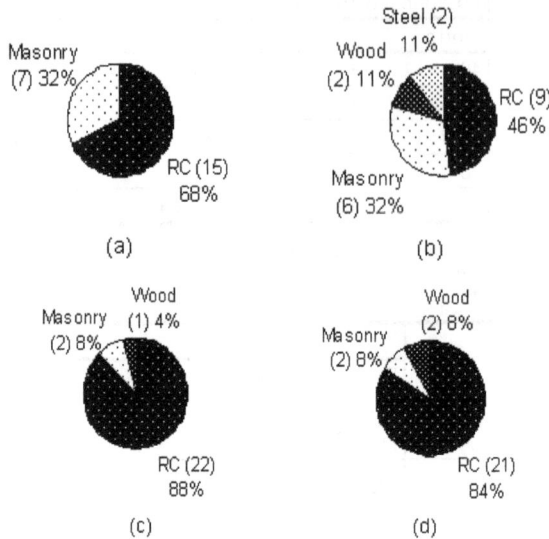

Figure 5. Type of construction for (a) mosques, (b) minarets in Bolu; and (c) mosques, and (d) minarets in Duzce and Kaynasli (number of mosques/minarets is given in parenthesis)

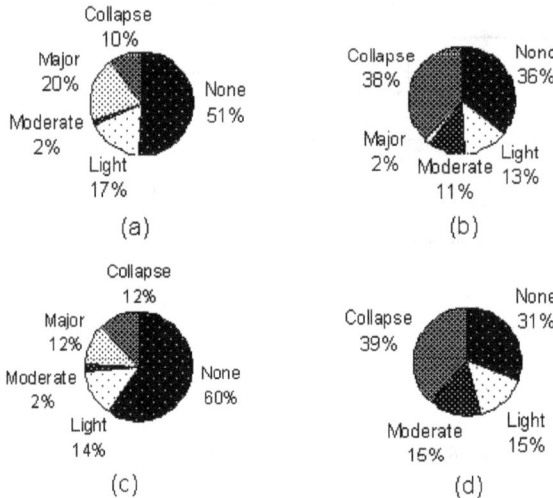

Figure 6. Damage distribution for (a) all mosques and (b) all minarets, and damage distribution for (c) RC mosques and (d) RC minarets

Comparison of Figures 6a and 6c indicates that the damage in RC mosques was less as compared to other structural systems. More than ten percent of the mosques surveyed collapsed. Three of the 26 mosques surveyed in Duzce and Kaynasli region collapsed. There was no collapsed mosque in Bolu, but five of the 21 mosques surveyed had to be closed after the October 12, 1999 earthquake. Closed Mosques in Bolu were Semsi Ahmet Pasa (Imaret),

Yildirim Bayezid, Kadi, Camli and Yesil. Note that the first three of these mosques are at least 200 years old.

As illustrated in Figures 6b and 6d, percentage of all damaged minarets and only RC minarets are similar. Almost forty percent of the minarets collapsed, and approximately one third of the minarets were undamaged. Failure plane for almost all collapsed RC minarets was within 1.5 meter long region above the minaret base or pyramid-shaped transition segment (Figure 7a), where the longitudinal reinforcing bars were usually spliced. Horizontal circumferential cracks and spalling of concrete were commonly observed at the bottom of the cylindrical body of RC minarets (Figure 7a). Frequently, flexural cracks, concrete crushing or spalling was observed in this region of the damaged RC minarets. As shown in Figure 7, less frequently collapse or damage occurred within the transition segment and middle of the cylindrical body or near the top of the minaret.

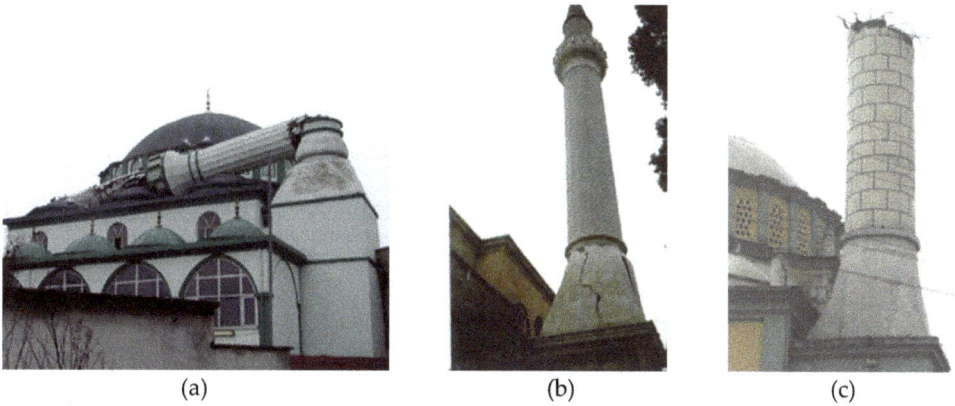

(a)   (b)   (c)

Figure 7. Collapse of a minaret near the bottom of cylinder (left), minaret damage within transition segment (middle), and collapsed minaret at mid-height of cylinder (right)

The ratio of collapsed or damaged unreinforced solid brick or stone masonry minarets was much larger than that of RC minarets. As listed in Table 4, majority of the visited masonry minarets collapsed (Sezen et al. 2003, and Firat 1999). Note that most of these minarets were either very new or few hundred years old (Tables 2 and 3). A minaret may either have an independent foundation (referred to as Type I minaret hereafter) or the base of the minaret may be attached to the roof of the mosque (referred to as Type II minaret). The minarets in Figures 2 and 7 are Type I and II minarets, respectively. In Type II minarets, the minaret and mosque structure respond independently. Large deformatiosn or failure in one structure does not affect the other one.

Lateral laod resisting mechanism of minarets are quite different from that of other structures. The height of center of minaret's mass can be very high above ground, resulting in large bending moments and shear forces. Masonry minarets without reinforcement or clamps (Figures 3 and 4) and with weak mortat are the most vulnerable against earthquakes. Masonry minarets typically either collapsed or suffered minor or no damage. This suggests that these minarets had no ductility or very limited deformation capacity. If the demand due to lateral earthquake forces is less than minaret's ultimate strength, minaret behaves elastically and suffers no damage. However, as soon as lateral demand exceeds the elastic capacity, the minaret collapses. Unlike RC minarets, masonry minarets failed at different locations along the height of the minaret inconsistently (Figure 8). The Imaret and Kadi mosques and possibly their minarets were 600 and 500 year old, respectively. As shown in Figures 8a and 8b, their minarets collapsed at the bottom of the cylindrical body. On the other hand, the Oksuztekke minaret collapsed at its mid-height (Figure 8c). Oksuztekke mosque was constructed only six years before the 1999 earthquakes (Table 2).

Name	City	Location or coordinates	Type	Observed damage
Aziziye Merkez	Düzce	40.50N-31.08E	II	Failed at the bottom of cylinders and collapsed
Cedidiye Merkez	Düzce	40.50N-31.09E	I	Failed at the bottom of cylinders and collapsed
Oksuztekke	Bolu	40.44.488N-31.35.851E	I	Minaret segment above the 2nd balcony level collapsed
Semsi Ahmet Pasa (Imaret)	Bolu	40.43.852N-31.36.635E	I	Failed near the bottom of the cylinder and collapsed
Sarachane	Bolu	40.43.935N-31.36.513E	I	Cracks in stone blocks in the mosque – no observed minaret damage
Yildirim Bayezid	Bolu	40.44.040N-31.36.576E	II	Dislocation of stone blocks in the mosque – no obserevd damage in minarets
Kadi	Bolu	40.43.901N-31.36.495E	I	Severe damage to mosque, minaret collapsed

Table 4. Masonry minarets surveyed after the 1999 earthquakes

4.1. Earthquake damage in historical mosques

Among the historical mosques surveyed, Düzce Merkez, 600-year-old Imaret, 500-year-old Kadi, 300-year-old Sarachane, and 200-year-old Yildirim Bayezid mosques observed to suffer significant structural damage after the 1999 earthquakes (Dognagun and Sezen, 2012). The Yıldırım Bayezid mosque was originally built in 1382 and was burned down in the 19th century. A new structure was constructed after the fire, and it was severely damaged during a 7.3 magnitude earthquake in 1944. The mosque was damaged during the 1999 earthquakes and was closed for a period of time. On the south side of the mosque, portion of the walls above and below the windows were subjected to larger shear stresses (compared to solid wall sections) during the strong ground shaking. Higher shear demand in those parts of the relatively thick walls created serious cracks and openings between the stone blocks (Figure 9). As shown in Figure 9 (middle photo), portion of the walls above and below the windows act, in a sense, like short columns during the earthquake. Larger shear demand in those parts of the relatively thick walls created wide cracks between the stones.

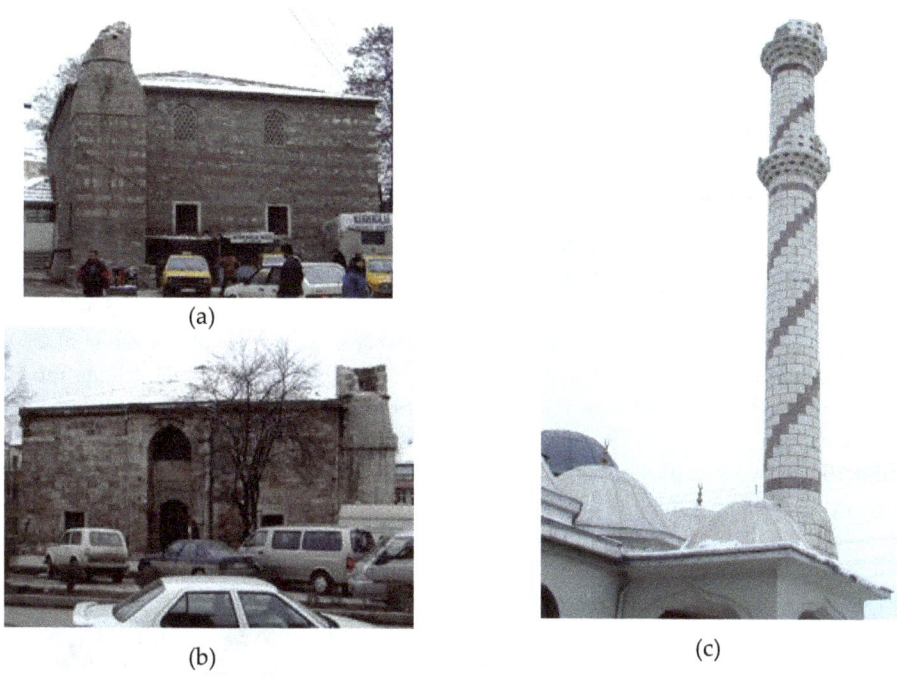

Figure 8. Failures near the bottom and mid-height of the masonry cylindrical minaret body: (a) Imaret, (b) Kadi, and c) Oksuztekke mosques in Bolu

Figure 9. Wall damage and plan view of historical Yildirim Bayezid mosque in Bolu

Another historical mosque, Kadi (Figures 8b and 10a) sustained substantial damage including large cracks around historical entrance door, severe cracks and stone dislocations in the 1.5 m wide unreinforced stone masonry perimeter walls. Damage was mostly concentrated below or above the windows. The missing two key keystones and dislocated stones on top of the upper window are visible in Figure 10a. The reduced wall area along the vertical section through the windows was stressed more, causing considerable damage in those wall areas. The stone masonry minaret collapsed right above its base because the minaret base was integral with minaret walls (Figure 8b) and was quite stiff compared to the cylindrical minaret body.

The Imaret mosque, one of the oldest structures in the region, and its minaret were built using stones and small solid bricks bounded by a thick layer of mortar. The mortar between the bricks is typically as thick as the bricks. Old brick masonry minaret collapsed and Imaret mosque was closed after the 1999 earthquakes due to cracks in the walls (Sezen et al. 2003). The 1999 earthquakes caused very limited damage to the 300-year-old Sarachane mosque and there was no visible damage to its minaret. Some large cracks were observed in the walls at locations similar to those observed in other mosques discussed in this paper. Figure 10b and 10c show examples of couple such cracks; one immediately above a window and another in a corner near the roof.

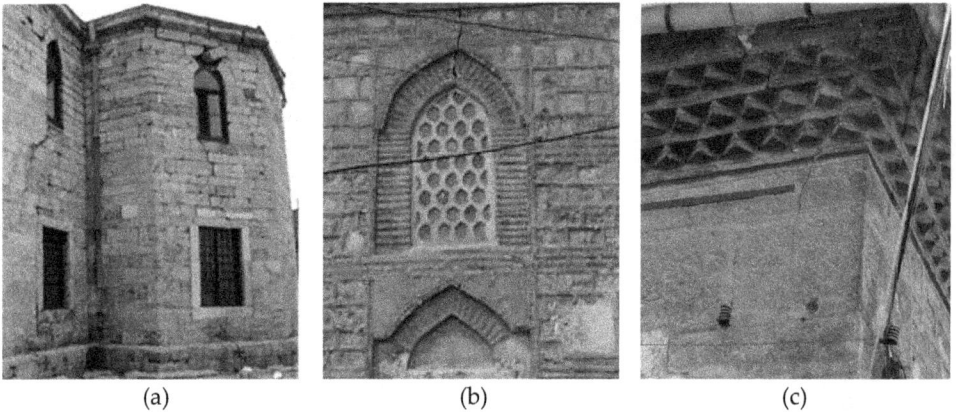

(a)   (b)   (c)

Figure 10. a) cracks developed in the walls and above windows of Kadi; and b, c) damage in in the walls of Sarachane mosque

5. Dynamic analysis of representative masonry minarets

Although numerous collapses and structural damage to minarets and mosques were documented after strong earthquakes, only few researchers investigated the seismic behavior and performance of minarets (Sezen et al. 2008, Portioli et al. 2011, Turk and

Cosgun 2012, Altunisik 2012, Oliveira et al. 2012). It is essential to understand the dynamic behavior of these structures to improve the life safety and to preserve and strengthen the historical monumental structures. In this research, as an example, dynamic modeling and seismic analysis of unreinforced masonry minarets is presented.

Recently, a study was conducted by the authors to investigate the seismic response of reinforced concrete (RC) and masonry minarets (Dogangun et al. 2008, and Sezen et al. 2008). Modal analysis and dynamic time history analyses were conducted. Strength and deformation capacities of the RC and unreinforced masonry models are influenced by unique differences in material properties, including weight, modulus elasticity, and Poisson's ratio. For example, the flexural strength of an unreinforced masonry model is significantly lower as the reinforcing steel in a similar RC model provides tensile resistance. While Sezen et al. 2008 investigates distinct parameters affecting the behavior of RC minarets using a single model, the example presented below examines the effect of masonry material properties and minaret height on the minaret response using three different models. Detailed modeling properties and analysis results are presented in Dogangun et al. (2008).

Although the height of a minaret varies greatly depending on many factors such as the location or construction materials used, the height of a typical minaret with a single balcony over its height is about 20 to 25 m. In this research, two generic minarets with a height of 25 and 30 m and double balconies (Minaret I and Minaret II) and another 20 m tall minaret with a single balcony (Minaret III) were modeled and analyzed. As shown in Figures 11 and 12, only the height of the cylindrical minaret body was varied while the geometrical properties of the base or boot, transition segment, balconies and conical cap were kept the same in all three models. It should be noted that not all conical caps at the top of the masonry minarets are constructed using masonry materials. For uniformity, all minarets including the conical caps are modeled with the same masonry material in this study.

The finite element models of the three minarets are shown in Figure 12, and significant mode shapes of a typical minaret are shown in Figure 13 (Sezen et al. 2008). The computer program, ANSYS was used to model and analyze the minarets (see Dogangun et al. 2008 for details). In the models, the modulus of elasticity of uncracked section, Poisson's ratio and unit weight of masonry material (ordinary limestone) are taken as 3000 MPa, 0.2, and 20 kN/m^3, respectively. In the models, linear elastic material models and five percent damping ratio is used in all dynamic time history analyses. It is assumed that the minaret is located in a high seismic region Zone 1 in Turkish Earthquake Code (TEC, 2007). Considering the TEC, a structural behavior factor, R of 3 and an importance factor, I of 1.2 can be used for such structures.

Dynamic time history analyses are carried out for each minaret model using two ground motions recorded during the 12 November, 1999 Duzce and 17 August, 1999 Kocaeli earthquakes (see Dogangun et al. 2008 for details). The input ground motion was applied only in one horizontal direction. The modal periods of the minaret models (calculated from the modal analysis) and their contribution to the total dynamic response are calculated. The modal periods are greatly affected by the height of the minaret which affects the total mass and flexural stiffness. The calculated first or fundamental periods were 1.41, 0.90, and 0.51 seconds for the minaret models I, II, and III, respectively. The calculated modal response quantities indicate that the contribution of higher mode effects to total dynamic response is significant. The first mode contribution to the total response is about 47 or 49 percent. The torsional or the fifth mode has virtually no effect on the total response of the almost symmetrical minaret structures.

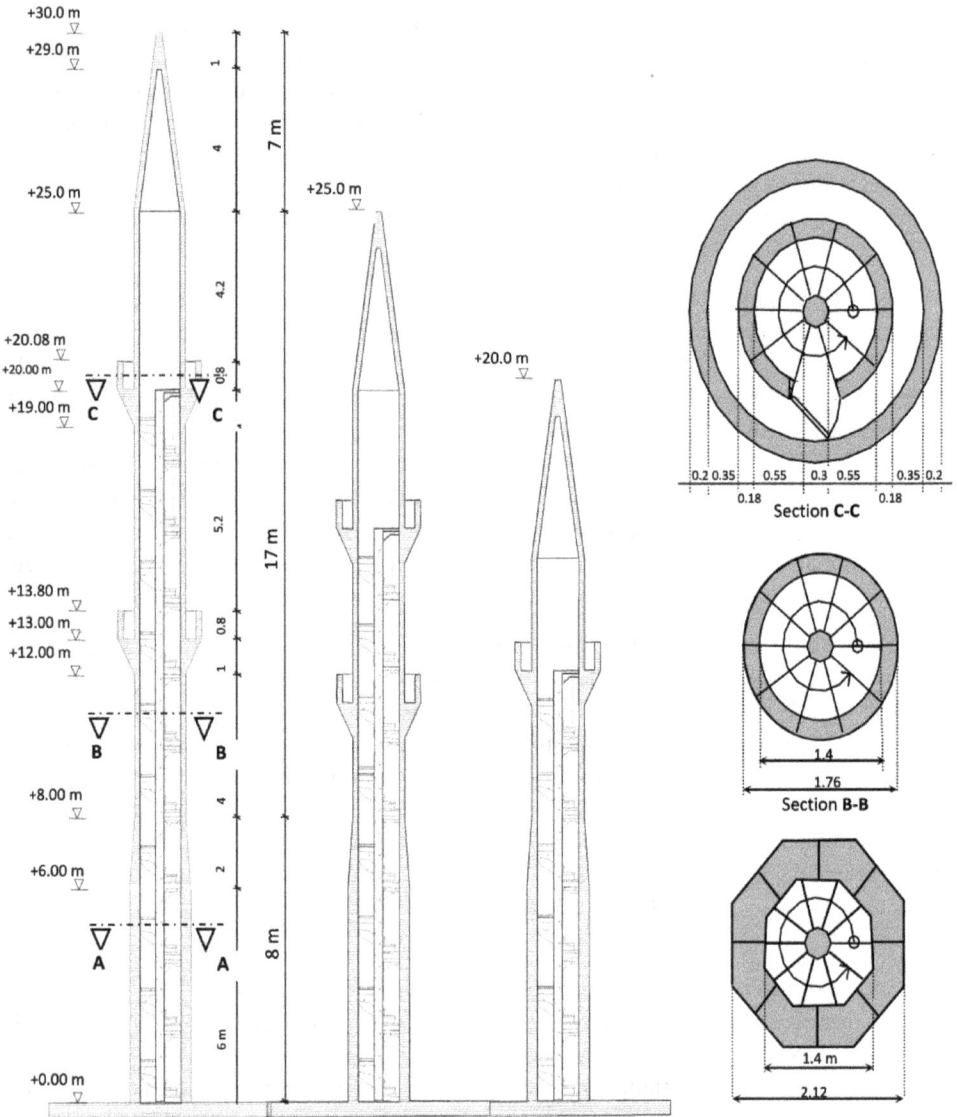

Figure 11. Three minaret models and their geometrical and cross-sectional properties

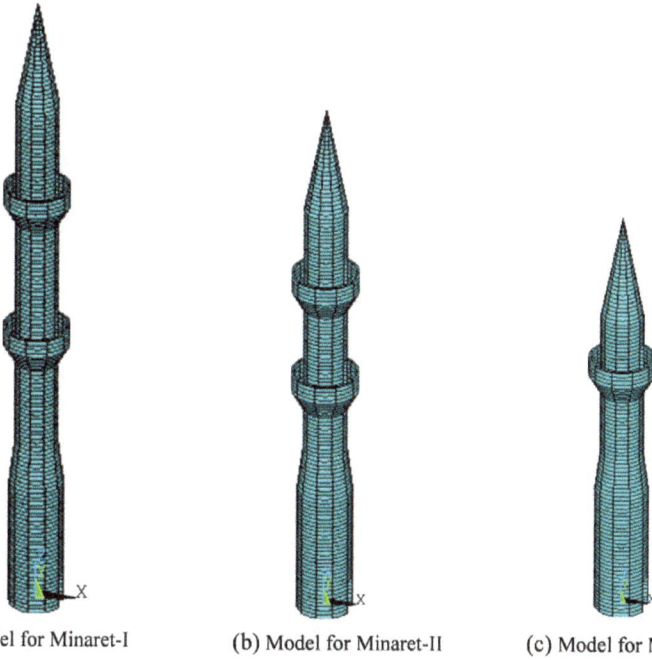

(a) Model for Minaret-I   (b) Model for Minaret-II   (c) Model for Minaret-III

Figure 12. Finite element models of the three minarets

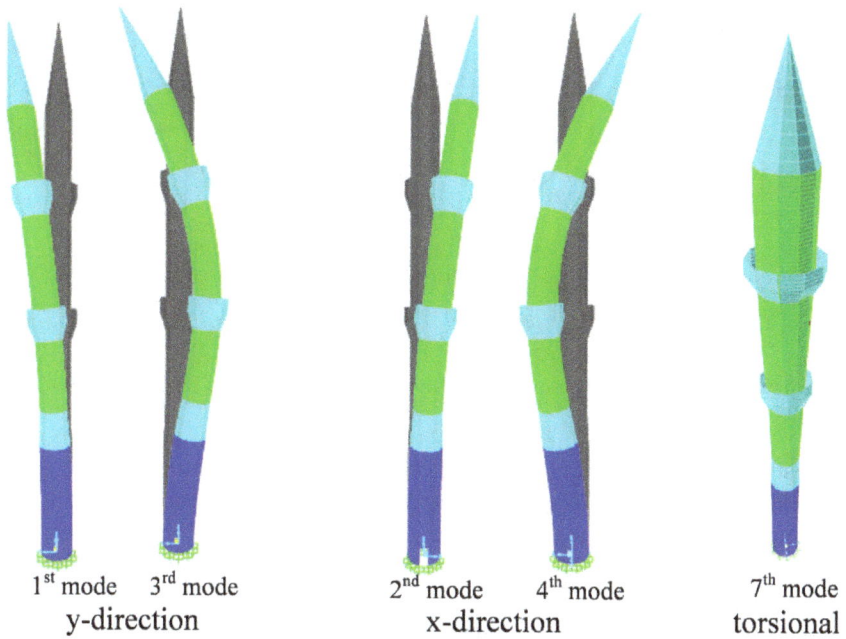

1st mode 3rd mode 2nd mode 4th mode 7th mode

y-direction   x-direction   torsional

Figure 13. First two translational and the seventh mode shapes for a typical model

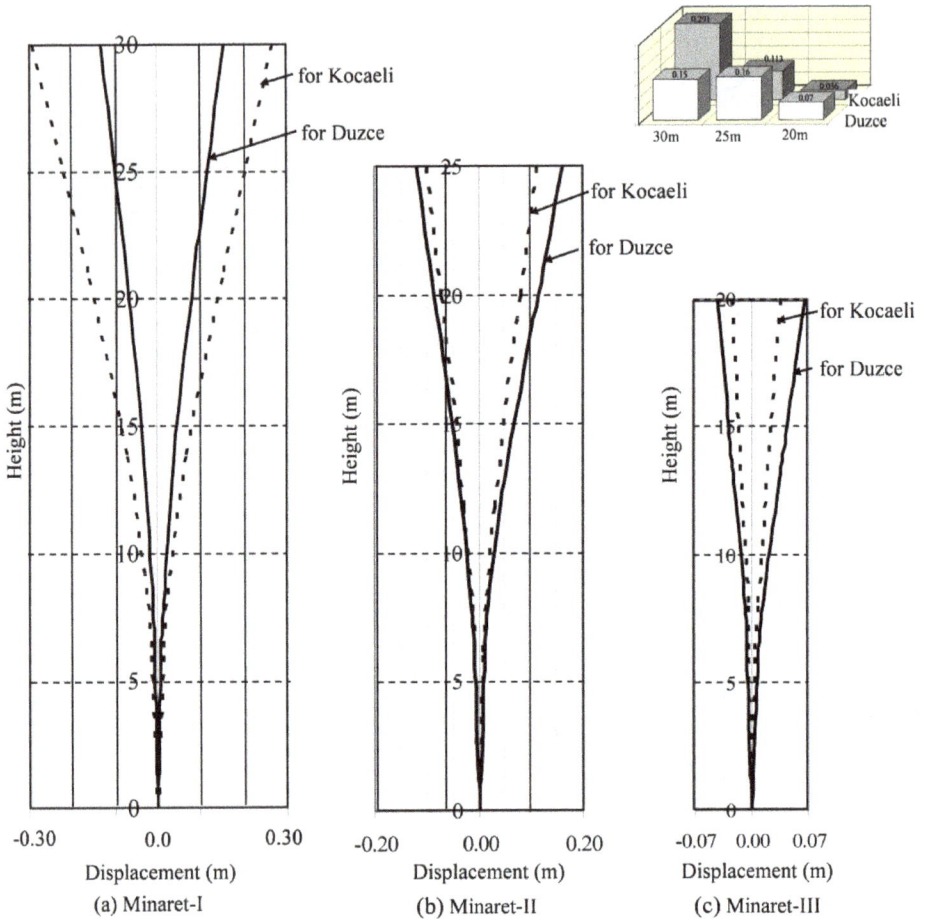

Figure 14. Lateral displacement distribution over the height of minarets for Duzce and Kocaeli ground motions

Figure 14 shows the calculated maximum elastic lateral displacement distribution along the length of three minaret models subjected to 1999 Duzce and Kocaeli earthquake ground motions. As shown in Figure 14 the deflected shapes of the minarets appear to show flexure dominated response with the largest displacements calculated at the top. Although the minarets act as cantilevers, the deformations are much smaller over the height of relatively stiff 6 m high base or boot. The displacements start to increase above the transition segment at about 8 m from ground level. The maximum dynamic displacement of Minaret I is 88%

larger for the Kocaeli input motion (0.291 m) than that for the Duzce ground motion (0.154 m), although the maximum acceleration of the Duzce motion was larger. For the other two shorter minarets Minaret I and II, the maximum calculated displacements are larger for the Duzce ground motion. The maximum dynamic lateral displacement of 0.29 m calculated for the 30 m tall minaret under Kocaeli earthquake exceeds the TEC (2007) code limit of 0.20 m for the Minaret I model (Dogangun et al. 2008).

The most critical axial stress contours calculated during the dynamic analysis of minaret models are shown in Figure 15. The maximum tensile and compressive stresses occur at a height of 8 m, immediately above the transition zone (Figure 12). This is consistent with the most common minaret failure mode observed in the field after the earthquakes (Figures 7a, 8a and 8b). The minaret models used here were stand-alone fixed-ended cantilevers, which did have any lateral support over their height. Nevertheless, in some cases one side of the minaret boot or base (bottom 6 m of the minaret models) are supported by or attached to the mosque structure (Type II in Table 4, e.g., Figures 8a and 8b). Such a partial lateral support, even on one side, provides added stiffness near the bottom of the minaret. This will further increase failure potential at the top of the transition zone as shown in Figures 7a, 8a, 8b and 14.

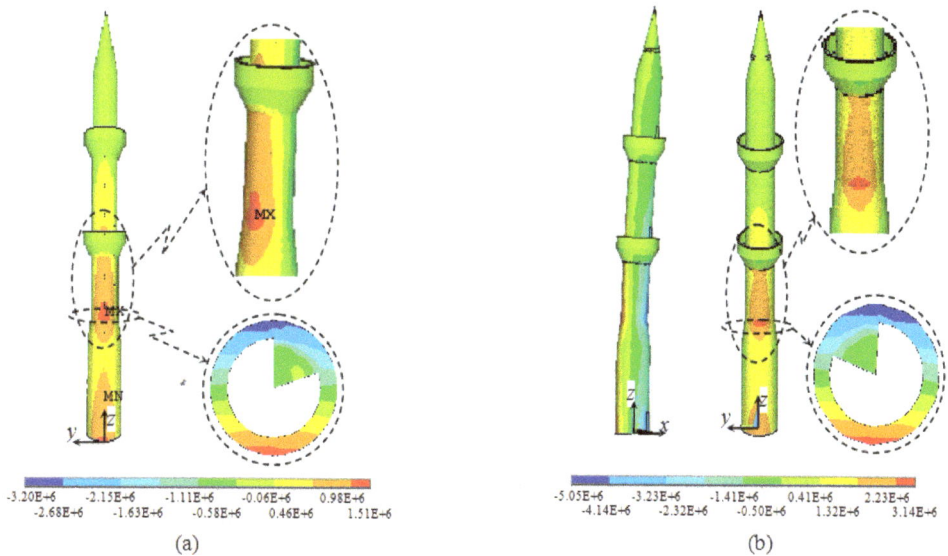

Figure 15. Maximum stress contours for Minaret-I under: a) Duzce, and b) Kocaeli records

6. Summary and conclusions

Structural damage and failures observed in 59 historical and monumental structures (mosques and/or minarets) were documented after the August 17 and November 12, 1999 earthquakes. The reported damage distribution in minarets in the cities of Düzce and Bolu gives an indication of the extent of damage. Ten percent of all visited mosques collapsed while 20% suffered severe damage. Of all visited minarets, 38 percent collapsed (Figure 6). Five out of seven masonry minarets (approximately 70%) in the cities of Düzce and Bolu reporeted in this chapter collapsed (Table 4).

Historical structures with heavy and stiff walls are subjected to larger lateral earthquake forces due to their small periods of vibration. The missing keystones in the upper-floor windows and wide cracks between the stones in the walls above and below window openings were the most common damage types.

The location of the failure in the minarets that collapsed during the 1999 earthquakes was generally at the region near the bottom of the cylinder where a transition was made from a square to a circular section. At the bottom of the cylinder body the lateral stiffness and strength are smaller compared with those of the transition region or the minaret base. Consistent with the observed damage, dynamic analysis results from three generic finite element minaret models showed the largest bending stresses in the same region of the minaret models. For minarets with the base or boot attached to the mosque structure (Type II), additional rigidity and stiffness provided by the mosque prevents deformation and damage within the base. In such cases the failure mostly occurs just above the transition region or base of the cylindrical body.

Author details

Halil Sezen
Department of Civil, Environmental, and Geodetic Engineering, The Ohio State University, Columbus, Ohio, USA

Adem Dogangun
Department of Civil Engineering, Uludag University, Bursa, Turkey

7. References

[1] Turkish Earthquake Code. (1998 and 2007).Specification for structures to be built in disaster areas; Part III—earthquake disaster prevention. Ministry of Public Works and Settlement, Government of Republic of Turkey.

[2] Turkish Building Code. TS-500 (1985). Building Code Requirements for Reinforced Concrete. Turkish Standards Institute, Ankara, Turkey (in Turkish).

[3] Sezen, H., Whittaker, A. S., Elwood, K. J., and Mosalam, K. M. (2003). Performance of Reinforced Concrete and Wall Buildings during the August 17, 1999 Kocaeli, Turkey

Earthquake, and Seismic Design and Construction Practice in Turkey. *Engineering Structures*, Vol. 25, pp. 103-114

[4]   Uniform Building Code. (1997). Structural Engineering Design Provisions, Volume 2. International Conference of Building Officials, ICBO. Whittier, California.

[5]   Çamlıbel, N. (1998). Sinan Mimarlığında Yapı Strüktürünün Analitik İncelenmesi, Doçentlik Tezi, İTÜ, Fen Bilimleri Enstitüsü, İstanbul (in Turkish).

[6]   Ural, A., Dogangun, A., Sezen, H. and Angin, S. (2012). Seismic Performance of Masonry Buildings during the 2007 Bala, Turkey Earthquakes. *Natural Hazards*, Vol. 60, No. 3, pp. 1013-1026

[7]   Sezen, H. and Doğangün, A. (2009). Vulnerability and conservation of historical masonry structures. Prohitech 2009, Protection of Historical Buildings, Rome, Italy, June 21-24

[8]   Oğuzmert, M. (2002). Dynamic behaviour of masonry minarets, M.Sc. Thesis, Istanbul Technical University, Turkey.

[9]   Dogangun, A., Sezen, H., Tuluk, O. I., Livaoglu, R., and Acar, R. (2007). Traditional Turkish Monumental Structures and their Earthquake Response. *International Journal of Architectural Heritage*, Vol. 1, No.3, pp. 251-271

[10] Sezen, H., Firat, G. Y., and Sozen, M. A. (2003). Investigation of the Performance of Monumental Structures during the 1999 Kocaeli and Duzce Earthquakes. Fifth National Conference on Earthquake Engineering, Istanbul, Turkey, 26-30 May

[11] Firat, Y. G. (2001). A Study of the Structural Response of Minarets in the 1999 Anatolian Earthquakes. M.S. Thesis. Purdue University. West Lafayette, Indiana.

[12] Dogangun, A., and Sezen, H. (2012). Seismic Vulnerability and Preservation of Historical Masonry Monumental Structures. *Earthquakes and Structures*, Vol. 3, No. 1, pp. 83-95

[13] Sezen, H., Acar, R., Dogangun, A., and Livaoglu, R. (2008). "Dynamic Analysis and Seismic Performance of Reinforced Concrete Minarets." *Engineering Structures*, Vol. 30, No. 8, pp. 2253-2264

[14] Portioli, F., Mammana, O., Landolfo, R., Mazzolani, F. M., Krstevska, L., Tashkov, L., and Gramatikov, K. (2011). Seismic Retrofitting of Mustafa Pasha Mosque in Skopje: Finite Element Analysis. *Journal of Earthquake Engineering*, Vol. 15, No. 4, pp. 620-639

[15] Turk, A. M., and Cosgun, C. (2012). Seismic Behaviour and Retrofit of Historic Masonry Minaret. *Gradevinar*, Vol. 64, No. 1

[16] Altunisik, A. C. (2012). Comparison of earthquake behavior of reinforced concrete minarets using fiber-reinforced polymer composite. *The Structural Design of Tall and Special Buildings*, DOI: 10.1002/tal.731

[17] Oliveira, C. S., Çaktı, E., Stengel, D., and Branco, M. (2012). Minaret behavior under earthquake loading: The case of historical Istanbul. *Earthquake Engineering and Structural Dynamics*, Vol. 4, No. 1, pp. 19-39

[18] Dogangun, A., Acar, R., Sezen, H., and Livaoglu, R. (2008). Investigation of Dynamic Response of Masonry Minaret Structures. *Bulletin of Earthquake Engineering*, Vol. 6, No. 3, pp. 505-517

Chapter 8

Dynamic Behaviour of the Confederation Bridge Under Seismic Loads

Lan Lin, Nove Naumoski and Murat Saatcioglu

Additional information is available at the end of the chapter

1. Introduction

The Confederation Bridge, which was opened for traffic in June 1997, is 12,910 m long and is one of the longest reinforced concrete bridges built over water in the world. The bridge crosses the Northumberland Strait in eastern Canada and connects the province of Prince Edward Island and the province of New Brunswick.

The bridge is located in a region known for very harsh environmental conditions. The Strait is covered by ice approximately three to four months in a year. Heavy storms with winds in excess of 100 km/h are often experienced at the bridge site. Given the importance of the Confederation Bridge, its length, and the environmental conditions, special criteria were imposed in the design and construction of the bridge in order to provide a high degree of safety during its operational life. The bridge was designed for a service life of 100 years, which is twice the service life considered in the Canadian codes for highway bridges that were in use during the design of the Confederation Bridge, i.e., the CSA Standard CAN/CSA-S6-88 [1], and the Ontario Highway Bridge Design Code (OHBDC) [2]. A safety index of 4.0 was used in the design, compared with 3.5 specified in CAN/CSA-S6-88 and OHBDC. Load combinations and load resistance factors were developed specifically for the design of the bridge, as described in [3]. A number of assumptions had to be made in the design, particularly for the long-term properties of the materials in the specific environmental conditions and for the effects of various dynamic loads on the performance of the bridge. Given these assumptions, a comprehensive research program was undertaken to monitor and study the behaviour of the bridge. As part of this program, a study was conducted to investigate the dynamic performance of the bridge under seismic loads. The objective of the study was to compare the responses of the bridge for seismic actions representative of the seismic hazard at the bridge location with those used in the design. There are two major reasons for undertaking this study. First, significant advancements in the understanding of the eastern Canadian seismicity and in the methods for seismic hazard computations have been made since the design of the bridge in the mid 1990s, and therefore, a more accurate estimate of the seismic hazard at the bridge location can now be made. Second, recorded vibrations of the bridge are available which enable the development of an accurate analysis model of the as-built bridge.

This paper describes the main findings from the study. It includes: (i) a brief description of the bridge; (ii) an overview of the seismic parameters used in the design of the bridge; (iii) development of a finite element model of the bridge for use in the seismic analysis; (iv) selection of seismic ground motions representative of the seismic hazard at the bridge location; and (v) dynamic analysis of the bridge model and comparison of the analytical results with the design values.

2. Description of the bridge

The Confederation Bridge consists of two approach bridges at its ends and a main bridge between them (Fig. 1). The approach bridge at the Prince Edward Island end (i.e., the east end) is 555 m long and has 7 piers, and that at the New Brunswick end (i.e., the west end) is 1,275 m long and has 14 piers. The longest span of the approach bridges is 93 m. The main bridge is 11,080 m long and has 44 piers, designated P1 to P44 in Fig. 1. Of the 45 spans of the main bridge, 43 spans are 250 m long and the two end spans are 165 m long. The cross section of the bridge girder is a single-cell trapezoidal box. The depth of the girder of the main bridge varies from 4.5 m at mid spans to 14 m at piers. The width of the bridge deck is 11 m.

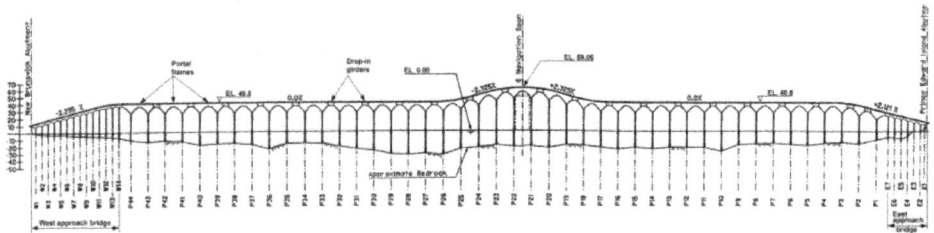

Figure 1. Elevation of the Confederation Bridge.

As shown in Fig. 1, the bridge deck of most of the main bridge is at elevation of 40.8 m above mean sea level (MSL). The height of the columns of this part of the bridge ranges from 38 to 62 m. In the middle portion of the main bridge, between piers P17 and P26, the elevation of the deck increases from 40.8 m at P17 and P26 to the highest elevation of 59.06 m at the central span P21-P22. This span is called the navigation span. The elevation of 59.06 m above MSL provides a 49 m vertical clearance for marine vessel traffic. The height of the piers of the navigation span is approximately 75 m.

Both the approach bridges and the main bridge were built of precast concrete segments which were assembled using post-tensioned tendons. A detailed description of the bridge and the construction methods is given in [4]. Because this study is associated with typical spans of the main bridge, the discussion in the rest of this section will be focussed on structural features of the main bridge.

The structural system of the main bridge consists of a series of rigid portal frames connected by simply supported girders, which are called drop-in girders (Fig. 1). Every second span is constructed as a portal frame, and all other spans are constructed using drop-in girders. In total, there are 21 portal frames in the main bridge. This structural system was selected to prevent progressive collapse of the bridge due to extreme effects of wind, ice, seismic, and traffic loads, and ship collisions.

Figure 2 shows a typical portal frame of the main bridge. The girder consists of two 192.5 m double cantilevers and a 55 m long segment between them. The connections between this segment and the cantilevers are detailed to behave as rigid joints. The drop-in girders that connect the frames are also shown in Fig. 2, in the spans adjacent to the portal frame span. The length of the drop-in girders is 60 m. Each of the drop-in girders sits on the overhangs of the two adjacent portal frames. Four specially designed elastomeric bearings are used as supports. One of the bearings is fixed against translations and the remaining three allow translations of the girder only in the longitudinal direction. All four bearings allow rotations about all axes. This configuration of the bearings provides a hinge connection at one end, and longitudinal sliding connection at the other end of the drop-in girder.

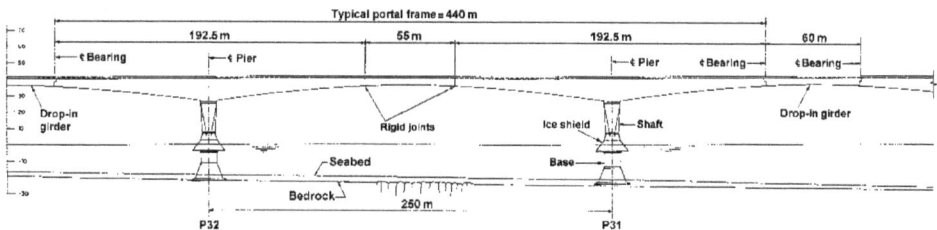

Figure 2. Typical portal frame.

The piers are constructed of two precast concrete units each, i.e., the pier base and the pier shaft (Fig. 2). The pier base is a hollow unit and has a circular cross section in plan with an outer diameter of 8 m at the top and 22 m at the footing. The pier shaft is also a hollow unit and consists of a shaft at the upper portion and an ice shield at the bottom portion of the pier. The cross section of the pier shaft varies from a rectangular section at the top to an octagonal section at the bottom of the shaft. Both the pier base and the pier shaft have very complex shapes. Detailed explanations for these and the geometrical properties of the piers can be found in [4].

3. Seismic design parameters and seismic hazard for the bridge

3.1. Seismic design parameters

The design life of 100 years and the safety index of 4.0 were the basic design requirements for the Confederation Bridge. These requirements were much higher than those prescribed in the highway bridge design codes available at the time when the bridge was designed. The specified design life and safety index for the Confederation Bridge required special studies in order to determine the seismic ground motion parameters at the bridge location.

The seismic ground motion parameters used in the design of the bridge were given in the design criteria specified by J. Muller International – Stanley Joint Venture Inc. [5]. These included the peak ground acceleration, the peak ground velocity, the peak ground displacement, and the seismic design spectrum for the bridge location. The methods for determining these parameters were described by [6]. Two methods were used for the estimation of the peak ground acceleration of the expected seismic motions at the bridge location. The first method was based entirely on probabilistic considerations. According to this method, the peak ground acceleration for the design service life of 100 years and the design safety index of 4.0 corresponded to an annual probability of exceedance of 0.00027. The value of the peak ground acceleration for this probability of exceedance was found to be A=0.136 g.

The second method was primarily based on engineering considerations. In this method, first, the peak ground acceleration was determined for a probability of exceedance of 10% during the design service life of 100 years. The background for this was to keep the same probability of exceedance during the service life as that required by the 1990 edition of the National Building Code of Canada (NBCC) [7]. Then, the acceleration value corresponding to 10% in 100 years probability of exceedance was increased by applying a factor of 1.43 representing the product of the commonly used importance factor of 1.3, and an additional importance factor of 1.1 because of the unusual importance of the bridge. The resulting peak ground acceleration was 0.12 g, and this value was adopted for the design. Using the same approach, the peak ground velocity was found to be 10.8 cm/s. Having the values for the peak ground acceleration (A) and the peak ground velocity (V), a value for the peak ground displacement (D) of 5.9 cm was obtained using the relationship between A, V, and D, proposed by [8].

The 5% damped elastic seismic design spectrum for horizontal seismic motions was developed using the foregoing values for the peak ground acceleration, velocity and displacement, and applying the corresponding spectral amplification factors proposed by [8] for the mean plus one standard deviation level. This level corresponds to a probability of 84% that the spectral amplification factors will not be exceeded. The parameters for the construction of the horizontal design spectrum are given in Table 1, adopted from the design criteria. It can be seen that the spectrum was defined assuming a constant spectral acceleration in the short period range (T<0.5 s), a constant spectral velocity in the intermediate period range (0.5 s < T< 3.0 s), and a constant spectral displacement in the long period range (T > 3.0 s), which is a common approach for constructing design spectra based on peak ground motions and spectral amplification factors [8]. The vertical design spectrum was taken as 2/3 of the horizontal spectrum [5], which is also a common practice for defining vertical design spectra, based on the findings reported in [9].

Period, T(s)	Governing parameter	Spectral acceleration (g)
< 0.5	Acceleration = 0.326 g	0.326
0.5 – 3.0	Velocity = 24.8 cm/s	0.1589 / T
> 3.0	Displacement = 11.8 cm	0.48 / T^2

Table 1. Parameters of the design spectrum for horizontal seismic motion; 5% damping [5].

Figure 3 shows the horizontal seismic design spectrum. The other spectrum in the figure, designated "uniform hazard spectrum" is discussed below.

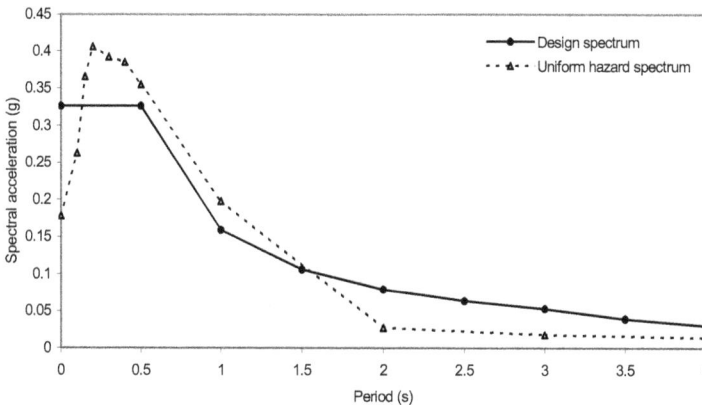

Figure 3. Design and uniform hazard spectra; 5% damping.

3.2. Seismic hazard for the bridge location

Since the development of the design parameters for the Confederation Bridge in early 1990s, there have been significant advances in the understanding of the seismic hazard in Canada. New source models, and most updated software have been used for the assessment of the seismic hazard. It should be mentioned, however, that there are still significant uncertainties in the estimation of seismic hazard. As pointed out by [10], the ground motion attenuation relations for eastern Canada are the major source of uncertainty in the seismic hazard estimations. This is because of lack of recordings of ground motions from strong earthquakes in eastern Canada for use in the calibration of the attenuation relations. It is noted that the ground motion attenuation relations for eastern Canada may change significantly as new events are recorded as reported in [10].

The seismic hazard in Canada is currently represented by uniform hazard spectra rather than by peak ground motions. A uniform hazard spectrum represents an acceleration spectrum with spectral ordinates that have the same probability of exceedance. Uniform hazard spectra can be computed for different probabilities and different confidence levels. Confidence levels of 50% (median) and 84% are typically used for uniform hazard spectra. These levels represent the confidence (in %) that the spectral values will not be exceeded for the specified probability.

For the purpose of this study, Geological Survey of Canada (GSC) computed the uniform hazard spectrum for the bridge location for an annual probability of exceedance of 0.00027 and confidence levels of 50% and 84%. Among the two confidence levels, the uniform hazard spectrum at the 84% confidence level was used in this study. The 84% (rather than 50%) level was chosen since the spectral amplification factors used in the development of the design spectrum are for that level. The 84% level uniform hazard spectrum (UHS) is shown in Fig. 3. The spectral values for periods below 2.0 s were provided by GSC. For periods between 2.0 s and 4.0 s, the spectrum was extended assuming a constant spectral velocity with the same value as that at 2.0 s. This is the same as assumed in the defining of the spectral values in the intermediate period range of the design spectrum.

It can be seen in Fig. 3 that the uniform hazard spectrum is somewhat higher than the design spectrum for periods below 1.5 s. As will be discussed later, this difference does not have significant effects on the seismic response of the bridge.

3.3. Scenario earthquakes for the bridge location

The seismic hazard at a given site represents the sum of the hazard contributions of different earthquakes at different distances from the site. For each site, however, there are a few earthquakes that have dominant contributions to the hazard. These earthquakes are normally referred to as scenario or predominant earthquakes. The shape of the uniform hazard spectrum for a given site, representing the seismic hazard for the site, depends on the magnitudes of the scenario earthquakes and the distances of these earthquakes from the site. In general, the dominant contribution to the short period ground motion hazard is from small to moderate earthquakes at small distances, whereas larger earthquakes at greater distance contribute most strongly to the long period ground motion hazard.

For the purpose of the selection of earthquake ground motions for use in the seismic analyses, it is necessary to determine the scenario earthquakes for the Confederation Bridge. This can be done by computing the seismic hazard contributions of selected magnitude-distance ranges that cover all possible magnitude-distance combinations. Figure 4, provided by Geological Survey of Canada, shows the magnitude-distance contributions for the Confederation Bridge for annual probability of exceedance of 0.000404 (i.e., 2% in 50 years). Such graph could not be produced for a probability of exceedance of 0.00027 because of the uncertainties in the hazard analysis due to the extrapolations relative to the current hazard models. However, it was reported by [11] that the predominant magnitude increases very slowly as probability decreases. Also, results reported in [12] indicated that the lowering of the probability has small effects on the predominant magnitude and distance values. Given this, the magnitude-distance contributions shown in Fig. 4 were considered to be representative of those for probability of exceedance of 0.00027.

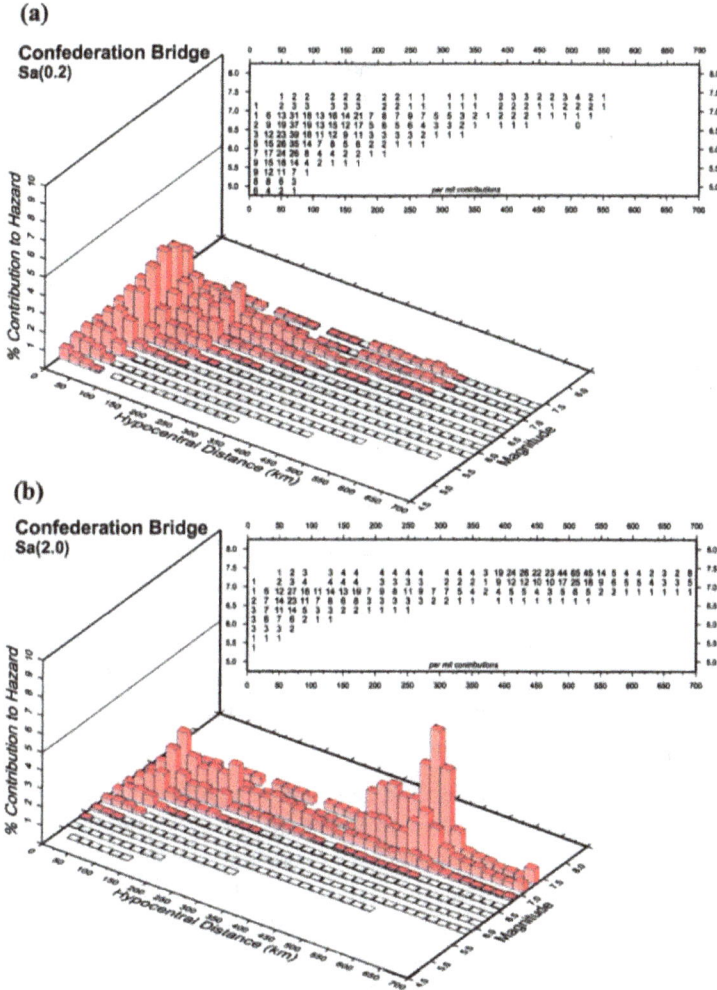

Figure 4. Magnitude-distance contributions to the seismic hazard of the Confederation Bridge, (a) for spectral acceleration at period of 0.2 s, and (b)for spectral acceleration at period of 2.0 s.

Figure 4(a) shows the contributions to the seismic hazard for period of 0.2 s, representing the short period ground motion hazard, while Fig. 4(b) shows the contributions for period of 2.0 s, representing the long period ground motion hazard. The contributions are computed for magnitude increments of 0.25, and distance increments of 20 km. It can be seen in Fig. 4(a) that the scenario earthquakes that have predominant contributions to the short period ground motion hazard are with magnitude ranging from 6 to 6.75 at distances of 60 km to 80 km. Similarly, Fig. 4(b) shows that the scenario earthquakes that have predominant contributions to the long period ground motion hazard are with magnitudes ranging from 7.25 to 7.5 at distances of approximately 500 km.

4. Modelling of the bridge

The structural system of the bridge allows the development of a model of a selected segment of the bridge rather than modelling the entire bridge. Because of the repetitiveness of the units of the structural system (i.e., portal frames and drop-in girders) along the bridge, a proper model of a selected segment would be quite representative of the whole bridge.

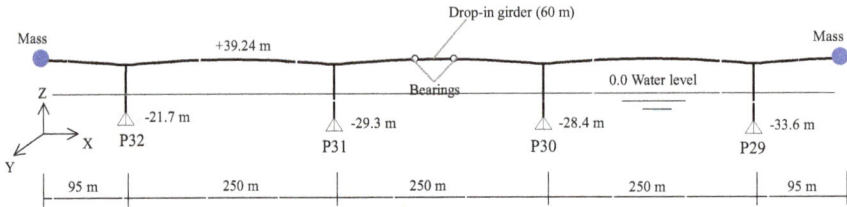

Figure 5. Model of two portal frames and one drop-in span using 3-D beam elements.

Figure 5 shows the model used in this study. It is a three-span frame model consisting of 3-D beam elements. The modelling was conducted using the computer program SAP 2000 [13]. The model represents the bridge segment between piers P29 and P32 (Fig. 1), which consists of two rigid portal frames (P29-P30 and P31-P32), and one drop-in span (P30-P31). This segment was modelled since it is the instrumented portion of the bridge, and recorded data is available for use in the calibration of the model. Also, the height of the piers of this segment is quite representative of the main bridge.

The model consists of 179 beam elements and 180 joints. The bridge girder is modelled by 123 elements, and each pier is modelled by 14 elements. The interaction with the adjacent drop-in girders (left of P32, and right of P29) was modelled by adding masses at the ends of the overhangs, as shown in Fig. 5. A half the mass of each drop-in girder was added at the end of the supporting overhang in transverse and vertical directions, full mass was added in the longitudinal direction for a hinge connection, and no mass was added in the longitudinal direction for a sliding connection. Similarly, vertical forces from a half the weight of each drop-in girder were applied at the ends of the overhangs.

In addition to the three-span model (Fig. 5), a single-span model consisting of a single portal frame (P31-P32), and a five-span model with three portal frames and two spans with drop-in girders (between P29 and P34; Fig. 1) were also considered. While the natural periods and mode shapes of these three models were quite comparable, the three-span model was chosen for the analysis in this study because it provides results for both the portal frame spans and the spans with drop-in girders, and requires an acceptable computation time for the analysis. The single-span model does not provide results for the drop-in girder, and the five-span model requires an excessive computation time. Note that the segment shown in Fig. 5 is normally used as a typical segment in studies on the behaviour of the Confederation Bridge [e.g., 14,15].

5. Calibration of the model using data of full scale test

The model shown in Fig. 5 was calibrated using records of vibrations and tilts of the bridge obtained during a full scale tests of the bridge were conducted on April 14, 1997, about two months before the official opening of the bridge. The objectives of the tests were: (i) to measure the deflection of the bridge pier under static loads, and (ii) to measure the free vibrations of the pier due to a sudden release of the static load. The instrumentation of the bridge (Fig. 6) was used to measure the bridge response during the pull tests. It consists of 76 accelerometers and 2 tiltmeters. The accelerometers were used to measure acceleration time histories of the response of the bridge. The two tiltmeters installed at locations 3 and 4 of pier P31 were used to measure the tilts of the pier.

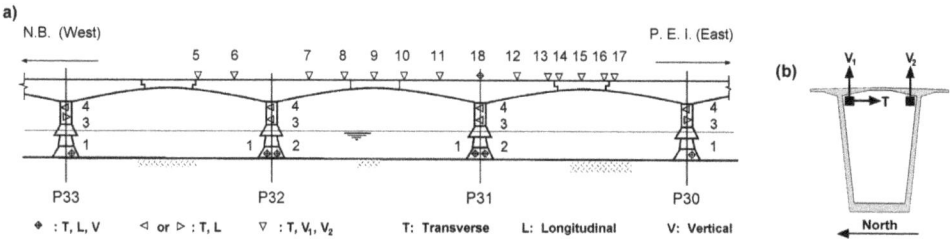

Figure 6. Locations of accelerometers: (a) instrumented sections of the bridge girder and piers, and (b) locations of accelerometers in the girder.

The first pull test was a static test. Using a steel cable, a powerful ship pulled pier P31 in the transverse direction of the bridge. The pulling was at the top of the ice shield, approximately 6 m above the mean sea level. The force was increased steadily up to 1.43 MN, and then released slowly.

The second pull test was a dynamic test. In this test, the load was applied at a slow rate up to 1.40 MN and then suddenly released. This triggered free vibrations of the bridge, which were recorded by several accelerometers. The acceleration time history of the transverse vibrations recorded at the middle of span P31-P32 (location 9 in Fig. 6) along with the recorded tilts at locations 3 and 4 were used in the calibration of the model.

The parameter that was varied in the calibration process was the foundation stiffness. Rotational springs in the longitudinal and transverse directions were introduced in the model, at the bases of the piers, to represent the foundation stiffness. A trial value of the stiffness of the springs was initially selected, and a number of iterations of static and dynamic elastic analyses were performed in order to determine the stiffness that provides a close match between the computed and the measured tilts and free vibrations of the bridge. In each iteration, the tilts and the response were computed by using a load function closely representing the actual loading during the test. A modulus of elasticity of the concrete of 40,000 MPa was used in the analyses. This value was based on experimental data for the bridge [14], and is representative of the modulus of elasticity at the time when the test was conducted.

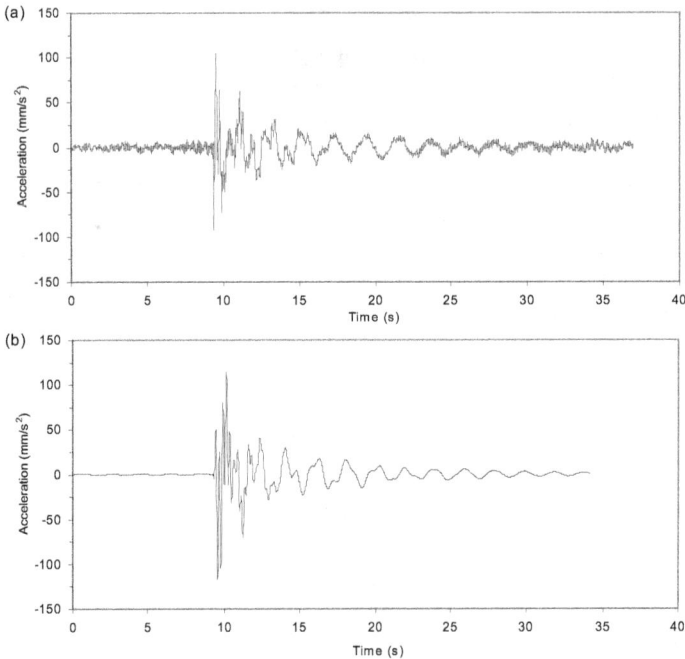

Figure 7. Acceleration time histories of transverse vibrations at midspan between piers P31 and P32 (a) measured, (b) computed.

It was found that the model with a rotational stiffness of 3.35×10^9 kN·m/rad provides the best matching of the computed and measured responses. Figure 7 shows the measured and the computed acceleration time histories of the transverse vibrations of the bridge girder at the mid-span between piers P31 and P32, and Fig. 8 shows the Fourier amplitude spectra of these time histories. It can be seen in Fig. 7 that the computed response of the bridge is very similar to the measured response. Also, Fig. 8 shows that the Fourier amplitude spectra of the computed and the measured responses are quite close. Note that the first two predominant frequencies of the computed response of 0.51 Hz and 1.28 Hz correspond respectively to the 7th and the 18th modes of the model.

Table 2 shows the natural periods of the first ten modes obtained from dynamic analysis of the model. For illustration, the vibrations of the first five modes are presented in Fig. 9. It is necessary to mention that a similar model was developed by Lau et al. [15] using the computer program COSMOS [16]. The natural periods and mode shapes of that model are very close to those of the model developed in this study.

It is useful to mention that certain variations of the dynamic properties of the model are expected due to different effects. For example, the modulus of elasticity increases with the age of concrete and varies due to temperature changes. Also, the responses used in the calibration of the model are substantially smaller than those from expected seismic motions at the bridge location. A comprehensive investigation of the possible variations of the dynamic properties due to the foregoing effects conducted by [17] showed that these variations are insignificant from practical point of view, therefore, the model developed as described above is considered appropriate for the seismic evaluation of the bridge.

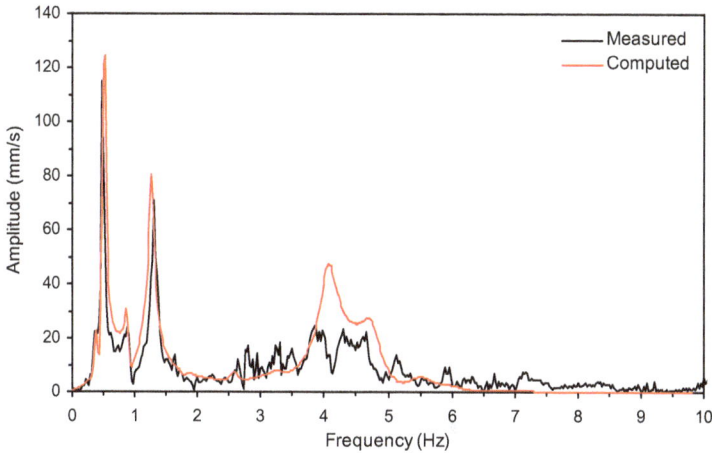

Figure 8. Fourier amplitude spectra of measured and computed acceleration time histories of vibrations at midspan between piers P31 and P32.

Mode No.	Period (s)	Mode type
1	3.13	Transverse
2	2.99	Transverse
3	2.72	Transverse
4	2.48	Transverse
5	2.22	Transverse
6	2.13	Longitudinal
7	2.08	Transverse
8	2.01	Longitudinal
9	1.54	Vertical
10	1.43	Vertical

Table 2. Natural periods of the first 10 modes of the bridge model.

6. Seismic excitations for time-history analysis

Given the uncertainties in the estimation of the seismic hazard for eastern Canada, a number of time-history analyses were conducted using excitation motions well beyond the scenario earthquake motions for the bridge location determined from the seismic hazard analysis as discussed in Section 3.3. In total, five groups of different seismic excitations were considered.

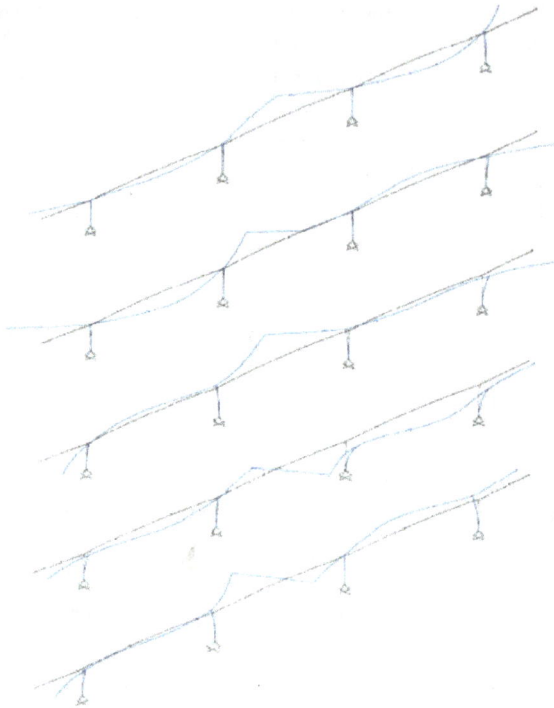

Figure 9. Mode shapes of the bridge model.

Because of lack of strong seismic motion records in eastern Canada, two ensembles of ground motion records obtained during strong earthquakes around the world were used in this study. The ensembles are described in [18, 19] and are characterized by different peak ground acceleration to peak ground velocity ratios (A/V ratios). The average A/V ratio (A in g, and V in m/s) of the records of one of the ensembles is 2.06, and that of the other ensemble is 0.48. Based on the A/V ratios of the records, the ensembles are referred to as the high and low A/V ensembles. In general, high A/V ratios are characteristics of seismic motions from small to moderate earthquakes at short distances, and low A/V ratios are characteristics of seismic motions from large earthquakes at large distances. Regarding the frequency content, high A/V motions normally have a high frequency content, and low A/V motions have a low frequency content. Seismic motions with a high frequency content are characterized by predominant frequencies higher than approximately 2 Hz (i.e., periods lower than 0.5 s), and seismic motions with a low frequency content are characterized by predominant frequencies lower than 2 Hz (i.e.,periods longer than 0.5 s).

In addition to the foregoing ensembles, ground motion records obtained during the 1988 Saguenay, Quebec earthquake, and the 1982 Miramichi, New Brunswick earthquake were used as excitation motions. Also, stochastic seismic motions generated for eastern Canada were used.

6.1. High A/V excitations

It is well known that seismic ground motions in eastern Canada are characterized by high frequency content and high A/V ratios [19, 20]. As discussed above, an ensemble of records with high A/V ratios from strong earthquakes around the world [19] was adopted for the analysis. The ensemble consisted of 13 pairs of horizontal and vertical records. The magnitudes of the earthquakes are between 5.25 to 6.9, the distances are between 4 km to 26 km. The average A/V ratio of the records is 2.06. It is necessary to mention that the magnitudes of these earthquakes cover the magnitude range of 6.0 to 6.75 of the scenario earthquakes for the short period ground motion hazard for the bridge location as discussed in Section 3.3.

The excitation motions for the time-history analysis were obtained by scaling the records to the peak ground velocity of 7.1 cm/s computed by GSC for an annual probability of exceedance of 0.00027. These excitations are referred to as high A/V excitations. Figure 10 shows the acceleration response spectra of the scaled horizontal records of the ensemble. For comparison, the design spectrum is superimposed on the figure. It can be seen that the spectra of the records exceed significantly the design spectrum for periods shorter than approximately 0.5 s, and the spectra are well below the design spectrum for periods longer than 0.5 s.

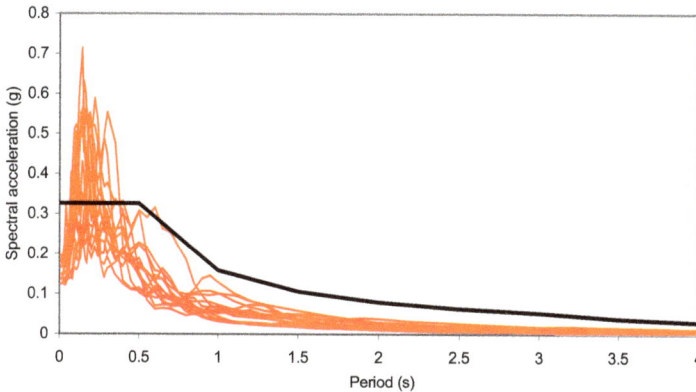

Figure 10. Design spectrum and scaled response spectra of high A/V excitations; 5% damping.

6.2. Low A/V excitations

The low A/V ensemble consisted of 15 pairs of horizontal and vertical records of seismic ground motions [18]. The records were taken during strong earthquakes around the world with magnitudes ranging from 6.3 to 8.1. The distances at which the records were taken were within the range from 38 km to 469 km. The average A/V ratio of the records is 0.48. Both the magnitudes and the distances cover the magnitude and distance ranges of the scenario earthquakes for short and long period ground motion hazards for the bridge location determined from the seismic hazard analysis (see Section 3.3).

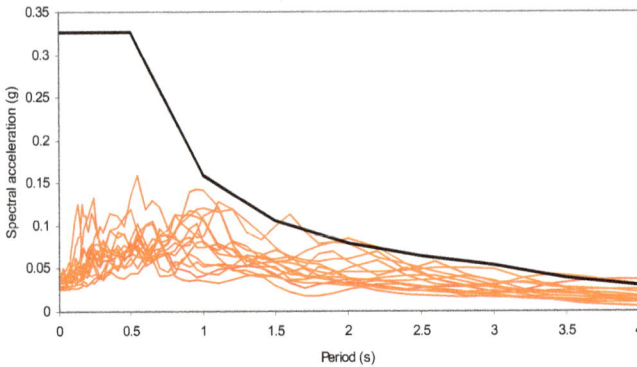

Figure 11. Design spectrum and scaled response spectra of low A/V excitations; 5% damping.

Figure 11 shows the acceleration response spectra of the horizontal records of the low A/V ensemble scaled to the peak ground velocity of 7.1 cm/s. The design spectrum is also included in the figure. It can be seen that the spectra for the low A/V records are all enveloped by the design spectrum. Given this, no time-history analyses were conducted for this ensemble.

6.3. Saguenay earthquake excitations

It was of special importance for this study to investigate the performance of the bridge when subjected to seismic motions from earthquakes in eastern Canada. On November 25, 1988, an earthquake of magnitude of 5.7 occurred in the Saguenay region of the province of Quebec. This was the most significant earthquake in the past 50 years in eastern North America. Ground motion records were obtained at 16 sites at distances ranging from 43 km to 525 km [21, 22]. The response spectra for all horizontal records were scaled to the peak ground velocity for the bridge location of 7.1 cm/s and were compared with the design spectrum. Based on the comparison, 5 horizontal records and the companion vertical records were selected for the analysis. The scaled spectra of the horizontal records together with the design spectrum are shown in Fig. 12.

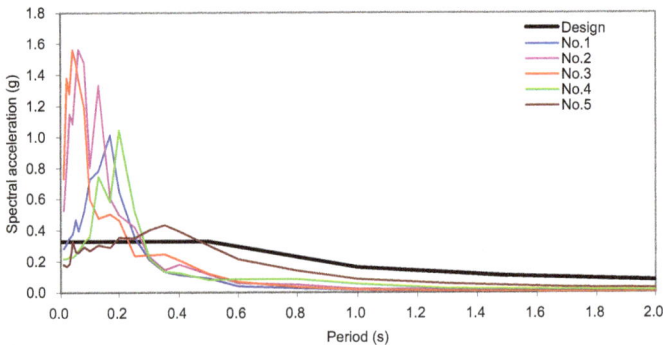

Figure 12. Design spectrum and scaled response spectra of Saguenay earthquake excitations; 5% damping.

It can be seen in the figure that the scaled spectra of the Saguenay earthquake motions are significantly higher than the design spectrum for periods below 0.25 s. The highest spectra (i.e., spectra of records No. 2 and No. 3) exceed the design spectrum by a factor of approximately 5.

6.4. Miramichi earthquake excitations

In 1982, several earthquakes occurred in the Miramichi region of the province of New Brunswick [23]. The epicentres of earthquakes were approximately 150 km from the bridge site. By considering the response spectra, three records representing the strongest motions during the earthquakes were selected for this study. It was found that the A/V ratios of the records are very high (about 11). Consequently, the ground motions from the Miramichi earthquakes are dominated by very short period (i.e., very high frequency) motions. The selected records were scaled to the peak ground velocity of 7.1 cm/s for the bridge location, and the scaled response spectra of the horizontal records are shown in Fig. 13. It can be seen clearly in Fig. 13 that the ground motions of the Miramich earthquakes are dominated by very short period (i.e., about 0.04 s). Figure 13 also shows that for the period of 0.04 s, the spectral acceleration for the strongest motion (i.e., record. No. 1) is approximately 9 times larger than the value of the design spectrum.

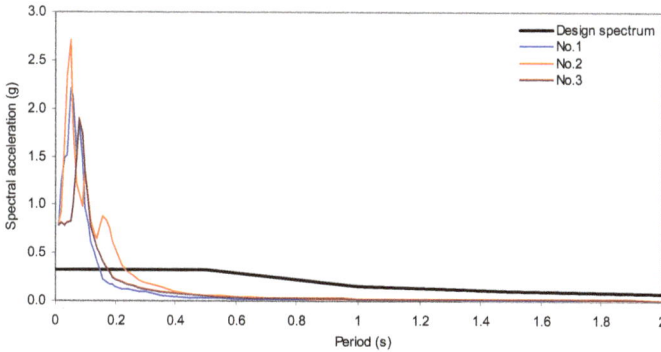

Figure 13. Design spectrum and scaled response spectra of Miramichi earthquake excitations; 5% damping.

6.5. Simulated excitations

In addition to the "real" records of seismic ground motions discussed above, "simulated" acceleration time histories (i.e., accelerograms) were also used as excitation motions. As reported by [11] seismic hazard for eastern Canadian sites can be approximated using a magnitude M=6.0 event to represent the short-period hazard, and M=7.0 event to represent the long-period hazard. They simulated ground motion accelerograms for eastern Canada for M=6.0 and M=7.0, and for different distances. For each distance, four accelerograms were simulated for a probability of exceedance of 2% in 50 years (i.e., annual probability of exceedance of 0.0004).

Since the seismic hazard based on the service life and the importance of the bridge corresponds to an annual probability of exceedance of 0.00027, it was necessary to scale the simulated accelerograms to be consistent with the uniform hazard spectrum (UHS) for a probability of exceedance of 0.00027 (Fig. 3). To determine the short-period hazard motions for the bridge, the simulated accelerograms for the M=6.0 event were scaled to have the same spectral values at the period of 0.2 s as that of the UHS for the bridge location. Similarly, the long-period hazard motions were obtained by scaling the simulated accelerograms for the M=7.0 event to have the same spectral values as that of the UHS at the period of 2.0 s.

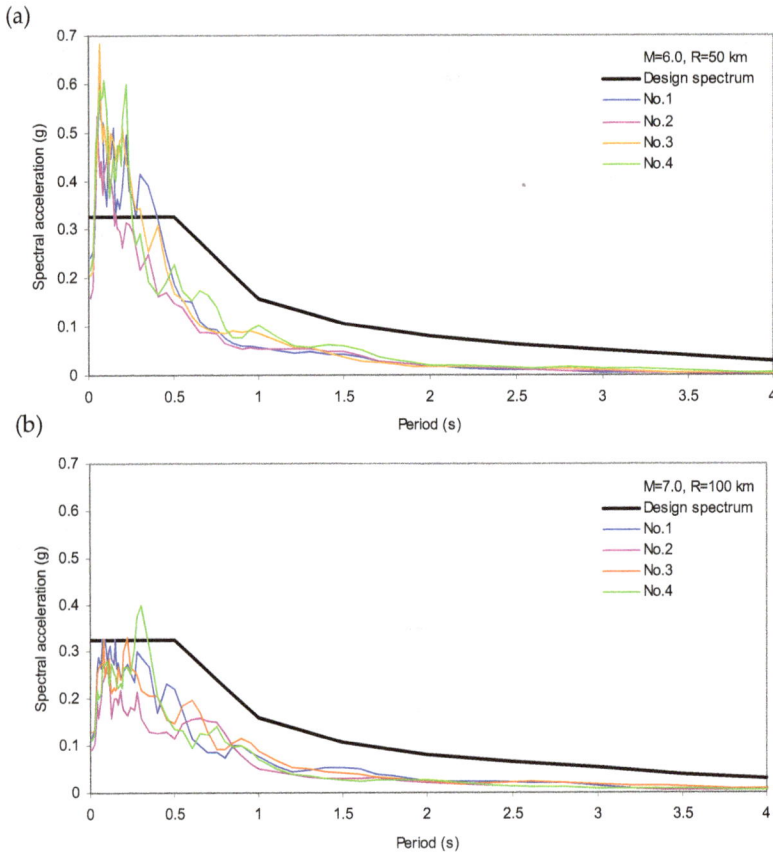

Figure 14. Design spectrum and scaled response spectra of simulated excitations; 5% damping (a) short-period hazard motions, (b) long-period hazard motions.

Trial time-history analyses showed that the largest responses of the bridge model are associated with the scaled accelerograms corresponding to the epicentral distances of R=50 km for the M=6.0 event and R=100 km for the M=7.0 event, and therefore, only these accelerograms were considered. The response spectra of the scaled short-period hazard accelerograms (R=50 km, M=6.0) and long-period hazard accelerograms (R=100 km, M=7.0) are shown in Figs. 14(a) and 14(b) respectively. It can be seen in Fig. 14(a) that the spectra of the short-period hazard accelerograms exceed the design spectrum by a factor of approximately 2.5 for periods below 0.2 s. On the other hand, the spectra of the long-period hazard accelerograms (Fig. 14(b)) are only about 20% higher than the design spectrum for periods below 0.3 s. Given these observations, only the short-period hazard accelerograms were used as excitation motions in the time-history analysis.

7. Dynamic analysis and results

For the purpose of the seismic evaluation of the bridge, dynamic analyses were conducted on the bridge model to determine the responses due to seismic actions represented by the uniform hazard spectrum and the selected sets of records. Elastic material properties of the model were assumed in the analyses. The dynamic analyses included both response-spectrum analyses and time-history analyses.

Response-spectrum analyses

Response-spectrum analyses were performed for seismic actions represented by the uniform hazard spectrum. Separate response-spectrum analyses were carried out for the following two cases of seismic actions: (i) seismic actions in the longitudinal and vertical directions of the model; and (ii) seismic actions in the transverse and vertical directions. These two cases were considered appropriate since the longitudinal and the transverse modes are well separated, and the vertical modes are combined mainly with the longitudinal modes. The horizontal and the vertical actions were applied simultaneously at the bases of the piers. The horizontal seismic actions were represented by the horizontal uniform hazard spectrum (UHS) (Fig. 3), and the vertical actions were represented by a spectrum obtained by multiplying the horizontal UHS by 2/3. The factor of 2/3 is commonly used for defining vertical design spectra relative to horizontal spectra [9].

The analyses included the first 100 modes, which covered all natural periods above 0.02 s. A modal damping of 5% was used for all the modes. The response maxima at each joint of the models were computed by combining the modal responses using the complete quadratic combination (CQC) rule.

As required by the Canadian Highway Bridge Design Codes [24], the mass participation of the modes considered in the analysis is larger than 90% in each of the three principal directions of the model. Namely, the amounts of the mass participation of the longitudinal, transverse and vertical modes used in the analysis are 95.3%, 95.5% and 93.6% respectively.

Time-history analyses

Time-history analyses were conducted to determine the responses of the model subjected to the records of the selected sets. As in the response-spectrum analysis, simultaneous seismic excitations in the longitudinal and vertical directions, and in the transverse and vertical directions of the model were used in the time-history analysis. In each analysis, the seismic excitations consisted of a pair of scaled horizontal and vertical acceleration time histories applied at the bases of the piers.

The mode-superposition method was used in the time-history analysis. As in the response-spectrum analysis, the first 100 modes and modal damping of 5% for all the modes were considered in the time-history analysis. The response time histories were obtained at equal time interval of 0.005 s.

8. Discussion of results

The response quantities obtained from both the response-spectrum analysis and the time-history analysis included bending moments, shear forces, axial forces, and displacements. A detailed review of the response results showed that the observations from the shear forces and the axial forces were the same as those from the bending moments. Given this, only the bending moments and the displacements were used for the evaluation of the seismic performance of the bridge. However only the results for bending moments are shown there, the results for deflections can be found in [17].

For simplicity in discussing the results, the simultaneous excitations in the longitudinal and vertical directions are referred to as excitations in the longitudinal direction (or longitudinal excitations), and those in the transverse and vertical directions are referred to as excitations in the transverse direction (or transverse excitations). This is the case for both the response-spectrum and the time-history analyses.

To assist in understanding the results from the analyses, it is useful to describe the convention for the moments, as used in this study. In reference to the coordinate system shown in Fig. 5, longitudinal moments in the bridge girder are those that act about the Y-axis, and transverse moments are those that act about the Z-axis. For the piers, the moments that result from longitudinal excitations and act about the Y-axis are referred to as "moments in the longitudinal direction", and those that result from transverse excitations and act about the X-axis are referred to as "moments in the transverse direction".

The moments at the joints of the model resulting from the response-spectrum analysis represent the maximum absolute values and by definition are positive. The time-history analysis provided a comprehensive set of results for each excitation motion. Time histories and maximum positive and negative values for the moments and displacements were obtained for the joints of the model. Moment and displacement envelopes for both the girder and the piers were determined using the largest *absolute* values of the computed (positive and negative) maxima for each of the selected sets of ground motions.

The comparisons of bending moments are shown in Figs. 15 and 16. Figure 15(a) shows the envelopes of the longitudinal moments in the bridge girder for seismic actions in the longitudinal direction, and Fig. 15(b) shows the envelopes of the transverse moments for seismic actions in the transverse direction. The moment envelopes are plotted using the corresponding values at selected sections along the bridge girder. Similarly, Figs. 16(a) and 16(b) present the moment envelopes for pier P31 for excitations in the longitudinal and transverse directions respectively. The moment envelopes for the other piers are similar to those for pier P31, and they are not shown here. The designation "Design" in Figs. 15 and 16 is for the design responses which were calculated by [7], and "UHS" is for the responses due to seismic actions represented by the uniform hazard spectrum. Furthermore, the designations "World-wide", "Saguenay", "Miramichi", and "Simulated" are respectively for the responses due to the selected world-wide records – short-period set (Fig. 10), the Saguenay records (Fig. 12), the Miramichi records (Fig. 13), and the simulated motions – short-period hazard set (Fig. 14(a)).

For the purpose of clarity, the results from the response-spectrum analysis (i.e., the "Design" and the "UHS" results) are discussed first. It can be seen from Fig. 15(a) that for the seismic actions in the longitudinal direction, the UHS envelope of the moments in the bridge girder is somewhat higher than the design envelope. Also, the values of the UHS envelope for the pier (Fig. 16(a)) resulting from the longitudinal seismic actions are larger than those of the design envelope in the upper 25 m of the pier. The largest differences are approximately 20%. These observations for the longitudinal seismic actions were expected because the periods of the predominant longitudinal and vertical modes of the bridge are shorter than 1.5 s, i.e., these are within the range in which the uniform hazard spectrum is higher than the design spectrum (Fig. 3). For seismic actions in the transverse direction, the UHS envelopes of the moments in the bridge girder and in the pier (Figs. 15(b) and 16(b), respectively) are all smaller than the design values. This is because the uniform hazard spectrum is lower than the design spectrum for the periods of the predominant transverse modes, i.e., periods longer than approximately 2.0 s (Fig. 3).

The 20% exceedance of the design responses by those from the UHS seismic actions in the longitudinal direction does not represent any concern regarding the seismic safety of the bridge. This is because of the following two reasons. First, conservative assumptions are involved in the design through the use of factored material strengths and specified safety factors, and therefore the actual *capacity* (i.e., resistance) of the bridge is substantially larger than the *demands* due to design loads. For example, considering only the resistance factors for concrete and reinforcing steel used in the design (i.e., ϕ_c=0.75 and ϕ_s=0.85, as specified in the Design Criteria [5]), the nominal flexural resistance of the bridge is about 20% larger than the design resistance. Other safety factors involved in the design, associated with the specified safety index [5], provide even larger resistances relative to the design resistance of the bridge.

The second reason is related to the conservatism of the response resulting from the uniform hazard spectrum. By definition, the uniform hazard spectrum at the bridge location represents the envelope of the spectral contributions of all possible earthquakes in the surrounding area that affect the seismic hazard at the location. This implies that the seismic response resulting from the uniform hazard spectrum represents the envelope of the response contributions from earthquakes with different magnitudes and at different distances from the bridge location, assuming that all the earthquakes occur at the same time. Obviously, the response from such combined earthquake actions is much larger than the responses from each of the earthquakes considered separately. These considerations clearly show that the response-spectrum analysis using the uniform hazard spectrum provides significantly larger responses than those from expected seismic ground motions represented by that spectrum.

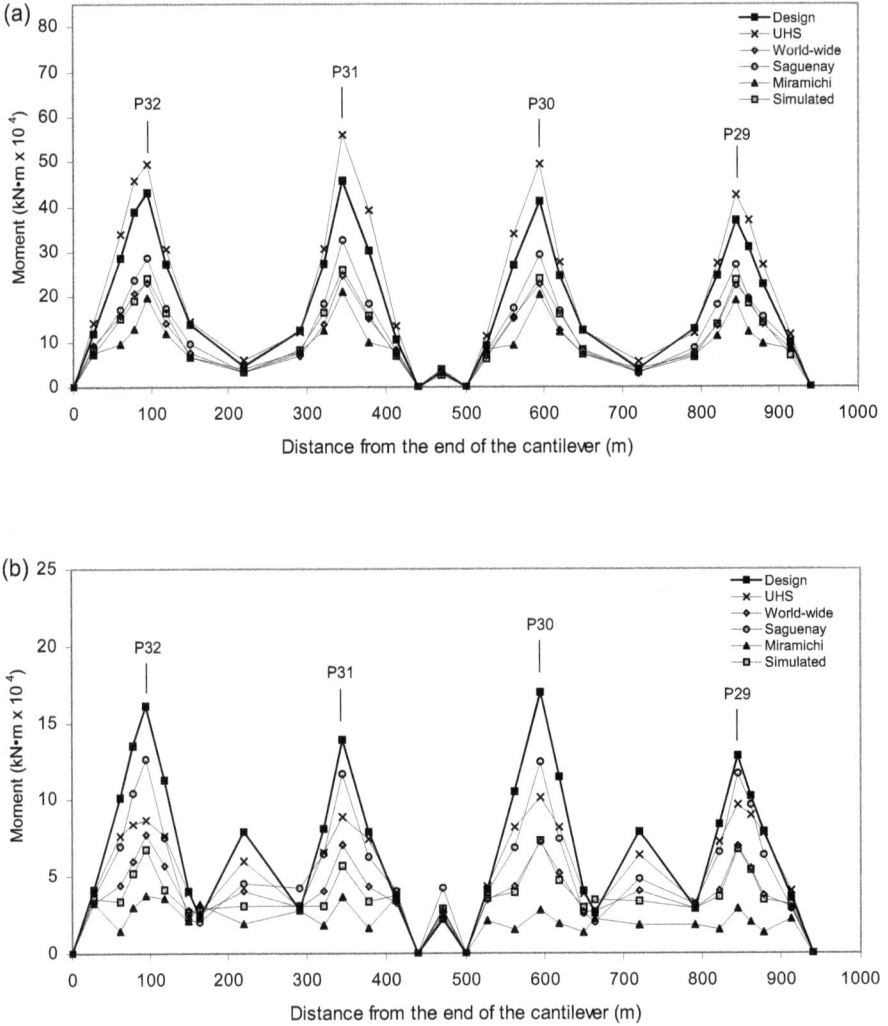

Figure 15. Moment envelopes for the bridge girder: (a) longitudinal moments, (b) transverse moments. Note: Piers P29 to P32 are indicated in the figures to identify the sections of the girder at the piers.

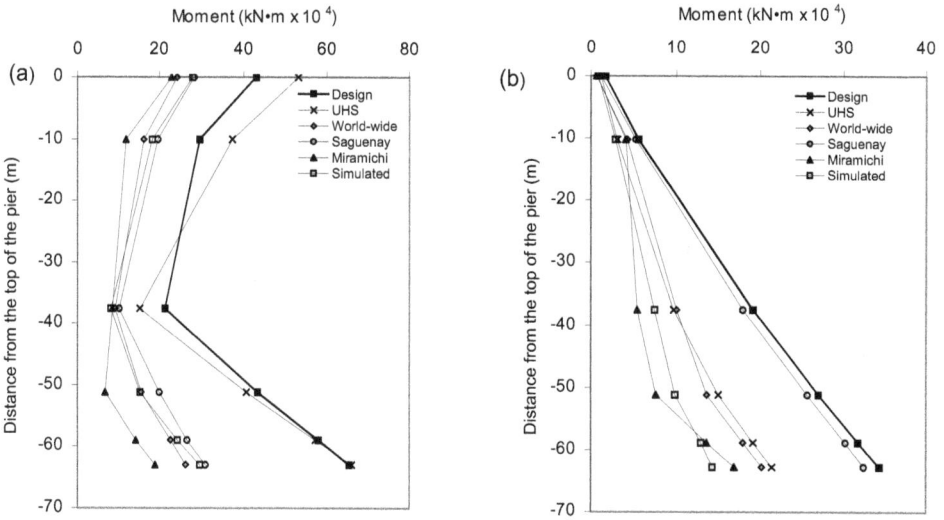

Figure 16. Moment envelopes for pier P31:(a) in longitudinal direction, (b) in transverse direction.

In regard to the response results obtained from the time-history analysis of the model subjected to the selected sets of excitations, it can be seen in Figs. 15 and 16 that the maximum moments are all smaller than the design responses for both the longitudinal and transverse excitations. This was expected based on the spectral characteristics of the excitation motions. As described earlier, the response spectra of the excitation motions used in the analysis (i.e., the World-wide short-period set, the Saguenay set, the Miramichi set, and the simulated short-period set) are all lower than the design spectrum for periods longer than approximately 0.5 s (Figs. 10, 12-14), i.e., within the period range of the longitudinal and transverse modes that produce almost the entire response. The contributions of the modes with periods below 0.5 s, where the spectra of the excitation motions exceed the design spectrum, are very small.

9. Conclusions

The objective of this study was to investigate the performance of the Confederation Bridge due to seismic excitations expected at the bridge location. A finite element model of a typical segment of the bridge was subjected to selected seismic motions representative of the seismic hazard for the bridge location. The response results obtained from the dynamic analysis of the model were compared with the seismic design parameters. The following are the main conclusions from this study:

- The responses from the linear time-history analyses (displacements and forces) were found to be smaller than those used in the design of the bridge.
- The longitudinal responses of some sections of the bridge obtained from the response spectrum analysis (i.e., for seismic actions represented by the horizontal and vertical

uniform hazard spectra) were found to be about 20% larger than the design values. Considering the conservatism in the design through the use of factored material strengths and specified safety factors, as well as the characteristics of the uniform hazard spectra, the exceedance of the design responses by 20% does not represent any concern regarding the safety of the bridge.

- The general conclusion is that the seismic effects considered in the design are appropriate for the required safety during the service life of the bridge.
- A finite element model consisting of 3D beam elements is suitable for the Confederation Bridge provided that the foundation flexibility is taken into account in the modeling.
- The modeling method used in this study is considered to be applicable to single-box girder bridges in general.

Author details

Lan Lin
Department of Building, Civil and Environmental Engineering,
Concordia University, Montreal, Canada

Nove Naumoski and Murat Saatcioglu
Department of Civil Engineering, University of Ottawa, Ottawa, Canada

10. References

[1] CSA. 1988. Design of highway bridges. Standard CAN/CSA-S6-88, Canadian Standards Association, Rexdale, Ontario.

[2] MTO. 1991. Ontario Highway Bridge Design Code. Ministry of Transportation of Ontario, Downsview, Ontario.

[3] MacGregor, J.G., Kenedy, D.J.L., Barlett, F.M., Chernenko, D., Maes, M.A., and Dunascegi, L. 1997. Design criteria and load and resistance factors for the Confederation Bridge. Canadian Journal of Civil Engineering, 24: 882-897.

[4] Tadros, G. 1997. The Confederation Bridge: an overview. Canadian Journal of Civil Engineering, 24: 850-866.

[5] JMS. 1996. Design criteria – Northumberland Strait Crossing Project. Revision 7.2. J. Muller International – Stanley Joint Venture Inc., San Diego, California.

[6] Jaeger, L.G., Mufti, A.A., Tadros, G., and Wong, P. 1997. Seismic design for the Confederation Bridge. Canadian Journal of Civil Engineering, 24: 922-933.

[7] NRC. 1990. National Building Code of Canada 1990. Institute for Research in Construction, National Research Council of Canada, Ottawa, Ontario.

[8] Newmark, N.M., and Hall, W.J. 1982. Earthquake spectra and design. Monograph, Earthquake Engineering Research Institute, Berkeley, California.

[9] Newmark, N.M., Blume, J.A., and Kapur, K.K. 1973. Seismic design spectra for nuclear power plants. Journal of the Power Division, Vol. 99, No. PO2, pp. 287-303.

[10] Adams, J., and Atkinson, G. 2003. Development of seismic hazard maps for the proposed 2005 edition of the National Building Code of Canada. Canadian Journal of Civil Engineering, 30: 255-271.

[11] Tremblay, R., and Atkinson, G.M. 2001. Comparative study of the inelastic seismic demand of eastern and western Canadian sites. Earthquake Spectra, Vol. 17, No. 2, pp. 333-358.

[12] Halchuk, S., and Adams, J. 2004. Deaggregation of seismic hazard for selected Canadian cities. Proceedings of the 13th World Conference on Earthquake Engineering, Vancouver, B.C., Canada, Paper No. 2470.

[13] CSI. 2000. SAP 2000 integrated software for structural analysis and design, Version 7. Computers and Structures Inc., Berkeley, California.

[14] Ghali, A., Elbadry, M., and Megally, S. 2000. Two-year deflections of the Confederation Bridge. Canadian Journal of Civil Engineering, 27: 1139-1149.

[15] Lau, D.T., Brown, T., Cheung, M.S., and Li, W.C. 2004. Dynamic modelling and behaviour of the Confederation Bridge. Canadian Journal of Civil Engineering, 31: 379-390.

[16] SRAC. 1994. COSMOS – Finite element analysis software. Structural Research and Analysis Corporation, Santa Monica, California.

[17] Lin, L. 2005. Seismic evaluation of the Confederation Bridge. M.A.Sc. thesis, Department of Civil Engineering, University of Ottawa, Ottawa, Ontario.

[18] Naumoski, N., Heidebrecht, A.C., and Rutenberg, A.V. 1993. Representative ensembles of strong motion earthquake records. EERG Report 93-1, Earthquake Engineering Research Group, McMaster University, Hamilton, Ontario.

[19] Naumoski, N., Tso, W.K., and Heidebrecht, A.C. 1988. A selection of representative strong motion earthquake records having different A/V ratios. EERG Report 88-01, Earthquake Engineering Research Group, McMaster University, Hamilton, Ontario.

[20] Adams, J., and Halchuk, S. 2003. Fourth generation seismic hazard maps of Canada: Values for over 650 Canadian localities intended for the 2005 National Building Code of Canada. Open File 4459, Geological Survey of Canada, Ottawa, Ontario.

[21] Munro, P.S., and Weichert, D. 1989. The Saguenay earthquake of November 25, 1988 – Processed strong motion records. Open File Report No. 1966, Geological Survey of Canada, Energy, Mines and Resources, Ottawa, Ontario.

[22] Friberg, P., Rusby, R., Dentrichia, D., Johnson, D., Jacob, K., and Simpson, D. 1988. The M=6 Chicoutimi earthquake of November 25, 1988, in the province of Quebec, Canada. Preliminary NCEER strong motion data report, Lamont-Doherty Geological Observatory of Columbia University, Palisades, N.Y.

[23] Weichert, D.H., Pomeroy, P.W., Munro, P.S., and Mork, P.N. 1982. Strong motion records from Miramichi, New Brunswick, 1982 aftershocks. Open File Report 82-31, Energy, Mines and Resources Canada, Ottawa, Ontario.

[24] CSA. 2006. Canadian Highway Bridge Design Code. Standard CAN/CSA-S6-06, Canadian Standard Association (CSA), Mississauga, Ontario.

Chapter 9

Seismic Performance Evaluation of Corroded Reinforced Concrete Structures by Using Default and User-Defined Plastic Hinge Properties

Hakan Yalçiner and Khaled Marar

Additional information is available at the end of the chapter

1. Introduction

There are several methods exist to define the seismic performance levels of reinforced concrete (*RC*) structures. Among these methods, the nonlinear dynamic and the static analyses in which both methods involve sophisticated computational procedures because of the non-linear behaviour of the *RC* composite materials. In order to simplify these analyses for engineers, different suggested guidelines such as *FEMA-356* (Federal emergency management agency [*FEMA-356*], 2000) and *ATC*-40 (Applied Technology Council [ATC-40, 1996]) were prepared to define the plastic hinges properties for *RC* structures in the United States, and thus they have been used by many computer programs (i.e., ETABS [CSI, 2003], SAP2000 [CSI, 2008]) as a default or ready plastic hinge documents. However, there are still contradictions exist in the available literature due to the use of these ready documents in which the buildings are not designed based on the earthquake code of United States. The assessment of seismic performance of structures under future earthquakes is an important problem in earthquake engineering (Abbas, 2011). The use of methods and assumptions to define the seismic performance levels of *RC* buildings become more and more important issue with time dependent effects of corrosion. Moreover, to the knowledge of the author, no any study has been performed up to date, which studies define the possible difference in the time-dependent seismic performance levels of *RC* buildings under the impact of corrosion by using default and user-defined plastic hinge properties.

The primary objectives of this study was to investigate the effects of default hinge properties based on *FEMA-356* (FEMA-356, 2000) and user-defined hinge properties on the time-dependent seismic performance levels of corroded *RC* buildings. An assumed corrosion rate was used to predict the capacity curve of the buildings by using default and user-defined plastic hinge properties as a function of time (*t*: 25 years, and *t*: 50 years). Two, four and seven stories of *RC* buildings were considered to represent the effects of default and user-defined hinge properties on story levels. For the modelling of user-defined hinge properties, the time-dependent moment-curvature relationships of structural members were predicted as a function of corrosion rate for two different time periods in order to perform push-over analyses, while default hinge properties were used for the other case based on the ready documents by *FEMA-356* (FEMA-356, 2000). Then, the nonlinear time-history analyses for both corroded and non-corroded buildings were performed by using 20 individual earthquake motion records. Seismic performance levels of non-corroded buildings and predicted time-dependent seismic performance levels of corroded buildings were compared based on their story levels as a result of user-defined and default hinge properties. Limit–states at each performance levels (e.i. immediate occupancy, life safety, collapse prevention and collapse) were obtained. The obtained results were summarized to compare the differences in the results of seismic response of the buildings due to user-defined and default hinge properties for both corroded and non-corroded cases.

2. Nonlinear material modelling

It is vital to accurately determine the effects of corrosion on the seismic analyses of *RC* buildings. Mainly, corrosion causes loss in the cross sectional area of the reinforcement bars and reduction in bond strength between reinforcement bars and concrete. A study done by Sezen and Moehle (Sezen & Moehle, 2006) indicated that, slip deformations contributed 25% to 40% of the total lateral displacement in the case of non-corroded reinforcement bars. These displacements might be more dramatic by lowering the bond strength due to corrosion. For the user-defined plastic hinge properties, time-dependent bond-slip relationships can be taken into account by modifying the moment-curvature relationships. Modified target post-yield stiffness of each structural member ensures the bond-slip relationships in nonlinear analyses. However, in the case of assigning the default hinge properties, available programs are not capable to consider the bond-slip relationships as a consequence of corrosion effects. Thus, it is inevitable to obtain huge difference in the result of time-dependent seismic performance levels of analysed buildings by using the default hinge properties. Therefore, in this study, the reduction in cross sectional area of reinforcement bars only considered as a function of time which can be also obtained by using available computer software programs.

A corrosion rate of 2.79 $\mu A/cm^2$ was assumed in order to predict the loss in the cross sectional area of reinforcement bars as a function of time where 0.0116 was used as a conversion factor of $\mu A/cm^2$ into mm/year for steel. For the user-defined plastic hinge properties, the obtained time dependent loss in cross sectional area of reinforcement bars were used to predict the moment-curvature relationships of *RC* sections. Developed model by Kent and Park (Kent & Park 1971) was used to model the stress-strain relationships of confined columns. Fig. 1 shows the well known developed model by Kent and Park (Kent & Park 1971) which was adapted for modelling the stress-strain relationships of *RC* sections in this study. Basically, the developed model by Kent and Park (Kent & Park 1971), has two branches. For the ascending branch (A-B), the curve reaches to maximum stress level at a strain of 0.002. After reaching maximum stress, two other different braches occurs (B-C, B-D) where two straight lines indicate different behaviour of concrete for confined and unconfined concrete. For the descending branch of the curve assumed to be linear and its slope specified by determining the strain when the concrete stress is decreased to half of its stress value as suggested by Park et al. (Park et al, 1982).

Mander's (Mander, 1984) model was used for each time periods (i.e., t: 25 years, and t: 50 years) for modelling the stress-strain relationships of steel as can be seen from Fig. 2. The developed model by Mander (Mander, 1984) includes linear elastic region up to yield, elastic-perfect-plastic region, and strain hardening region. The Mander's model (Mander, 1984) has control on both strength and ductility where the descending branch of the curve that first branch increases linearly until yield point then the curve continues as constant. In order to model the material properties, the following required assumptions were made. The modulus of elasticity of concrete $E_c = 3250\sqrt{f'_c} + 14000$ MPa was calculated according to TS500 (TSI, 2000). The mechanical properties of steel in the analyses were selected according to TS500 (TSI, 2000), where the minimum rupture strength (f_{su}) was equal to 500 MPa, the yield strain (ε_y) was equal to 0.0021, the strain hardening (ε_{sh}) was equal to 0.008, the minimum rupture extension (ε_{su}) was equal to 0.12% and the modulus of elasticity of steel (E_s) was taken as 200,000 MPa.

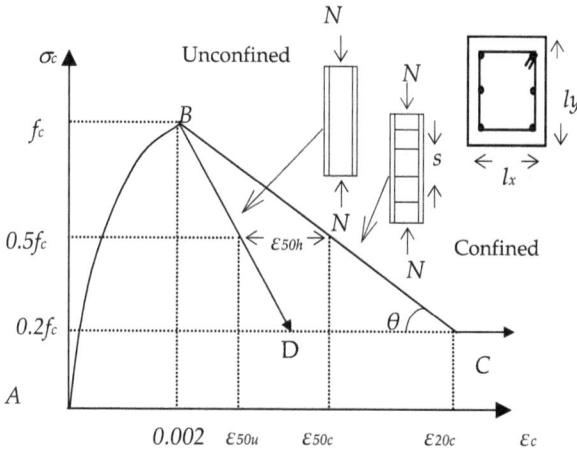

Figure 1. Used stress-strain relationship of concrete (Kent & Park 1971).

3. Description of structures

Three *RC* buildings having two, four and seven stories were considered in this study. The assessed three *RC* buildings were selected among the typical constructed *RC* buildings in North Cyprus where the buildings were designed according to Turkish earthquake code (TEQ, 1997). The soil classes were classified as soft clay (group D), the building importance factor was taken as 1, and the effective ground acceleration coefficient (A_0) was taken as 0.3g (seismic zone 2) according to Turkish earthquake code (TEQ, 1997). The buildings were remodelled to select the most critical frames by using the existing plans of the buildings. Fig.

3 shows the three dimensional modelling of a two story of *RC* building. In Fig. 3, the total height of the building is 6 m where the typical floor height is identical and equal to 3 m. The slab thicknesses of the building are same and equal to 0.15 m. The dead (*G*) and live (*Q*) loads of the slabs were designed to be 5.15 kN/m² and 1.96 kN/m², respectively. Additional wall load on the beams were designed to be 3.19 kN/m².

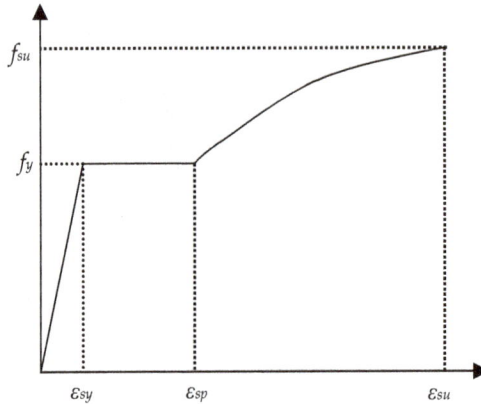

Figure 2. Stress-strain relationship of steel (Mander, 1984).

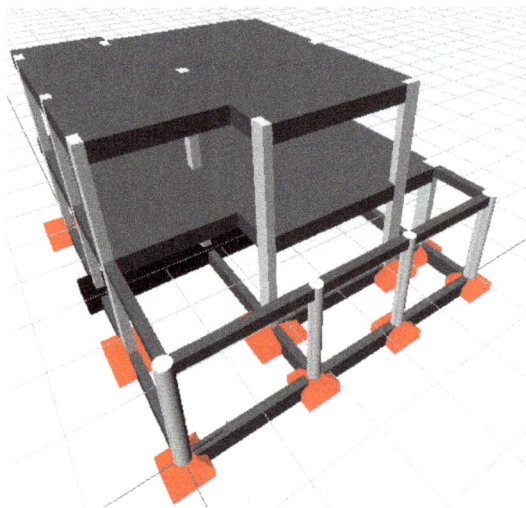

Figure 3. Three dimensional view of two story reinforced concrete building.

Fig. 4 shows two dimensional view of the selected frame from the two story of *RC* building. In Fig. 4, the member names and sectional dimensions of columns and beams with the amount of longitudinal reinforcement bars are also represented. The vertical distributed loads that were used in the analyses are also depicted in Fig. 4. The frame has a first-mode period of T_1: 0.40 seconds having a total weight of 19.69 tons. For the second case study, a four story *RC* building was selected to be analysed which represents a typical apartment buildings in North Cyprus. Figs. 5 and 6 show three dimensional view and the selected frame of the building, respectively. In Fig. 5, the total height of the building is equal to 12 m where the typical floor height is identical and equal to 3 m. The slab thicknesses of the building are same and equal to 0.17 m. The dead and live loads on the slabs are 5.64 kN/m² and 1.96 kN/m², respectively. Additional wall load on the beams are identical and equal to 3.19 kN/m². In Fig. 6, the sectional dimensions of all beams are identical and equal to 0.25 m by 0.60 m, with the same details of reinforcement bars. The frame has a first-mode period of T_1: 1.09 seconds having a total weight of 55.53 tons.

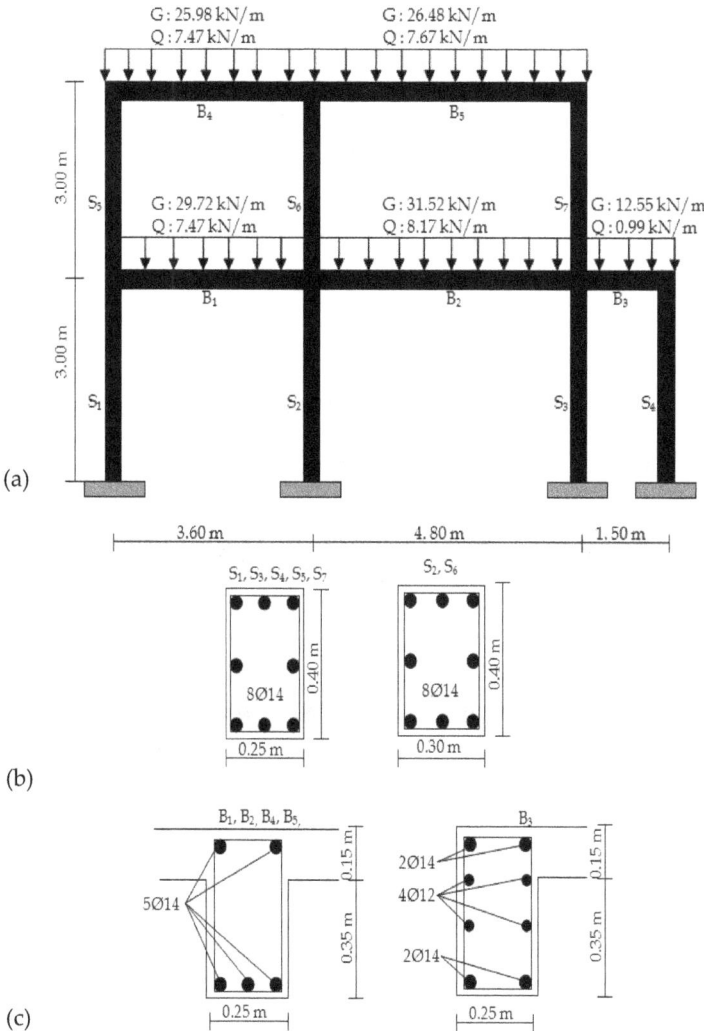

Figure 4. Dimensional and reinforcement details of a two story frame: (a) Used vertical loads in the analyses, (b) Reinforcement details of columns, (c) Reinforcement details of beams.

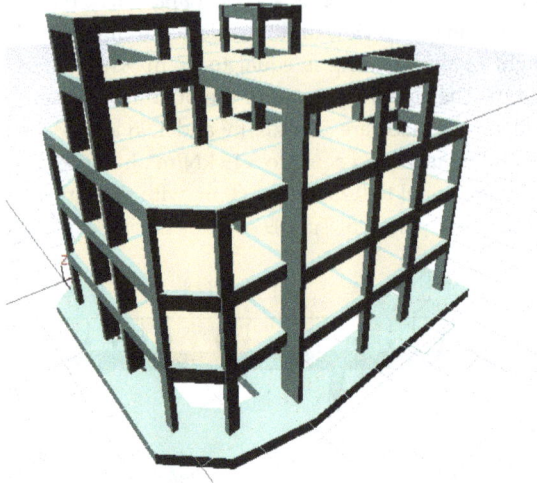

Figure 5. Three dimensional view of a four story *RC* building.

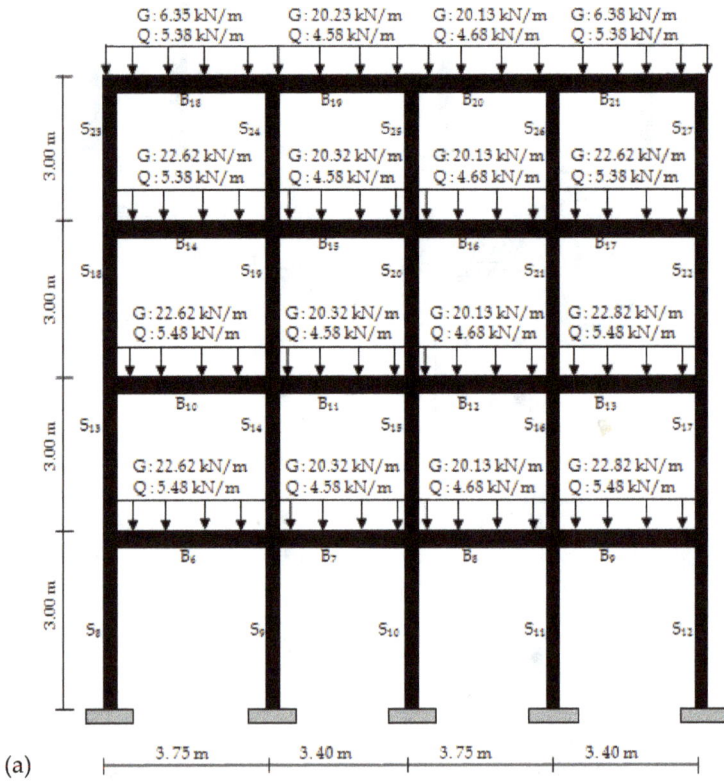

(a)

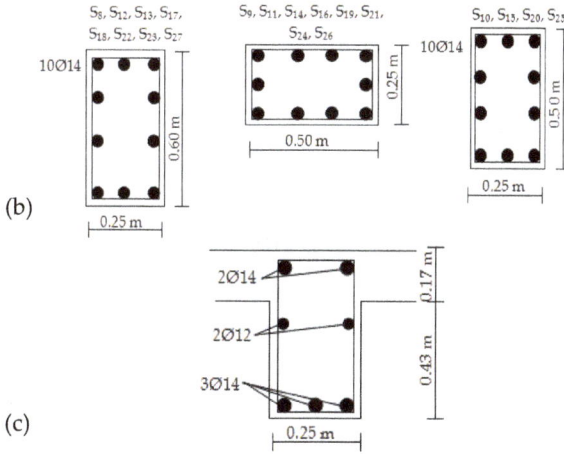

Figure 6. Dimensional and reinforcement details of a four story frame: (a) Used vertical loads in the analyses, (b) Reinforcement details of columns, (c) Reinforcement details of beams.

The third case study deals with an existing seven story of a RC building. Figs. 7 and 8 show three dimensional view and the selected frame of the analysed building, respectively. In Fig. 7, the total height of the building is equal to 27 m where the typical floor height is identical and equal to 4.50 m. The slab thicknesses of the building are same and equal to 0.17 m. The dead and live loads on the slabs were 6.25 kN/m² and 4.90 kN/m², respectively. Additional wall load on the beams were identical and equal to 3.19 kN/m². Because of having more than twenty different reinforcement details of beams, only reinforcement details of columns are shown in Fig. 8. In Fig. 8, the depicted vertical distributed loads of seventh floors are same for other floors. The frame has a first-mode period of T_1: 4.27 seconds having a total weight of 151.56 tons.

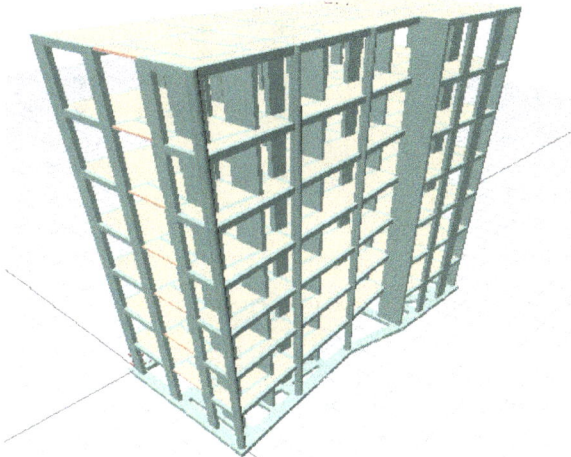

Figure 7. Three dimensional view of a seven story of RC building.

G: 14.93 kN/m   G: 20.8 kN/m   G: 18.63 kN/m   G: 17.02 kN/m   G: 16.34 kN/m
Q: 6.68 kN/m   Q: 11.36 kN/m   Q: 9.57 kN/m   Q: 8.37 kN/m   Q: 7.77 kN/m

(a)

(b)

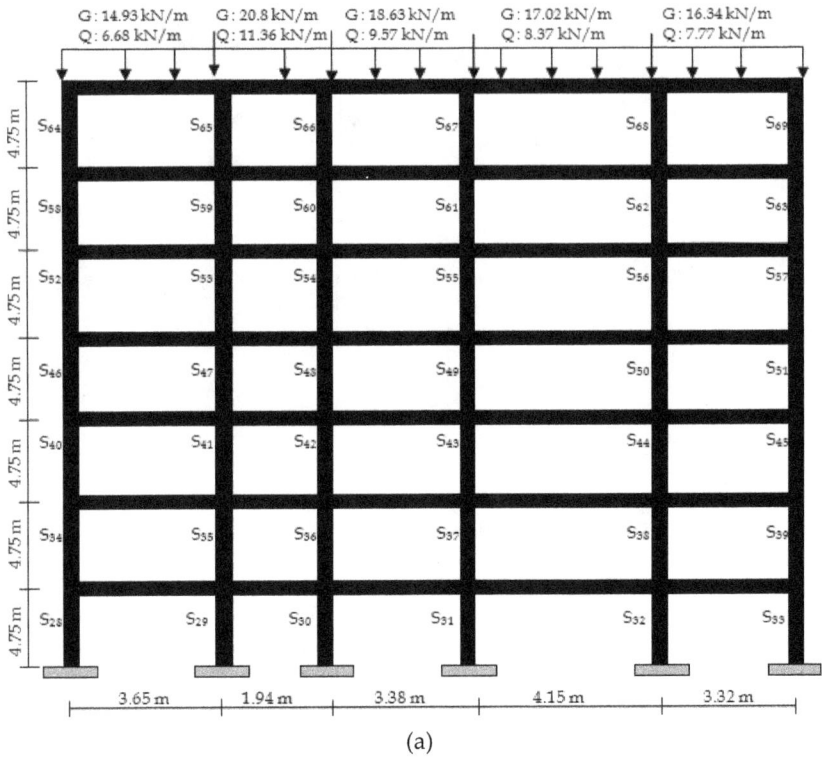

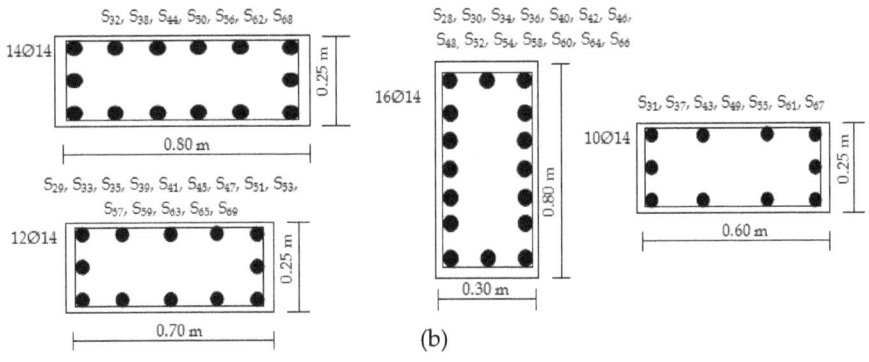

Figure 8. Dimensional and reinforcement details of seven story of frame: (a) Used vertical loads in the analyses, (b) Reinforcement details of columns.

4. Moment-curvature relationships

Moment-curvature relationships were predicted in order to define the user-defined plastic hinge properties as a function of time. Moment-curvature relationships of columns were carried out from the calculated section properties and constant axial forces acting on the elements. Axial loads on the beams were assumed to be zero. A total of 210 plastic hinge properties as a function of time (t: 0 (non-corroded), t: 25 years, t: 50 years) were defined to be used in the nonlinear static push-over analyses. In order to predict the moment-curvature relationships, a new developed software program SEMAp (Inel et al., 2009) was used. SEMAp (Inel et al., 2009) models the stress-strain relationships of steel and concrete by the user. Fig. 9 shows the predicted moment-curvature relationships of randomly selected RC columns and beams as a function of time for different story levels. In Fig. 9, time dependent moment-curvature relationships of the assessed RC members basically indicates three segments; the elastic region prior to cracking, the post-cracking branch between the cracking and yield points and the post-yield segment beyond yielding, respectively. As shown in Fig. 9, premature yielding occurs due to the loss in cross sectional area of the reinforcement bars. For instance, for the same story level, the premature yielding moments of the S_1 column corresponding to time periods of 25 and 50 years were 18% and 39%, respectively. As shown in Fig. 9, at the same moment values, curvature of a structural section increases as a function of time which affects the demand capacity of the frame by the defined plastic hinge properties.

In Fig. 9, the area under moment-curvature represents the storage energy capacity of a section in inelastic behaviour. As shown, in Fig. 9, the area under the curvature decreases due to premature yielding of reinforcement bars which causes cracking of concrete at early stages. The results of time period of 50 years showed that concrete crushes before the reinforcing bars exceed the strain hardening region with increased corrosion level.

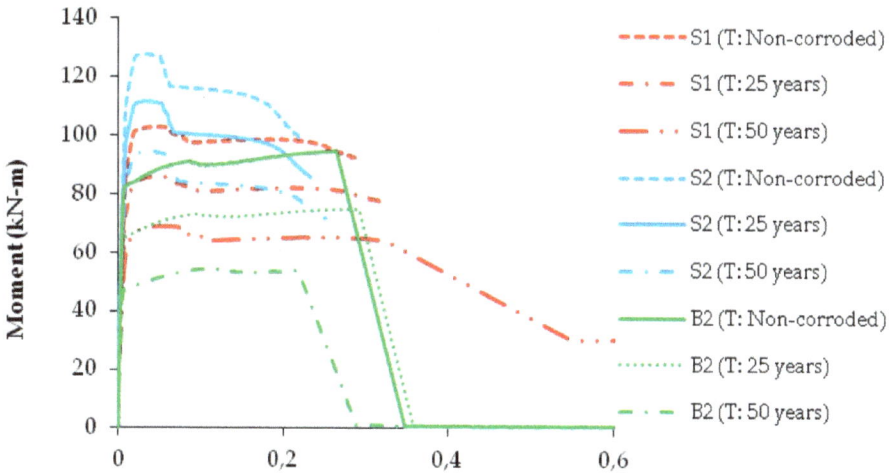

Curvature (rad/m)

(a)

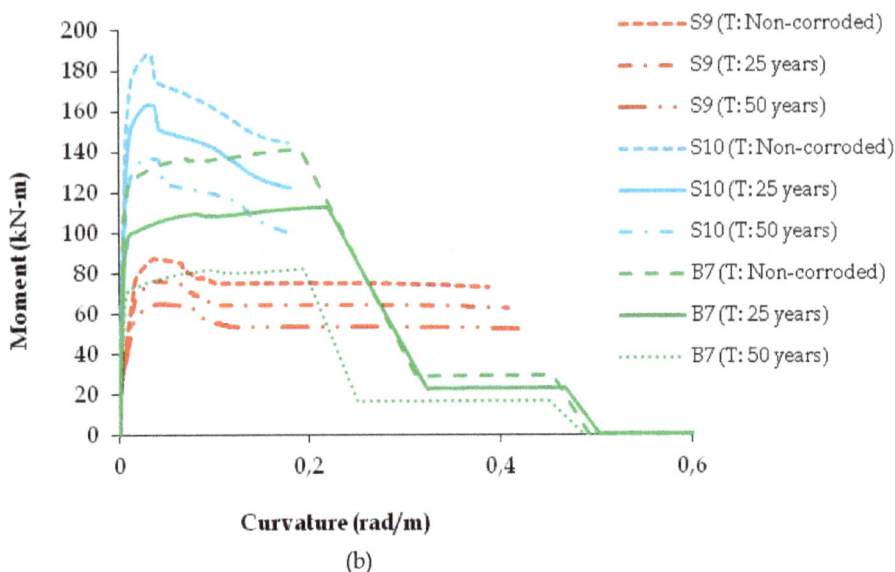

(b)

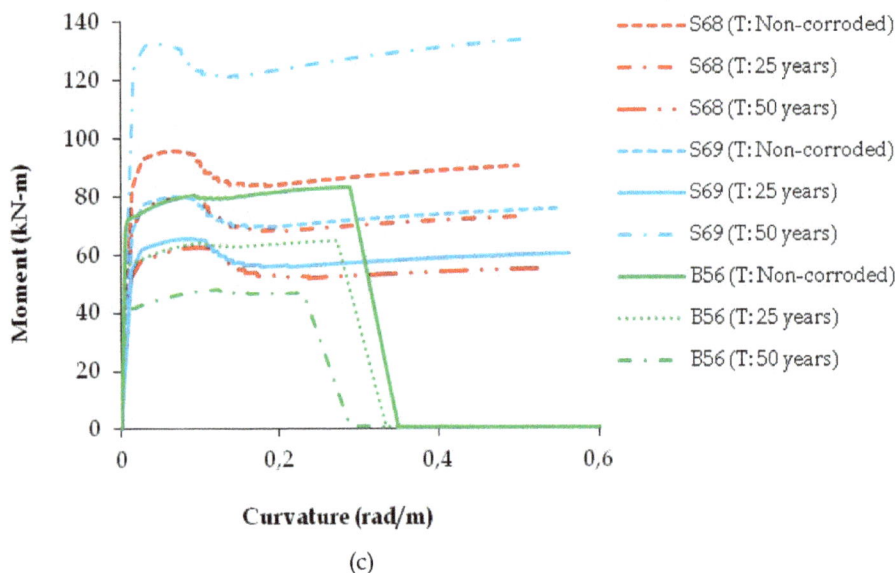

(c)

Figure 9. Moment-curvature relationships of *RC* members as a function of time: (a) Two story, (b) Four story, (c) Seven story.

5. Nonlinear static analysis

SAP2000 (CSI, 2008) computer program was used to analyse the selected frames as a function of time. For the user-defined plastic hinge properties, the force-deformation behaviour needs to be plotted to define the behaviour of plastic hinges. Fig. 10 shows a typical force-deformation relationship to define the behaviour of plastic hinges by *FEMA-356* (FEMA-356, 2000) and also the required acceptance criteria of immediate occupancy (*IO*), life safety (*LS*), collapse prevention (*CP*) and collapse (*C*).

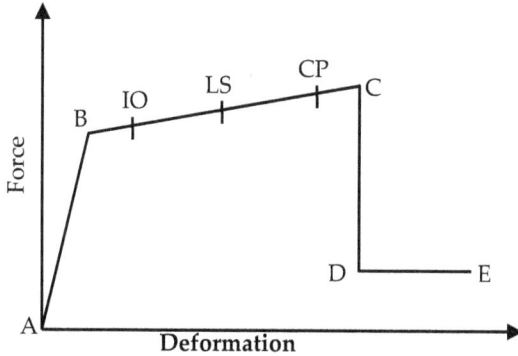

Figure 10. Force-deformation relationship of a plastic hinge.

In Fig. 10, point *A* corresponds to unloaded condition of hinge deformation. Point *B* represents yielding of structural elements that controlled by moment-curvature relationships. Hinge deformation shows strength degradation at point *D* where the structure might show sudden failure after this point. The failure of the structure can be defined by reaching the point *D* and *E*. In this study, the locations of the hinges of the selected frames were located according to the study done by Inel and Ozmen (Inel & Ozmen, 2006). The lengths of the plastic hinges were used to calculate the moment-rotation instead of moment-curvature given by Eq. 1 (Varghese, 2006):

$$\int \varphi \, ds : \varphi \, L_p : \theta \tag{1}$$

where θ is the rotation of plastic hinge, L_p is the plastic hinge length, and φ is the curvature at a point.

There are different proposed models available in the literature to calculate the length of the plastic hinges. Since the mechanical properties of reinforcement bars play an important role for the user-defined plastic hinge properties, proposed model by Paulay and Priestley (Paulay & Priestley, 1992) to calculate the length of plastic hinges was used according to the given Eq. 2.

$$L_p = 0.08 \, L + 0.022 \, d_b \, f_y \tag{2}$$

where *L* is the critical distance from the critical section of the plastic hinge to the point of contra flexure, f_y and d_b are the yield strength and the diameter of longitudinal reinforcement bar, respectively.

As shown in Eq. 2, the proposed model by Paulay and Priestley (Paulay & Priestley, 1992) is important to ensure the effect of corrosion on the length of plastic hinges as a function of time. Shear strength hinge properties were calculated by using Eq. 3 according to ACI 318 code (ACI 318, 2005):

$$V_c = 0.17x\sqrt{f_c}x\,b\,x\,d\,x\left(1+\frac{N}{14\,A_c}\right) \tag{3}$$

where V_c is the shear strengths provided by concrete, b is the section width, d is the effective depth, f_c is the unconfined concrete compressive strength, N is the axial load on the section, and A_c is the concrete area.

The calculated plastic hinge properties were assigned to each floor at both ends of the beams and columns of the assessed frames according to the corresponding time periods. Triangular lateral load pattern was applied to the frames to perform nonlinear push-over analyses. There are different options are available in *SAP2000* (CSI, 2008) to define the loading of the hinge properties. In this study, unload entire structure option was selected for the method of hinge unloading. When the hinges reach point *C* in Fig. 10, the program continues to increase the base shear force. After point *D* the lateral displacement begins to reduce with the reduced base shear force and the structural elements starts to be unloaded. Fig. 11 shows the predicted time-dependent push-over analyses of the selected frames as a function of time for both of user-defined and default hinge properties.

As can be seen from Figs. 11a-c, the collapse mechanisms of non-corroded frames were affected by corrosion as a function of time. For instance, by using the user-defined plastic hinge properties, the collapse mechanism of the non-corroded frame of the two story of *RC* building started at a top displacement of 0.2633 m when the base shear force was 206 kN (see Fig. 11a). However, for the time periods of 25 and 50 years, collapse mechanism started at top displacements of 0.2608 m and 0.2612 m when the base shear forces were 170 kN and 130 kN, respectively. Same behaviour can be also observed for other performed frames. When the results were compared for the default hinge properties, the effect of corrosion can

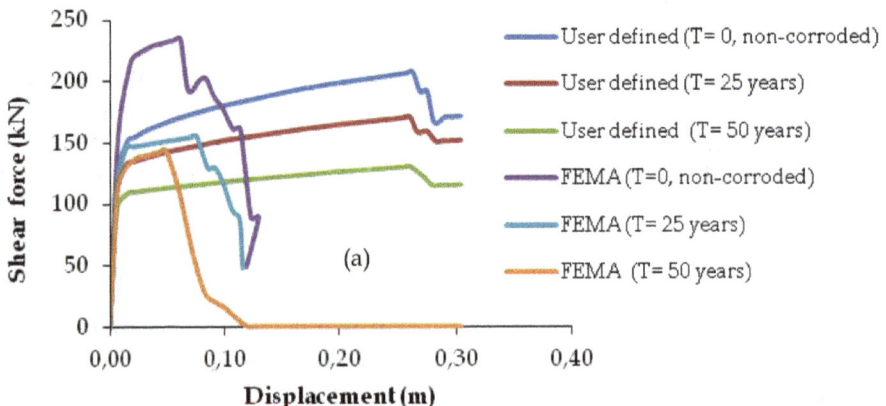

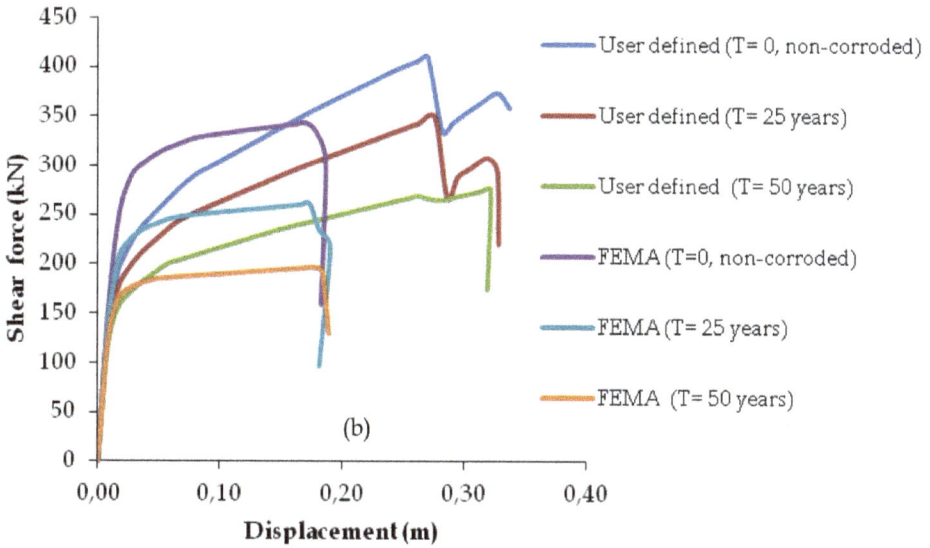

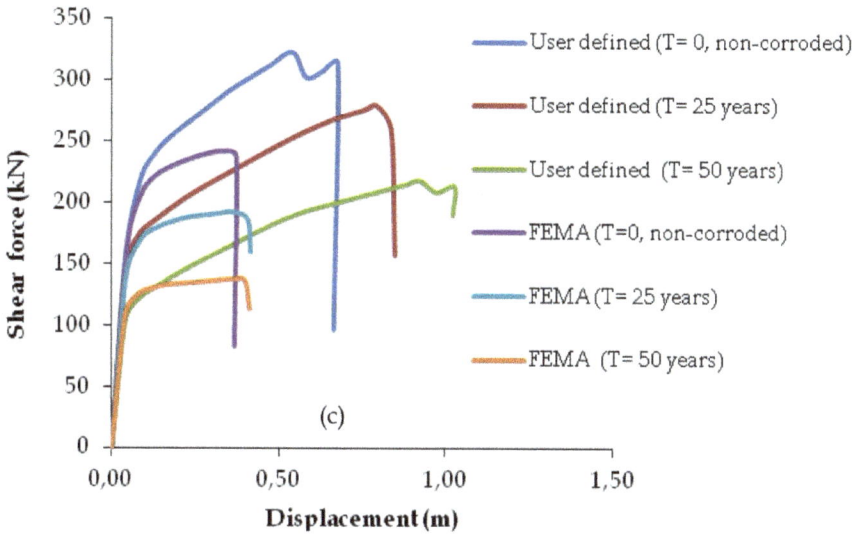

Figure 11. Time dependent load-displacement relationships by using default and user defined plastic hinge properties: (a) Two story, (b) Four story, (c) Seven story.

be also observed. However, there is a huge difference for the collapse mechanism of the assessed frames by default hinge properties. For instance, the time period of 50 years of the seven story of a *RC* building (see Fig. 11c), the recorded top displacement by user-defined plastic hinge properties was 0.92 m when the base shear force was 217 kN. For the same

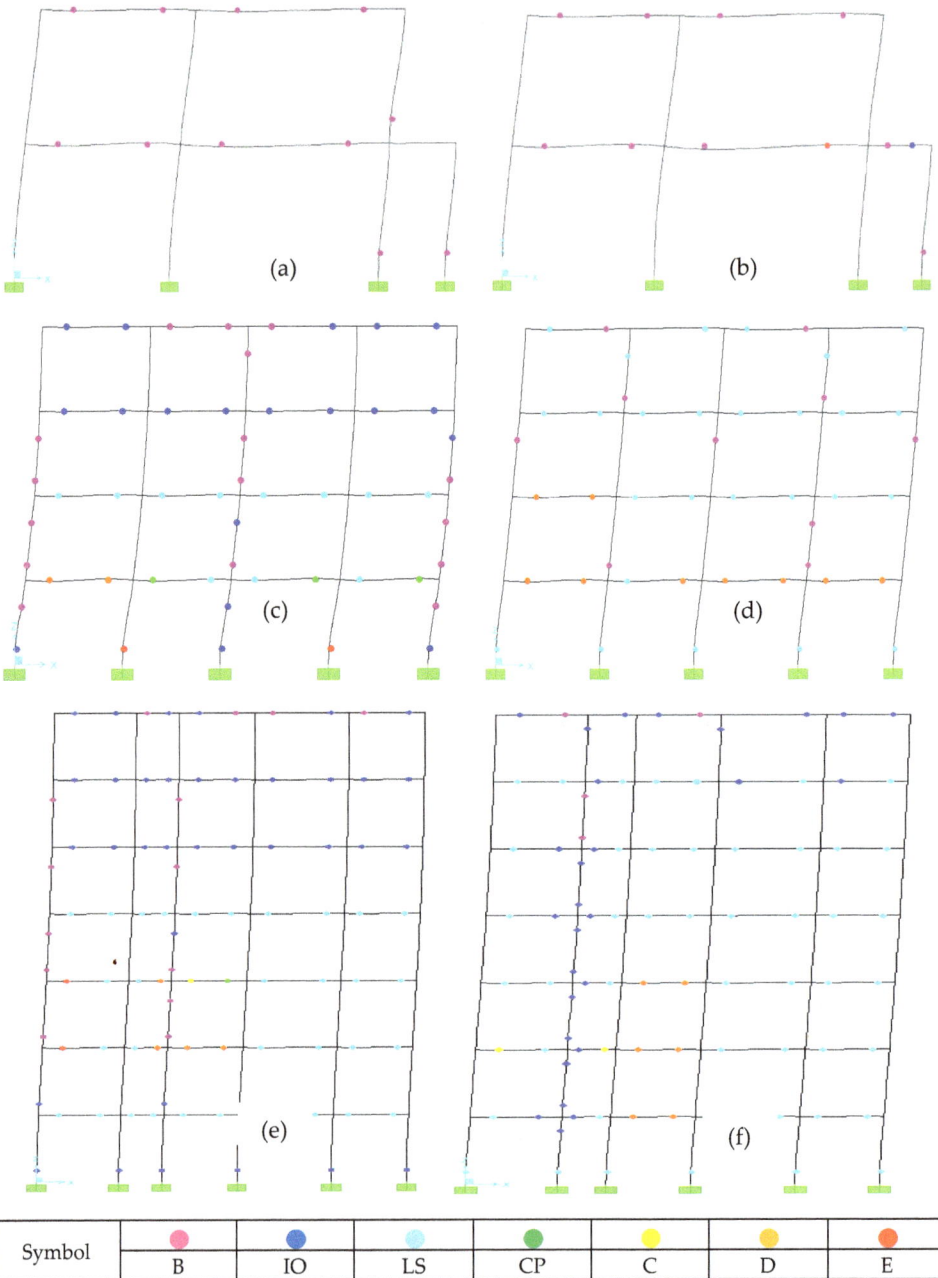

Symbol	●	●	●	●	●	●	●
	B	IO	LS	CP	C	D	E

Figure 12. Plastic hinge patterns by using default and user defined plastic hinge properties: (a) Two story user-defined, (b) Two story *FEMA-356*, (c) Four story user-defined, (d) Four story *FEMA-356*, (e) Seven story user-defined, (f) Seven story *FEMA-356*.

time period of the seven story building (see Fig. 11c), the recorded top displacement by default hinge properties was 0.36 m when the base shear force was 137 kN. Thus, it is clear that the shear capacities obtained from default hinge properties gave underestimate results when compared with the user-defined hinge properties for each case and time periods. For the time period of 25 years, the hinge patterns of two, four and seven stories of frames are plotted in Fig. 12.

As can be seen from Fig. 12, significant differences in hinging by the user-defined and default hinge properties. By increasing the number of stories, the differences become more significant. In both plastic hinge properties, plastic hinge formations at both columns and beams show almost similar behaviour for a two story of RC frame. However, for upper stories, hinge formations especially in columns show significant differences. Non-linear time history analyses were performed in the following section to define the effects of both plastic hinges modelling on performance levels.

6. Seismic performance analyses

Incremental dynamic analyses (IDA) were performed to predict the performance levels of the assessed frames as a function of corrosion rate by the using user-defined and default hinge properties. For IDA, the 5% damped first-mode spectral acceleration (S_a (T_1, 5%)) was selected. Twenty ground motion records were used to predict the performance levels of the building as a function of time. For the current study, the associated roof drift ratios corresponding to performance levels, IO, LS and CP were adopted from the study done by Stanish et al. (Stanish et al., 1999) and reduced drift values of 0.5%, 1%, and 2% were used for IO, LS, and CP, respectively. In order to perform IDA, NONLIN (Charney, 1998) a software computer program was used. By using the NONLIN (Charney, 1998), the material nonlinearity could be taken into account by specifying the yield strength and initial and post yield stiffness, which were calculated from the time-dependent load-displacement relationships (see Fig. 11). Twenty ground motion records were used to predict the performance levels of the buildings as a function of time, where the randomly selected motions records of pseudo velocity versus to period in seconds are shown in Fig. 13., where earthquake moment magnitudes (M) ranged from 4.7 to 7.51, PGA varied from 0.016 to 0.875g, and peak ground velocity (PGV) ranged between 1.65 to 117 cm/sec.

Figs. 14a-c, 15a-c and 16a-c show fragility curves of two, four and seven stories of RC buildings, respectively. In Figs. 14, 15 and 16, the obtained time dependent fragility curves which were in terms of PGA, compare the differences in the results of performance levels of the buildings as consequences of user-defined and default hinge properties.

The obtained fragility curves indicated that the performance levels of RC structures obtained by the default hinge properties based on FEMA may under-estimate or over-estimate results. Moreover, in the case of corroded conditions, the response of the buildings obtained by the default hinge properties does not represent the actual behaviour of the structures due to ductility problems of the structural members. Although the collapse mechanism of structures were affected by corrosion; directly reduced cross sectional area of reinforcement bars to perform ready documents hinge properties based on *FEMA* might provide more ductile structural members which might also over-estimate results in the performance levels of *RC* structures. For instance in Fig. 14b, when the *PGA* is equal to 0.4g, the probability of exceeding the limit state corresponding to *LS* is 11% for user-defined plastic hinge properties while this probability is 2% based on *FEMA* ready documents plastic hinge properties. Such differences can be also observed in the case of non-corroded conditions. From Fig. 15a, it can be seen that, when the *PGA* is equal to 0.4g, the probability of exceeding the limit state corresponding to *IO* is 43% for the user-defined plastic hinge properties while this probability is 23% based on *FEMA* ready documents plastic hinge properties. It should be noted that for any story level, the maximum story displacements thus roof drift ratios occurred at different times according to user-defined or ready documents plastic hinge properties. Moreover, the results clearly showed that, the percentage of errors (i.e., *IO, LS, CP*) occurred due to use ready document plastic hinge properties were not proportional with story levels.

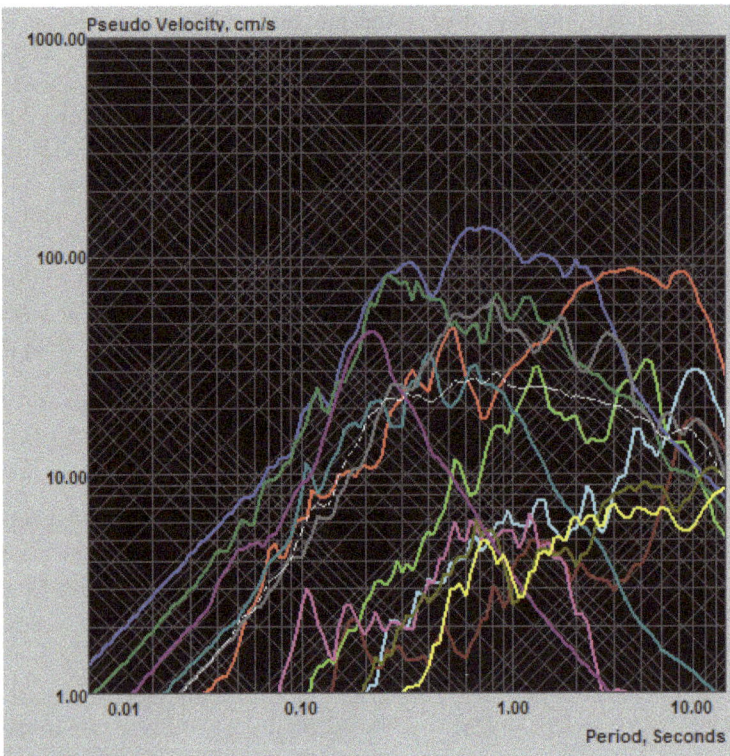

Figure 13. Pseudo velocity spectrum for used ground motion records.

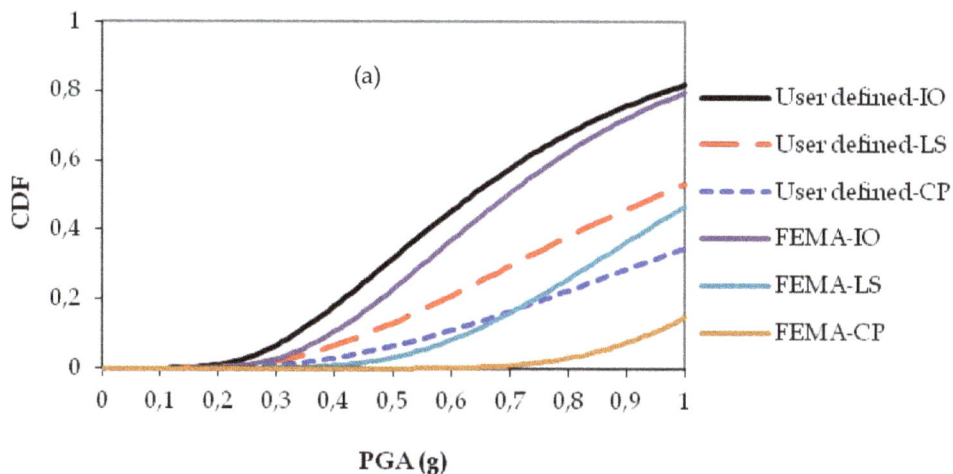

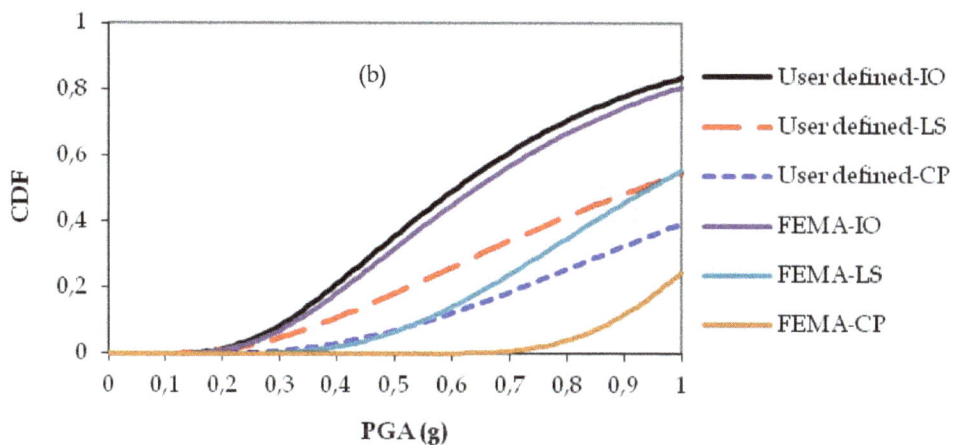

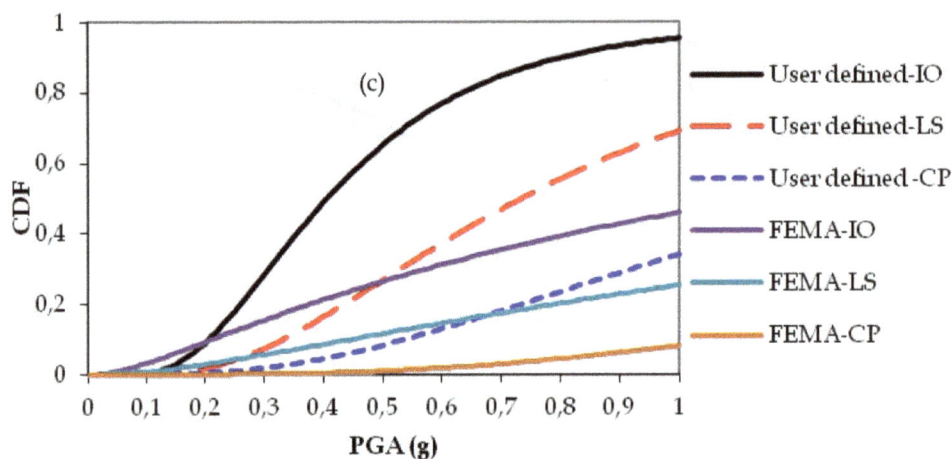

Figure 14. Fragility curves of two story *RC* building: (a) Non-corroded, (b) *T*: 25 years, (c) *T*: 50 years.

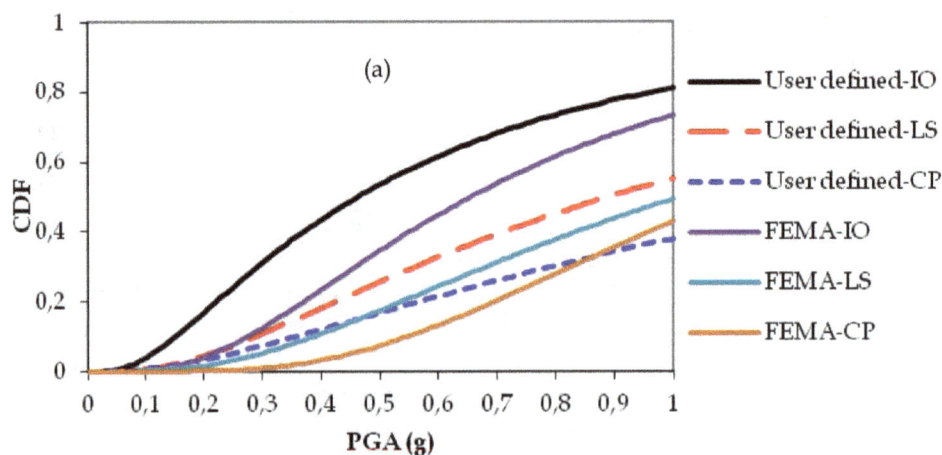

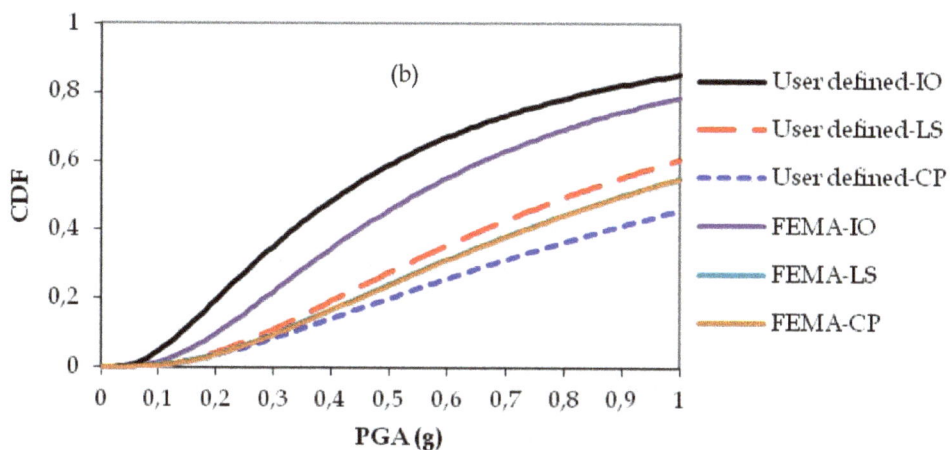

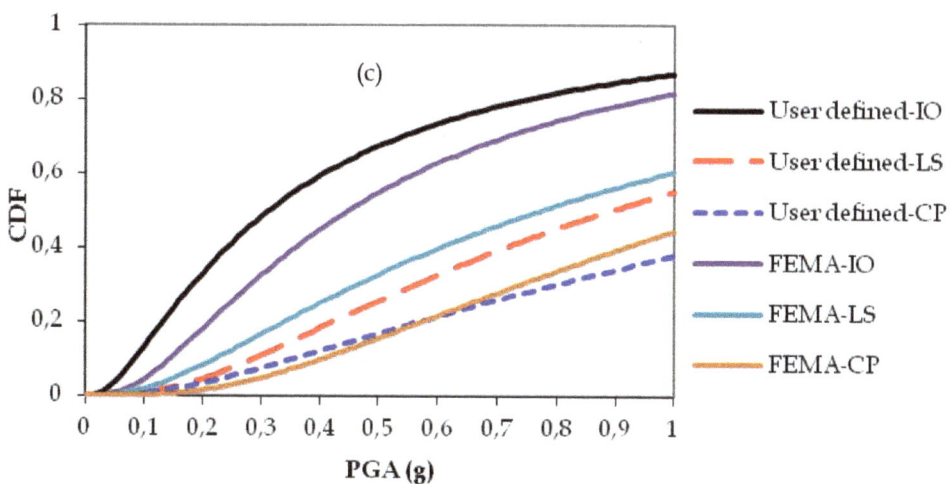

Figure 15. Fragility curves of four story *RC* building: (a) Non-corroded, (b) *T*: 25 years, (c) *T*: 50 years.

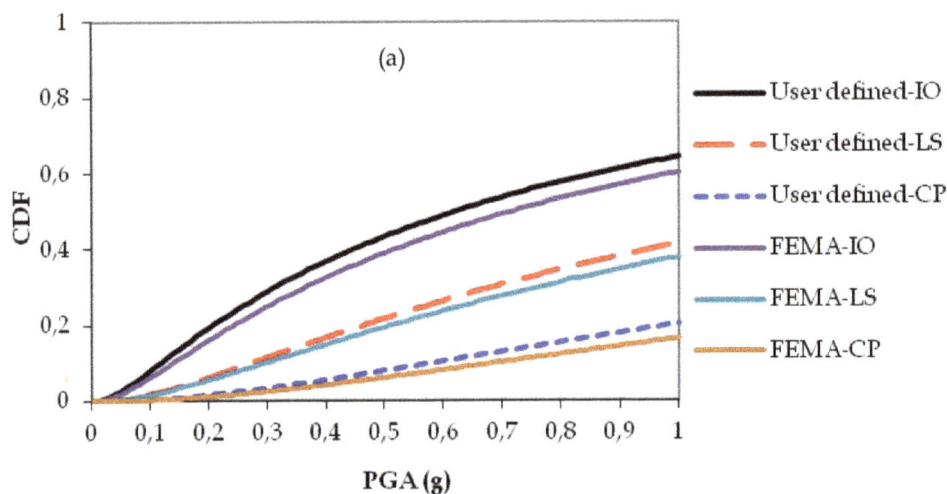

(a)

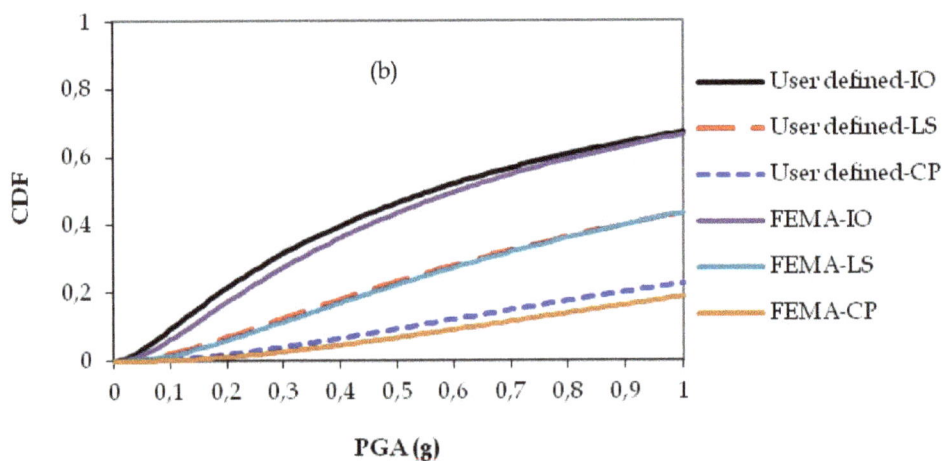

(b)

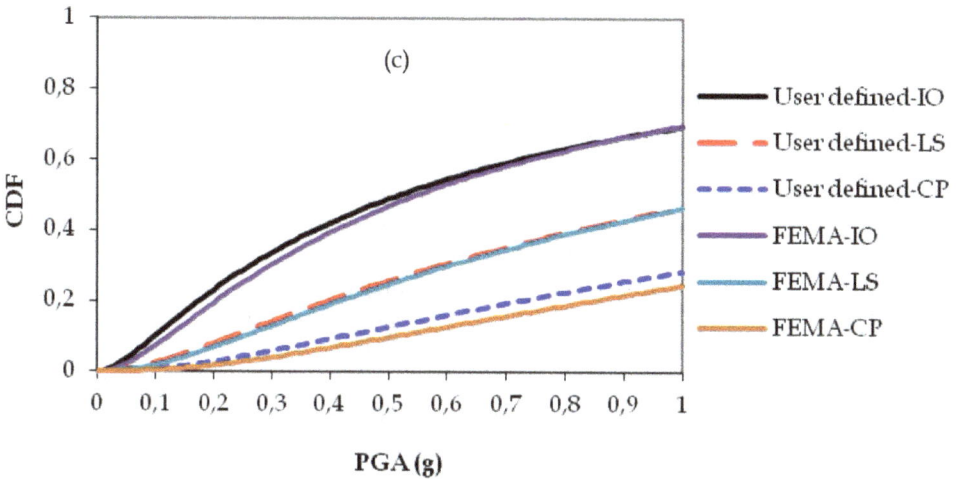

Figure 16. Fragility curves of seven story *RC* building: (a) Non-corroded, (b) *T*:25 years, (c) *T*:50 years.

7. Conclusion

Incremental dynamic analyses for three *RC* buildings having 2, 3 and 7 stories were carried out as a function of time. The performed push-over analyses and *IDA* clearly showed that there were important differences due to the use of the plastic hinge properties based on ready documents and user defined hinge properties. If the user knows the capability of the program where *SAP2000* (CSI, 2008) automatically stops the analysis when a plastic hinge reaches its curvature capacity, ready document plastic hinge properties might be used for rapid and preliminary assessment of *RC* buildings. However, the obtained time dependent results clearly showed that the user defined plastic hinge properties give better and correct results than default hinge properties. Additional studies are also required for accurate performance assessment of multi-degree-of-freedom systems. Bond-slip relationships and cover cracking of concrete due to corrosion need to be taken into account in seismic analyses where the effect of additional displacement due to slippage of reinforcement bars can be provided by the modification of plastic hinge properties. When the effects of corrosion on seismic performance levels and economical impacts in construction industry are considered, time-dependent nonlinear models rather than walk-down surveys are required for better decision making of strengthening of *RC* buildings to prevent serious damage under the expected seismic motions.

Author details

Hakan Yalçiner and Khaled Marar
European University of Lefke, Department of Civil Engineering, Mersin, Turkey

8. References

Abbas, Moustafa (2011). Damage-Based Design Earthquake Loads for Single-Degree-Of Freedom Inelastic Structures. *American Society of Civil Engineers, Journal of Structural Engineering*, Vol. 137, No. 3, pp.456–467.

ACI Committee 318 (2005). Building Code Requirements for Reinforced Concrete and Commentary. *American Concrete Institute*, Detroit, Michigan, pp. 423.

Charney, F.A. (1998). NONLIN V-7: Nonlinear Dynamic Time History Analysis of Single Degree of Freedom Systems, Blacksburg, Virginia, Advanced Structural Concepts.

CSI, ETABS (2003): Integrated design and analysis software for building systems, California, USA, Computers and Structures Inc.

CSI, SAP2000 V-12 (2008): Integrated finite element analysis and design of structures basic analysis reference manual, Berkeley, Computers and Structures Inc.

Inel, M., Ozmen, H.B. (2006). Effects of Plastic Hinge Properties in Nonlinear Analysis of Reinforced Concrete Buildings. *Engineering Structures*, Vol.28, No.11,pp. 1494-1502,

Inel, M., Ozmen, H.B., Bilgin, H. (2009). SEMAp: Modelling and Analysing of Confined and Unconfined Concrete Sections. *Scientific and Technical Research Council of Turkey*, Project No. 105M024.

Kent, D.C., Park, R. (1971). Flexural members with confined concrete. *American Society of Civil Engineers, Journal of the Structural Division*, Vol.97, No.7,pp. 1969-1990.

Mander, J.B. (1984). Seismic Design of Bridge Piers, Ph.D. Thesis, Department of civil engineering, University of Canterbury, New Zealand.

Park, R., Priestly, M.J.N., and Gill, W.D.(1982). Ductility of Square Confined Concrete Columns. *ASCE Journal of Structural Engineering*, Vol. 108, No. 11,pp. 929-951.

Sezen, H., Moehle, J.P. (2006). Seismic test of concrete columns with light transverse reinforcement. *American Concrete Institute Structural Journal*, Vol.103, No.6,pp. 824-849, ISSN

Stanish, K., Hooton, R.D., Pantazopoulou, S.J. (1999). Corrosion effects on bond strength in reinforced concrete. *American Concrete Institute Structural Journal*, Vol.96, No.6, (November-December 1999), pp. 915-922.

Turkish Earthquake Code (TEQ) (2007). Ministry of Public Works and Settlement Government of Republic of Turkey, Specification for Structures to be Built in Disaster Areas, Earthquake Disaster Prevention, Ankara, Turkey.

Turkish Standards Institute (TSI), TS500 (2000). Requirements for Design and Construction of Reinforced Concrete Structures, Ankara, Turkey.

Varghese, PC (2006). Allowable rotation for collapse load analysis, In: *Advanced reinforced concrete design 2nd edition*, pp. 399-402, Prentice-Hall press, 81-203-2787-X, India.)

Chapter 10

Bridge Embankments – Seismic Risk Assessment and Ranking

Wael A. Zatar and Issam E. Harik

Additional information is available at the end of the chapter

1. Introduction

Seismic stability analysis and retrofit of earth embankments, including site remediation, has been, to date primarily, focused on embankment dams and earth retaining structures [1]. If a bridge embankment on a priority route is at a high failure risk, soil stabilization may be required, depending on the importance of the bridge. The Seismic Retrofit Manual for Highway Bridges [3] stipulates techniques for assessing the seismic vulnerability of bridges with regard to technical and socio-economic issues. The seismic retrofit manual stipulates that for bridges near unstable slopes, detailed geotechnical investigations should be carried out to assess the potential for slope instability under seismic excitations. The required detailed investigations include material testing, borehole examination, and trenching to check for unstable layers and vertical fissures. However, for the preliminary evaluation of bridges on priority routes the use of detailed geo-technical investigations and sophisticated models are typically limited because of the associated cost and effort.

There is current interest in a careful assessment of the "most critical" embankments along priority routes. In order to achieve this goal, a means of assessing the embankments that qualify as "most critical" is required. Other than the work reported by the authors, almost no complete studies have been reported to identify and prioritize highway embankments that are susceptible to seismic failure. Data regarding soil types and depth of bedrock required for detailed seismic analysis and risk assessment are not available for the majority of bridge embankments. For instance, while the total number of bridges located on both I-24 and the Parkways in western Kentucky is 519 bridges, soil data is only available for few bridge sites. Therefore, the objective of this study is to provide a methodology to conduct seismic evaluations of bridge embankments in order to identify, rank, and prioritize the embankments that are susceptible to seismic failure and are in need of detailed analysis.

This Chapter addresses the technical component of embankment prioritization and is well-suited to a reliability-based model for seismic risk assessment.

In order to achieve the objective of this study, a flowchart is generated to assess the seismic vulnerability of multiple bridge embankments simultaneously. The embankment geometry, material, type of underlying soil, elevation of natural ground line, upper level of bedrock, and expected seismic event in accordance with associated seismic zone maps constitute the variables for each embankment. This methodology results in calculating the seismic slope stability capacity/demand (C/D) ratio, estimated displacement, and liquefaction potential of each bridge embankment for the respective expected seismic event. Seismic vulnerability ranking and prioritization of embankments are conducted by using the "Kentucky Embankment Stability Rating" (*KESR*) model. Three categories are identified in the KESR model to represent the failure risk of the embankments. A priority list of the embankments with the highest seismic risk can be generated for any set of embankments.

2. Seismic vulnerability and ranking of bridge embankments

In general, data regarding soil types and depth of bedrock are not available for many existing bridge embankments to allow for detailed seismic analysis and risk assessment. This Chapter provides a methodology that enables identifying the embankments that are susceptible to failure during a seismic event. Having categorized the embankments in a designated region according to the respective failure risk, a priority list that includes the most critical bridge embankments can then be highlighted. When site-specific data for a bridge embankment is available, it can be used to obtain the list of seismically deficient embankments. When site-specific data for a bridge embankment is not available, the proposed methodology outlines an approach to estimate the information that is required to obtain the priority list. It is understood that the resulting seismic risk of a specific embankment may not be very accurate due to limited available data or lack thereof. However, the estimated data and strength parameters that are available for utilization shall be assessed by a qualified geo-technical engineer in order to ensure valid results. In order to facilitate the application of the proposed ranking methodology, assumptions, calculations, and required checks are presented along with the parameters of each embankment. The parameters of each embankment include the respective geometry, material, seismic event, upper level of bedrock, level of natural ground line, soil type, and anticipated failure types. The following sections: input variables, embankment vulnerability analysis, ranking parameters, category identification, and ranking and prioritization are provided to outline all of the necessary steps to achieve the study objective.

3. Input variables

The geometry, material, level of natural ground line, soil type, seismic event, and upper level of bedrock constitute the required input for each embankment and are addressed in the following sub-sections.

Geometry: The ideal case for obtaining the geometry of a given embankment is to carry out an on-site inspection. Should there be difficulties encountered in gathering such on-site information, however, the embankment geometry may be taken from the bridge plans. It is assumed that utilizing data from a finalized set of bridge plans will not affect the accuracy of the final seismic ranking and priority list for a given embankment case. Embankment slopes are assumed to be free of any evidence of impending failure, swampy conditions, or other terrain conditions that might be relevant to their stability. For a typically irregular slope, an idealization of the slope has to be performed in such a way that results in the lowest seismic slope stability C/D ratio. It is assumed that the material that might have been used for erosion protection of the slope will not greatly influence the resulting seismic slope stability, and therefore is not considered as an input parameter. The embankment slope geometry is identified by its height (H) and the idealized inclination (b) (Figure 1). The water table is assumed to be located below the embankment base in order to obtain the most critical seismic stability conditions. Analysis shall be carried out on both ends of each bridge and the most critical embankment slope at either end, which results in the lower seismic slope stability C/D ratio, shall be considered in the ranking analysis and priority list.

Materials, Natural Ground Line and Soil Properties: The soil profile at a bridge site is often composed of naturally deposited soils rather than controlled fill. The profile usually consists of multiple layers of different soils and the contact between softer foundations and stiffer bedrock soils is typically irregular. Defining the soil conditions at a site requires detailed site-specific sub-surface exploration that is not available at the majority of existing bridge embankment sites. Therefore, another approach is employed herein to specify the soil types and properties of applicable sites. It is assumed that any soil outside the embankment zone at a bridge site has uniform un-drained shear strength. The soil is considered to be in continuous contact with the bedrock layer, where the bedrock acts as a layer possessing high strength at some depth below the embankment.

Soil data is dependent on the level of the Natural Ground Line (*NGL*), shown in Figure 1. Both the *"Geologic Quadrant Maps of the United States"* that are provided in *"United States Geologic Survey (USGS)"* maps [2] and the *"Soil Conservation Service, Soil Survey"* maps that are reported by *"United States Department of Agriculture (USDA)"* [3] are used to identify the soil type underneath an embankment. The way by which either map is chosen is based on the level of the *NGL* as compared to the embankment base. Whenever the level of the *NGL* is above the level of the embankment base by more than 1.5 m. (5 ft), the analysis is solely based on the soil data obtained from the *"Geologic Quadrangle Maps of the United States"*, provided by *USGS* [2]. Otherwise, the soil data is derived from the *"Soil Conservation Service, Soil Survey"*, provided by *USDA* [3].

The dependency on the *USDA* maps in this case can be attributed to the fact that the top 1.5 m. (5 ft) soil can be accurately obtained from these maps. Shear strengths are assigned as done so by [4] for non-cohesive soil materials, which were derived from analysis of standard penetration tests (Table 1). When a range of values is given for the shear strength of a given soil, the lowest value is assigned to accommodate for the anticipated liquefaction potential at many bridge sites [5]. The shear strength assigned for cohesive soils in Table 1 is chosen after examining commensurately accurate un-confined compression data. Shear strengths assigned to the embankment fill are adjusted to reflect the cyclic loading effects between un-drained failure for both cohesive and saturated cohesion-less soils, in addition to the intermediate behavior between drained and un-drained for dry and partially saturated soils. The density and shear strength of the embankment soils are conservatively estimated by assuming that marginal compaction may have occurred during construction. Should there be more accurate soil properties, they may replace those provided in Table 1.

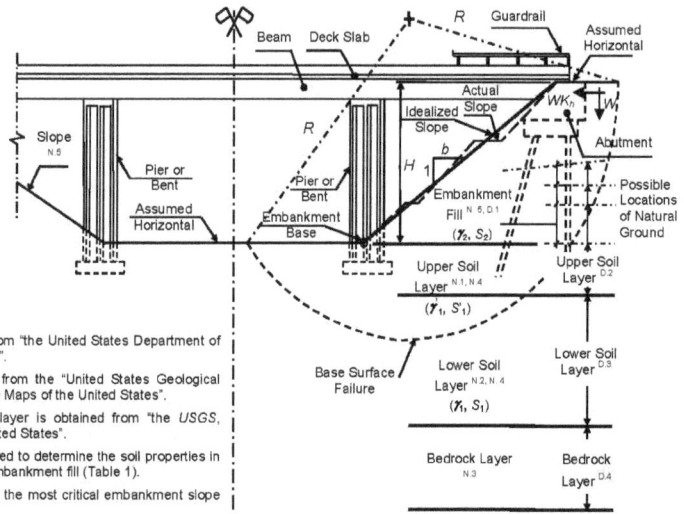

Definitions

D.1- Embankment: portion of the slope facing a pier in case of a multi-span bridge or an abutment in case of a single span bridge.

D.2- Upper soil layer: A layer bounded by the natural ground line and the lower soil layer.

D.3- Lower soil layer: A layer bounded by the upper soil layer and the bedrock layer.

D.4- Bedrock layer: A high strength layer beneath the lower soil layer.

Notes

N.1- Type of upper soil layer is obtained from "the United States Department of Agriculture, Soil Conservation Service".

N.2- Type of lower soil layer is obtained from the "United States Geological Survey (USGS), Geologic Quadrangle Maps of the United States".

N.3- Location of the top of the bedrock layer is obtained from "the USGS, Geologic Quadrangle Maps of the United States".

N.4- The soil types identified by [10] are used to determine the soil properties in the upper and lower soil layers and embankment fill (Table 1).

N.5- The ranking analysis is carried out on the most critical embankment slope at either end of the bridge.

Figure 1. Bridge embankment representation for seismic ranking

Upper Level of Bedrock Layer: Data regarding the level under which a hard stratum, stiff bedrock layer, exists is not available for the majority of existing embankment sites; especially for small bridges. An initial assumption of the upper level of this hard stratum is estimated from the *"Geologic Quadrant Maps of the United States"* [2]. The actual upper level of the stiff bedrock layer specifically falls within the range from the level of the embankment base down to the top level of the hard stratum. For the sake of seismic risk assessment of a bridge embankment, few upper levels of the bedrock layer within that range are considered. Wherever the upper level of the bedrock layer is not known at a bridge site, the following three assumptions of this level are made, and the most critical case is considered in the ranking analysis: (1) at the same level of the embankment base; (2) at the bottom level of the lower soil layer, which is also the upper level of the hard stratum; and (3) at mid-height of the lower soil layer. Other assumptions of the top level of the bedrock layer may be considered if those assumptions yield a lower seismic slope stability C/D ratio. The top level of the bedrock layer, adopted in the ranking analysis, is the assumed elevation that results in the lowest seismic slope stability C/D ratio.

Geologic Formation	Mass Density (γ)		Shear Strength (S)	
	(g/cm^3)	(lb/ft^3)	(kg/cm^2)	(lb/ft^2)
Alluvium	1.92	120	0.20	410
Weathered loess	1.84	115	0.35	717
Continental deposits	2.00	125	0.75	1536
Residuum	2.08	130	1.00	2048
Embankment	2.00	125	0.50	1024

Table 1. Density and strength of soils and embankments

Seismic Event: The input Peak Ground Acceleration (*PGA*), which is the maximum bedrock acceleration at a designated embankment site, is obtained from seismic maps that are generated for specific seismic events. The choice of the seismic event is based on the importance and anticipated performance of the bridge as well as its geographic location on the seismic maps. The seismic maps to define the acceleration coefficient based on a uniform risk method of seismic hazard can be used. The probability that the acceleration coefficient will not be exceeded for a 50-year event is estimated to be 90%, with an expected return period is of 475 years [6]. Alternatively, seismic maps that may have been generated by State Departments of Transportation can be used. For the Commonwealth of Kentucky 50-year, 250-year, and 500-year seismic events were developed [7]. These events have a 90% probability of not being exceeded in 50 years, 250 years, and 500 years, respectively. All but four of the bridges and their embankments on priority routes in western Kentucky are required to withstand the 50-year and 250-year seismic events. The four other bridges are required to resist the 500-year seismic event.

4. Embankment vulnerability analysis

The potential for slope displacement to occur during an earthquake is assessed using a two-dimensional limit equilibrium stability analysis. Sutterer et al. [8] summarized the stability analysis using numerical formulation of both critical circular and wedge–shaped failures (Figure 2). Sutterer et al. [4] reported that pseudo-static analysis of homogeneous slopes showed that seismically loaded embankments with uniform foundation soils, and slope inclinations flatter than 1 horizontal to 1 vertical and steeper than 4 horizontal to 1 vertical, most probably fail in a base failure mode. Steeper slopes may be subjected to a toe circle failure type in the embankment alone (Figure 2). Regardless, most highway bridge embankments fall within the range dominated by base failures. In assessing the seismic vulnerability of each embankment, both failure types are considered in the proposed methodology, and the one that results in a lower *C/D* ratio is considered. This Chapter defines a process to assign the seismic risk, rank and priority of a set of bridge embankments rather than providing only the required derivations and equations.

The horizontal earthquake acceleration in the seismic slope stability analysis often ranges from 50% to 100% of the *PGA* assigned for the embankment site. The *PGA* is often a single spike of motion of a very brief duration and causes little if any significant displacement. A horizontal earthquake acceleration (K_h) equals to two-thirds of the *PGA* is selected in the proposed methodology. This assumption accounts for those embankments in which the seismic acceleration either never exceeds the yield acceleration or very briefly exceeds the yield acceleration, and results in little or no displacement.

5. Ranking parameters

The embankment ranking and prioritization procedures in the seismic vulnerability methodology are based on three parameters that have to be derived for each embankment. They are the seismic slope stability *C/D* ratio, anticipated embankment displacement, and liquefaction potential at the embankment site. The way by which each parameter is calculated is described in the following sub-sections. After calculating the three ranking parameters, categorization of the embankment behavior during a specified seismic event is carried out using Table 2.

Capacity/Demand (C/D) Ratio: The seismic slope stability *C/D* ratio of a bridge embankment is calculated for two possible failure types, known as circular base failure and wedge type failure. For a circular base failure that is shown in Figure 2a, the factor of safety (FS_{cb}) is calculated from Eq. 1.

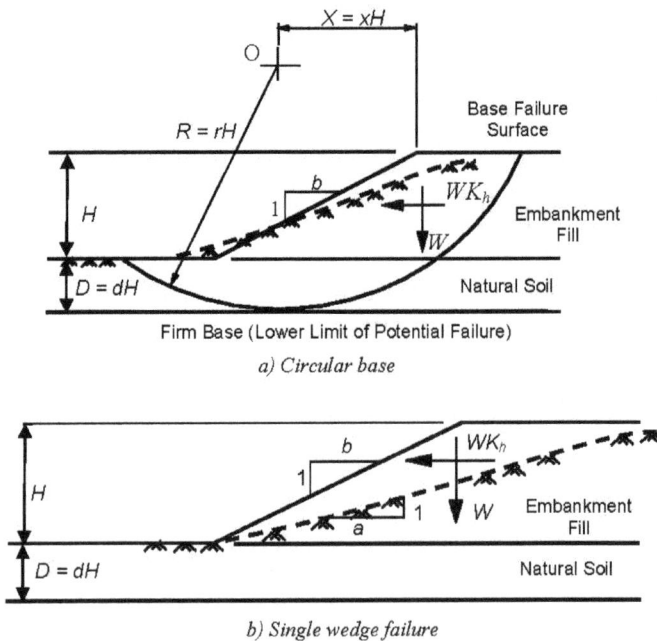

a) Circular base

b) Single wedge failure

Figure 2. Failure types of bridge embankments *(a) circular base failure and (b) single wedge failure*

Category	Category Description	Ranking Parameter	Embankment Displacement	Seismic Risk
A	1) High liquefaction potential, or 2) Displacement exceeds 10 centimeters	Displacement & liquefaction potential	Loss of embankment	High risk
B	1) Moderate liquefaction potential, or 2) Capacity/Demand (C/D)min. ratio is less than 1.0, and displacement is less than 10 centimeters	Displacement & liquefaction potential	Significant movement	Significant risk without loss of embankment
C	(C/D)min ratio is greater than or equal to 1.0	(C/D)min ratio	No significant movement	Low risk

Table 2. Categories of bridge embankment behavior during a seismic event

$$FS_{cb} = \left[\frac{R_1 - R_2}{D_1 + K_h \cdot D_2}\right] \cdot \frac{S_1}{\gamma_1 \cdot H} \tag{1}$$

where FS_{cb} is the factor of safety against circular base failure, S_1 is the un-drained shear strength of the soil beneath the embankment, H is the embankment height (Figure 1), and γ_1 is the density of the soil layer (Table 1). The parameters R_1, R_2, D_1, and D_2 are obtained from Equations (2), (3), (4), and (5), respectively.

$$R_1 = 40 \cdot \sqrt{\frac{d}{r}} \cdot r \cdot (d + 12 \cdot r) \cdot (\lambda - 2) \tag{2}$$

$$R_2 = \sqrt{\frac{1+d}{r}} \cdot (9 \cdot (1+d)^2 + 40 \cdot (1+d) \cdot r + 480 \cdot r^2) \cdot \lambda \tag{3}$$

$$D_1 = 40 \cdot \sqrt{2} \cdot \left(1 + b^2 + 3d + 3d^2 - 3r - 6dr - 3bx + 3x^2\right) \tag{4}$$

$$D_2 = 40 \cdot \sqrt{2} \cdot (b + 3bd - 2 \cdot (-d \cdot (d - 2r))^{3/2} - 2 \cdot ((-1-d) \cdot (1+d-2r))^{3/2} - 3br - 3x - 6dx + 6rx) \tag{5}$$

where λ is the ratio of S_2/S_1, S_2 is the embankment soil un-drained shear strength and γ_2 is the embankment soil density (Table 1). For the values of x and r that result in the lowest factor of safety, designated x_c and r_c, the term in brackets of Eq. 1 has to be calculated and is called the stability number for the designated slope. The use of Eq. 1 in a spreadsheet with an optimization function provides reliable estimates of these parameters over the designated slope inclinations. Specifically, the "Solver®" function in "Microsoft Excel XP®" can be utilized to find r_c and x_c in order to minimize the factor of safety. By using pseudo-static analysis, assuming $FS_{cb} = 1.0$ in Eq. 1, and optimizing for r_c and x_c the horizontal earthquake acceleration factor (k_{hf}) is obtained for different assumed elevations of the upper level of the bedrock layer. The critical K_{hf} causing a circular base failure is obtained from Eq. 6.

$$K_{hf} = \frac{(R_1 - R_2) \cdot \dfrac{S_1}{\gamma_1 \cdot H} - D_1}{D_2}$$ (6)

Although a base failure predominates for the slope geometry typically encountered in highway embankments, a wedge failure extending upward from the toe of the embankment may be more critical for steeper slopes. The wedge type failure geometry is depicted in Figure 2b. For a wedge type failure, the factor of safety (FS_w) is obtained from Eq. 7.

$$FS_w = \frac{2 \cdot (1 + a^2)}{(a - b) \cdot (1 + a \cdot K_h)} \cdot \frac{S}{\gamma \cdot H}$$ (7)

where FS_w is the factor of safety against embankment wedge failure; S is selected as the estimated shear strength along the base of the failure wedge and the parameter a; shown in Figure 2b, is the parameter to be optimized. The horizontal earthquake acceleration factor (k_{hfw}) shall be obtained for different assumed elevations of the upper level of the bedrock layer by using pseudo-static analysis, assuming $FS_w = 1.0$ in Eq. 7, and optimizing by changing the parameter a. The critical K_{hfw} causing a wedge type failure of the embankment is obtained from Eq. 8.

$$K_{hfw} = \frac{1}{a} \cdot \left[\frac{2 \cdot (1 + a^2)}{(a - b)} \cdot \frac{S}{\gamma \cdot H} - 1 \right]$$ (8)

The lesser factor of safety for a circular base failure (FS_{cb}) and for a wedge type failure (FS_w) is then called the capacity/demand (C/D) ratio for the designated elevation of the upper level of the bedrock layer. Similar processes are followed for other elevations of the upper level of the bedrock layer in order to obtain the overall least C/D ratio, which is called the minimum capacity/demand ratio, $(C/D)_{min.}$. The considered horizontal earthquake acceleration (K_{hf}) is the one that corresponds to the $(C/D)_{min.}$ from all of the failure cases.

Embankment Displacement: For an embankment with $C/D_{min.} < 1.0$, it is important to estimate how far the mass actually displaces during the seismic event. This is carried out by calculating the anticipated embankment displacement (u). For a designated embankment, the PGA is identified for a specified seismic event; this parameter is also known as the maximum acceleration ($A_{max.}$). For the embankment to displace, the maximum acceleration has to exceed the acceleration causing embankment yielding. Assuming that the yield acceleration is equal to the K_{hf}, that corresponds to the $(C/D)_{min.}$ from all the failure cases, the yield factor (Y) is estimated as the ratio of A_y/A_{max}, where A_y is the yield acceleration, and

$A_{max.}$ equals to the *PGA*. By utilizing the site geometry and the specified sub-surface conditions, it is possible to use a simple model to determine the approximate yield acceleration of a bridge embankment. A sliding block solution can then be applied to estimate the displacement of the slope for a specified *PGA*, exceeding A_y. As *Y* decreases, u increases correspondingly. For *Y*<1.0, embankment displacement is likely to occur. The displacement (u) can be estimated by the use of Eq. 9 [6].

$$\log_{10}(u) = \alpha + \beta_1 \log_{10}\left(1 - \frac{A_y}{A_{max}}\right) + \beta_2 \log_{10}\left(\frac{A_y}{A_{max}}\right) \tag{9}$$

where u is the displacement, in centimeters; α, β_1, β_2 are the bedrock coefficients that are required to calculate the embankment displacement. Dodds [8] reported the way by which the bedrock coefficients are calculated for both bedrock and soil sites based on the potential earthquake magnitude at the geographic location of the bridge site. The value of α for both bedrock and soil can be obtained by use of Eq. 10a and Eq. 10b. The parameter, β_1, can be calculated for both the bedrock and soil by the use of Eq. 11a and Eq. 11b, while β_2 can be calculated by the use of Eq. 12a and Eq. 12b.

$$
\begin{aligned}
(\alpha)_{bedrock} &= 0.735 \cdot M_{b,Lg} - 4.41 & \text{(a)} \\
(\alpha)_{soil} &= 1.025 \cdot M_{b,Lg} - 6.292 & \text{(b)}
\end{aligned}
\tag{10}
$$

$$
\begin{aligned}
(\beta_1)_{bedrock} &= 0.35 \cdot M_{b,Lg} + 1.94 & \text{(a)} \\
(\beta_1)_{soil} &= 3.58 - 0.174 \cdot M_{b,Lg} & \text{(b)}
\end{aligned}
\tag{11}
$$

$$
\begin{aligned}
(\beta_2)_{bedrock} &= 0.21 - 0.15 \cdot M_{b,Lg} & \text{(a)} \\
(\beta_2)_{soil} &= -0.794 - 0.056 \cdot M_{b,Lg} & \text{(b)}
\end{aligned}
\tag{12}
$$

where $M_{b,Lg}$ is the body-wave magnitude of the anticipated earthquake. As the seismic slope stability of an embankment decreases, a larger displacement is expected, providing a stronger indication of an at-risk embankment than that obtained from the $(C/D)_{min.}$ ratio. The analysis using this method eliminates the misleading condition of how to assess an embankment with $(C/D)_{min.}$ ratio<1.0. Instead, this method forces a consideration of the possible displacement that may be observed, a better prediction of the actual behavior of a given embankment during a seismic event.

Liquefaction Potential: The mechanical behavior, which includes the liquefaction potential during the seismic event, is another important parameter in the seismic vulnerability assessment and prioritization of bridge embankments. Cohesion-less soils, such as alluvium and sandy/gravelly continental deposits are susceptible to liquefaction, and alluvium is the most likely to experience liquefaction. Where the boring logs data is available, straight-forward steps are followed to define the liquefaction potential as reported by [9]. In order to overcome the difficulties encountered when such data is not available, an alternate approach to define the liquefaction potential is followed. The liquefaction potential has to be assessed in accordance with the following two sub-sections

Boring Logs Are Not Available: Where the boring log data of each embankment site is not available, the liquefaction potential can be addressed based on the Seismic Retrofit Manual for Highway Bridges [1]. The susceptibility of the embankment soil to liquefaction is classified as one of three possible types (Table 3).

Liquefaction Type	Liquefaction Susceptibility	Parameters and Signs
A	High	1) Associated with saturated loose sands, saturated silty sands, or non-plastic sands. 2) A bridge that crosses a waterway is often constructed on loose saturated cohesion-less deposits that are most susceptible to liquefaction.
B	Moderate	Associated with medium dense soils such as compacted sand soils.
C	Low	Associated with dense soils.

Table 3. Liquefaction Susceptibility at a bridge embankment site

The three liquefaction possibilities are: high susceptibility, moderate susceptibility, and low susceptibility. High susceptibility is associated with saturated loose sands, saturated silty sands, or non-plastic sands. A bridge that crosses a waterway where soils have been deposited over long periods of time by flowing water is often constructed on loose saturated cohesion-less deposits that are the most susceptible to liquefaction. Moderate susceptibility is associated with medium dense soils such as compacted sand soils. Low susceptibility is associated with dense soils.

Boring Logs Are Available: Where the boring log data is available, the liquefaction potential at the bridge site is determined by the method reported by [9]. To determine a reasonably accurate value of the cyclic stress ratio causing liquefaction and induced by the earthquake motion, a correlation between the liquefaction characteristics and standard penetration test (*SPT*) blow-count values (*N* values), described by [10] is used. The average cyclic shear induced by the seismic event is obtained from Eq. 13.

$$\frac{\tau_{h,avg}}{\sigma_0'} \cong 0.65 \frac{A_{max.}}{g} \cdot \frac{\sigma_0}{\sigma_0'} \cdot r_d \tag{13}$$

where $\tau_{h,avg}$ is the average cyclic shear stress during the time history of interest, σ_o' is the effective overburden stress at any depth, A_{max} is the maximum earthquake ground surface acceleration, and r_d is a stress reduction correction factor. The mean effective and total stresses (σ_o' and σ_o') are replaced with the effective and total vertical stresses. The stress reduction factor (r_d), defined by [10], is computed using the depth (z) in meters as shown in Eq. 14.

$$r_d = (1 - \frac{z}{91})$$ (14)

The soil penetration resistance is the corrected normalized standard penetration resistance, $N_{1,60}$, which is defined by [10] and [5] in Eq. 15.

$$N_{1,60} = C_N \cdot \frac{ER_m}{60} \cdot N_m$$ (15)

Where C_N is the correction coefficient, ER_m is rod energy ratio, and N_m is the measured *SPT* blow-count per foot. With the determination of both the cyclic stress ratio induced during the earthquake and the cyclic stress ratio required to cause liquefaction, the factor of safety against liquefaction (FS_l) is calculated as shown in Eq. 16.

$$FS_l = \frac{[\tau_{avg}/\sigma_0 \cdot]_{l,M=M}}{[\tau_{h,avg}/\sigma_0 \cdot]}$$ (16)

Where $[\tau_{avg}/\sigma_0 \cdot]_{l,M=M}$ is the cyclic stress ratio required to cause liquefaction at any magnitude M, and $[\tau_{h,avg}/\sigma_0 \cdot]$ is the cyclic stress ratio induced during an earthquake of the same magnitude. No liquefaction is predicted to occur for $FS_l > 1.0$.

6. Category identification

Ranking and prioritization of embankments is based on the input parameters including geometry, material, seismic event, upper level of bedrock layer, level of natural ground line and soil type. Seismic vulnerability ranking and prioritization is conducted using the 'Kentucky Embankment Stability Ranking' (*KESR*) model in which three categories are incorporated to specify the failure risk of each embankment [4]. Application of the proposed methodology results in obtaining the three aforementioned ranking parameters known as the $(C/D)_{min.}$ ratio, embankment displacement, and liquefaction potential. The *KESR* model assumes one of the following three possibilities (*A*, *B*, or *C*) of embankment behavior during a seismic event, as described in Table 2: (*A*) loss of embankment, (*B*) significant movement, and (*C*) no significant movement. High seismic risk is assigned to category *A*. Significant seismic risk without loss of the embankment is assigned to category *B*, while low seismic risk is assigned to category *C*. The embankment displacement and the liquefaction potential are the ranking parameters for category *A* and category *B*. Conversely, the ranking of embankments within category *C* is solely based on the anticipated $(C/D)_{min.}$ ratio. For an embankment to be assigned category *A*, either the displacement shall exceed 10 centimeters (4 inches) or a high liquefaction potential is probable during the specified seismic event.

An embankment in category B meets one of the following two criteria: (1) moderate liquefaction potential; or (2) an anticipated $(C/D)_{min.}$ ratio less than 1.0, along with a displacement of less than 10 centimeters (4 inches). An embankment in category C shall have $(C/D)_{min.}$ ratio greater than or equal to 1.0.

7. Ranking and prioritization

After classifying the bridge embankments to category A, category B, or category C in accordance with the criteria listed in Table 2, a prioritization within each category is carried out based on the significance of the three ranking parameters. For instance, the higher the displacement of an embankment in category A, the higher its seismic risk, and thus it is assigned a higher priority or ranking. The same applies for the prioritization of the embankments in category B. On the other hand, the lower the $(C/D)_{min.}$ ratio of an embankment in category C, the higher its seismic risk, and thus it is assigned a higher priority or ranking.

Having completed the classification and categorization of all embankments in a certain region due to an anticipated seismic event, the embankment prioritization in each category becomes a feasible task. This proposed ranking model is useful for a quick sensitivity assessment of the effect of various site conditions, earthquake magnitudes, and site geometry on possible movement of a designated embankment. Since the intent of the provided ranking model is to compare the seismic risk of the several embankments, regardless of the existence of highly accurate input data in the ranking model, it is the authors' recommendation to further conduct detailed assessments of the behavior of those at-risk embankments. In such detailed assessments, accurate data from sub-soil explorations is to be incorporated. Eventually, a priority list for the seismic risk of all the considered embankments can be prepared, which enables decision makers to take appropriate actions.

8. Step-by-step seismic risk identification of bridge embankments

In order to facilitate the application of the proposed ranking methodology to prioritize bridge embankments, a complete flowchart has been generated. The flowchart provides a useful tool that promotes achieving the final goal of the study. The flowchart in its current form and sequences ensures a minimal effort from the engineer/researcher to apply the specified ranking methodology. Parameters of each embankment including its geometry, material, seismic event, upper level of bedrock, level of natural ground line, soil type, and anticipated failure types are taken into consideration during the development of the flowchart. All considerations, assumptions, calculations and required checks are arranged in a defined order in the flowchart. The loops of the flowchart, shown in Figure 3, allow relative ranking of bridge embankments. Titles are provided to identify the different sections of the flowchart including geometry, materials, seismic event, soil type, analysis, ranking parameters, category identification, and final ranking/prioritization. Notes to explain the steps of the methodology are numbered consecutively, listed in Table 4, and need to be considered along with the flowchart during the seismic risk prioritization of bridge embankments in a designated region.

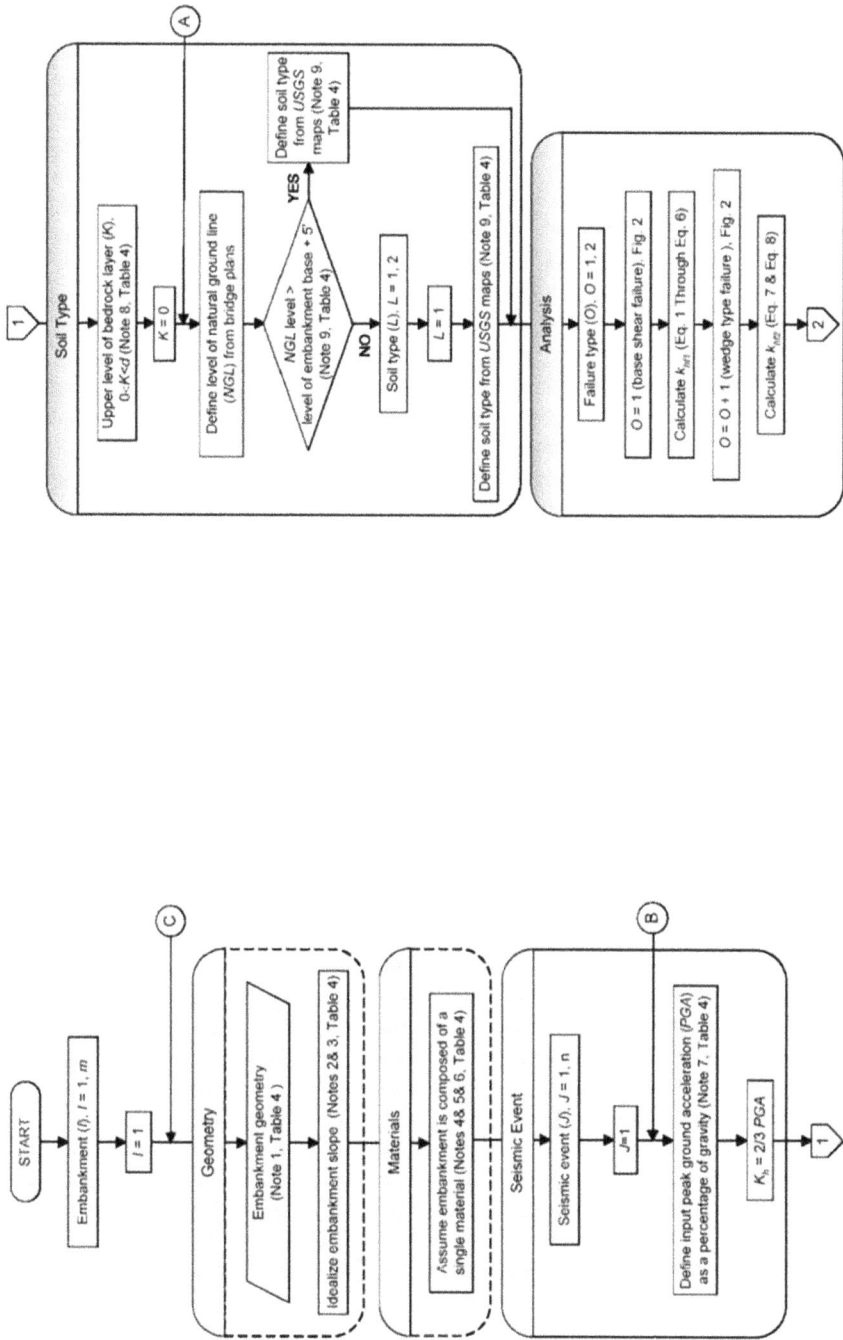

Soil Type

Upper level of bedrock layer (K). $0 < K < d$ (Note 8, Table 4)

$K = 0$

Define level of natural ground line (NGL) from bridge plans

NGL level > level of embankment base + 5' (Note 9, Table 4) — **YES** → Define soil type from $USGS$ maps (Note 9, Table 4)

NO

Soil type (L), $L = 1, 2$

$L = 1$

Define soil type from $USGS$ maps (Note 9, Table 4)

Analysis

Failure type (O), $O = 1, 2$

$O = 1$ (base shear failure), Fig. 2

Calculate k_{m1} (Eq. 1 Through Eq. 6)

$O = O + 1$ (wedge type failure), Fig. 2

Calculate k_{m2} (Eq. 7 & Eq. 8)

$START$

Embankment (I), $I = 1, m$

$I = 1$

Geometry

Embankment geometry (Note 1, Table 4)

Idealize embankment slope (Notes 2 & 3, Table 4)

Materials

Assume embankment is composed of a single material (Notes 4 & 5 & 6, Table 4)

Seismic Event

Seismic event (J), $J = 1, n$

$J = 1$

Define input peak ground acceleration (PGA) as a percentage of gravity (Note 7, Table 4)

$K_h = 2/3 PGA$

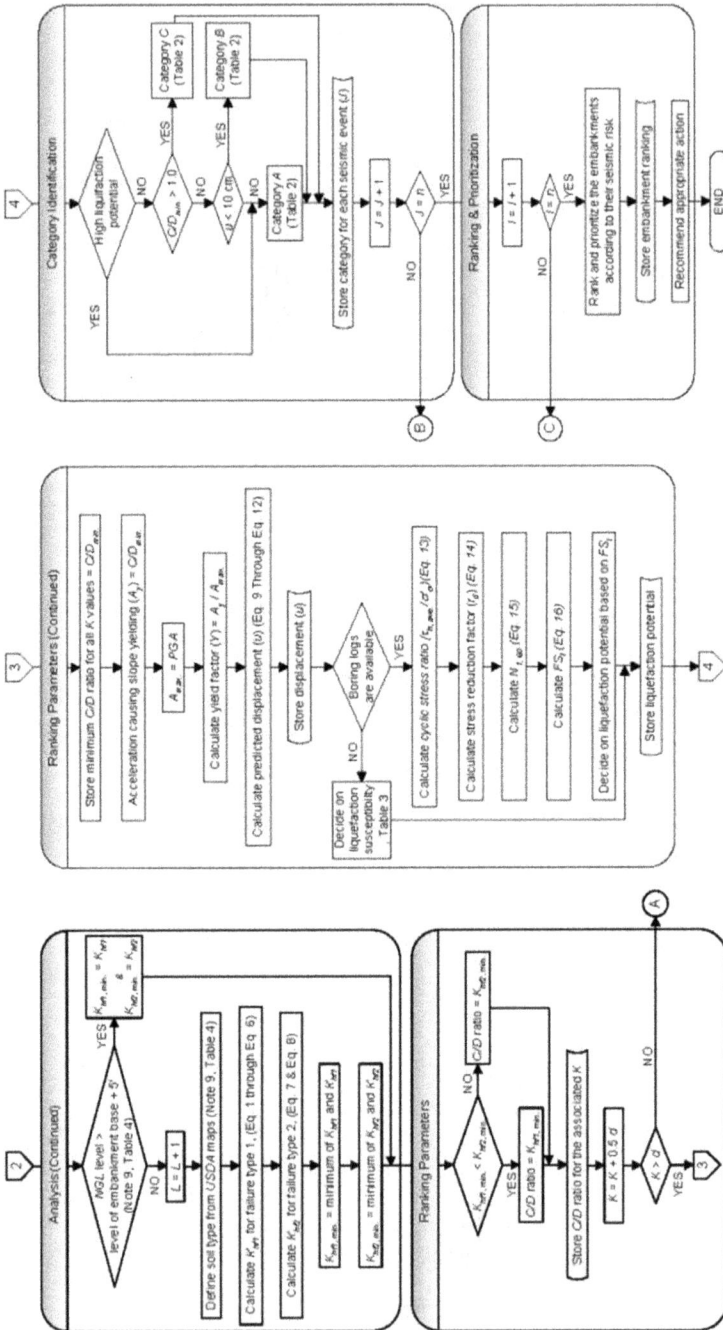

Figure 3. Flowchart for seismic risk assessment and ranking of bridge embankments

Note #	Description
Note 1	Embankment slope and geometry are taken from the bridge plans.
Note 2	For a typically irregular embankment slope, the highest idealized slope is used and is chosen in a way that results in the least $(C/D)_{min.}$ ratio.
Note 3	H = height of the embankment, b = slope inclination (Figure 1).
Note 4	Density and shear strengths of the embankment fill and soils underneath are identified in Table 1. A soil layer beneath the embankment fill has uniform un-drained shear strength different from that of the embankment fill.
Note 5	The analysis is carried out on the most critical embankment slope at either end of the bridge.
Note 6	The water table is assumed to be located below the embankment base.
Note 7	The input surface acceleration for a designated embankment site is obtained in accordance with the earthquake magnitude.
Note 8	The upper level of the bedrock layer is based on the "*Geologic Quadrangle Maps of the United States*" provided by "*United States Geological Survey (USGS)*", where the bedrock layer provides the lower bound for the soil layer (Figure 1).
Note 9	The choice whether to only consider soil data from the *USGS* maps or to consider an additional case from the *USDA* maps is dependent upon the level of the "Natural Ground Line" (*NGL*) as compared to the level of the embankment base.

Table 4. Complimentary notes to Figure 3 "Flowchart for seismic risk assessment and ranking of multiple bridge embankments"

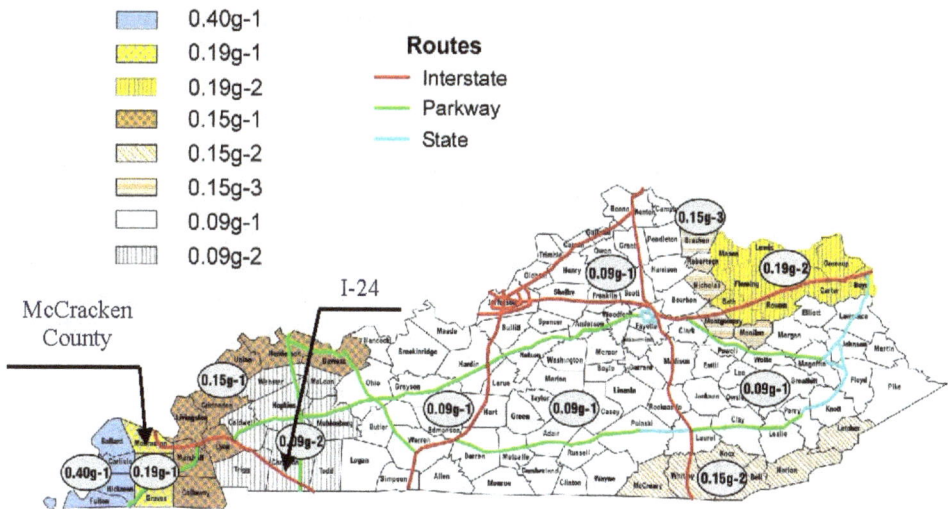

Figure 4. Predicted "Peak Ground Acceleration" (PGA) of all counties in the Commonwealth of Kentucky during a 250-year seismic event

9. Bridges in the commonwealth of Kentucky

Bridges in the western region of the Commonwealth of Kentucky are located near the New Madrid seismic zone, which is potentially one of the most destructive fault zones in the United States. It extends through the Mississippi River Valley and encompasses 26 counties in western Kentucky in the area of its strongest influence. Studies have shown that the probability of an earthquake with a 6.3 magnitude on the Richter scale to hit this area within the next 50 years exceeds 80%.

Passing through seven counties in western Kentucky, I-24 is considered a vital transportation link for the commonwealth of Kentucky. I-24 passes through McCracken, Livingston, Marshall, Lyon, Trigg, Caldwell, and Christian counties in western Kentucky (Figure 4). The objective of this part of the Chapter is to investigate the seismic risk of all bridge embankments on or over I-24 in western Kentucky.

In order to achieve the study objective, a means of accessing which embankments qualify as "most critical" is required. The methodology presented earlier in this Chapter is applied to assess the seismic vulnerability of I-24 bridge embankments. The embankment geometry, materials, type and properties of underlying soil, elevation of the natural ground line, and upper level of bedrock are estimated for each embankment. The minimum seismic slope stability capacity/demand, $(C/D)_{min}$ ratio, embankment displacement, and liquefaction potential of each bridge embankment are calculated. Bridge embankments along I-24 in western Kentucky are assigned one of three possible categories to represent their seismic failure risk. A final priority list of the embankments with the highest seismic risk is generated for the 127 bridges on or over I-24 in western Kentucky.

On-Site Inspection of I-24 Bridges in Western Kentucky: On-site inspection of the bridges, including photographing different structural components of each bridge, was carried out. The on-site inspection records form an invaluable source that assists in pre-earthquake evaluation studies as well as post-earthquake inspection.

I-24 Bridge Inventory in Western Kentucky: One objective of the on-site inspection is to have an informative source of accurate and updated bridge records, which are required for most assessment studies including the current study of seismic ranking and prioritization of I-24 bridge embankments in western Kentucky. Another objective of the on-site inspection is to provide engineers and transportation officials with information delineating the current bridges' conditions in order to facilitate future comparisons with post-earthquake conditions immediately after future earthquakes. Through these comparisons, significant changes can be reported and further studies can be carried out. All the bridges and embankments along I-24 in western Kentucky were visually inspected, photographed and the records were stored in a database. The on-site inspection represents a significant supplement to the "as-built" bridge plans. A comprehensive inventory of the bridges was compiled by review of the "as-built" bridge plans, construction and maintenance records, and on-site inspection forms. The inventory provides an essential data record, which is utilized for risk assessment of I-24 bridges and embankments in western Kentucky. A one-page sample of the I-24 bridge inventory for McCracken County is presented in Table 5. Similar inventories for Livingston, Marshall, Lyon, Trigg, Caldwell, and Christian counties are shown elsewhere [11].

Characteristics of I-24 Bridge Inventory in Western Kentucky: Eighty-one bridges are located on I-24 and 45 bridges are constructed over I-24, resulting in a total of 127 bridges either on or over the interstate in western Kentucky. Of the 127 bridges, many bridges were designed without following stringent seismic design guidelines, and may not withstand severe seismic events. Lyon and Marshall Counties are located approximately 115 Kilometers (72 miles) and 96 kilometers (60 miles) northeast of the center of the New Madrid seismic zone, respectively. McCracken County, located approximately 72 kilometers (45 miles) northeast of the center of the New Madrid seismic zone, has the largest number of bridges among all other counties with an average of two bridges per mile. The 127 bridges are categorized based on several characteristics, including: structural type, number of spans, maximum span length, skew angle, construction materials, and bearing types. Eighty three percent of the bridges are skewed, of which, 13% have a skew angle exceeding 40 degrees. McCracken County includes the largest number of bridges (38 bridges), followed by Lyon County (27 bridges), Marshall County (21 bridges), Christian County (20 bridges), Trigg County (11 bridges), Livingston County (seven bridges), and Caldwell County (three bridges).

10. Embankment properties

The geometry of each bridge embankment on or over I-24 in western Kentucky is taken from the bridge plans. The geometry of the 127 studied embankments is classified into five types (Figure 5a-5e). An embankment has either a single slope or double slopes separated by a perm. The inventory of I-24 bridge embankments in western Kentucky shows that a given slope has one of three possible inclinations (1:1, 2:1, or 3:1), where the first number of the ratio represents the horizontal unit and the second number represents the vertical unit. The drawings shown in Figure 5a-5d are for cases where the feature crossed by the bridge is either a highway or a railway. The drawing shown in Figure 5e is found when the bridge crosses a waterway. The embankment slope geometry is identified by its height (H) and the idealized inclination (b) (Figure 6). The analysis is carried out on both ends of each bridge and the most critical embankment slope at either end; whichever analysis results in a lower seismic slope stability C/D ratio is considered in the seismic vulnerability ranking.

Accurate identification of the soil characteristics requires detailed site-specific subsurface exploration. This approach is expensive, and such data is not available for the majority of the bridge embankments along I-24 in western Kentucky. Pflazer [14] reported on the use of existing geo-technical data to supplement site investigations. Another approach to specify the soil type and its properties is to use existing geological and agricultural maps. The source of soil data is dependent on the *NGL* (Figure 5f-5g). The *USGS* and the *USDA* are used to identify the soil type underneath an embankment. The way by which either

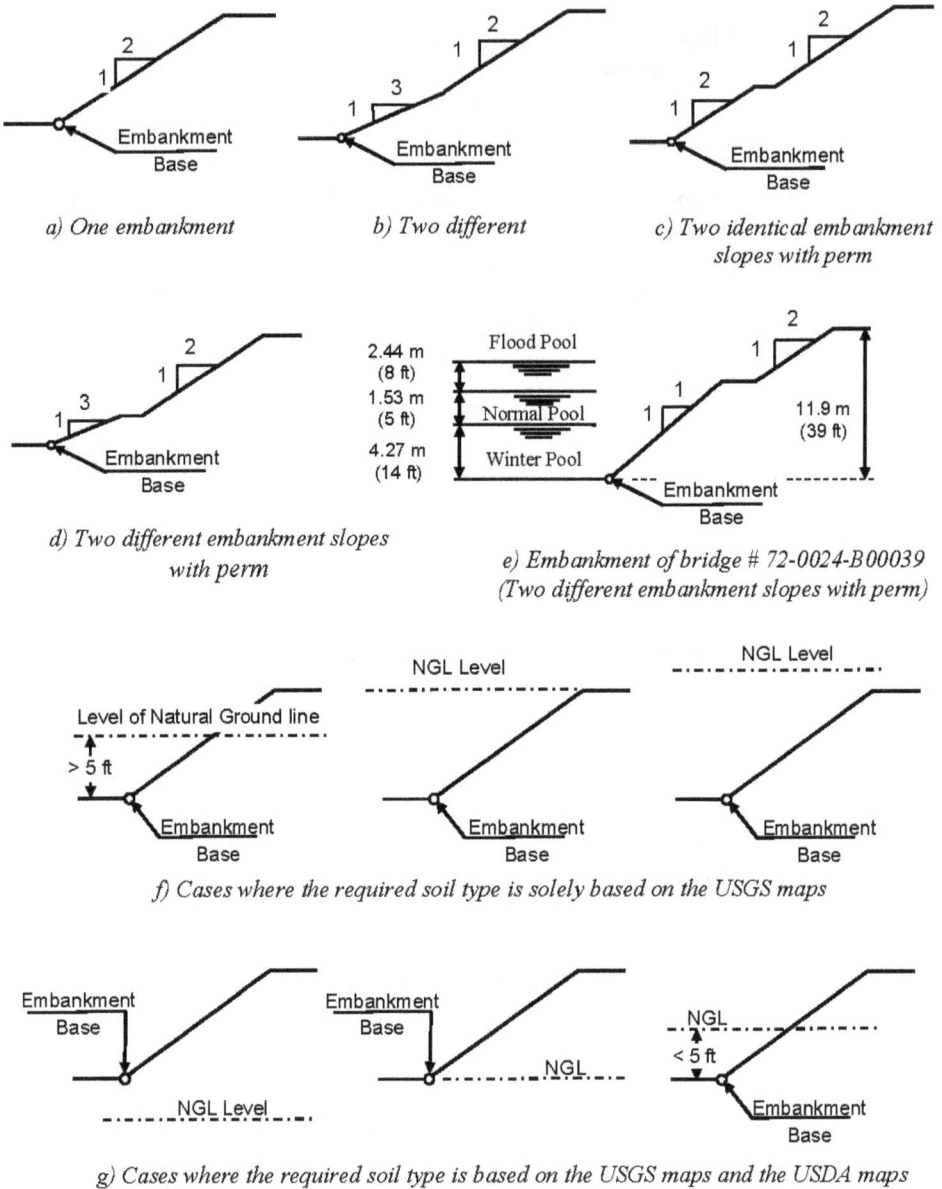

a) One embankment

b) Two different

c) Two identical embankment slopes with perm

d) Two different embankment slopes with perm

e) Embankment of bridge # 72-0024-B00039 (Two different embankment slopes with perm)

f) Cases where the required soil type is solely based on the USGS maps

g) Cases where the required soil type is based on the USGS maps and the USDA maps

Figure 5. Embankments along I-24 in western Kentucky: geometry classification (Figs. a, b, c, d, e), level of "Natural Ground Line" (NGL) and source of the soil data (Figs. f, g)

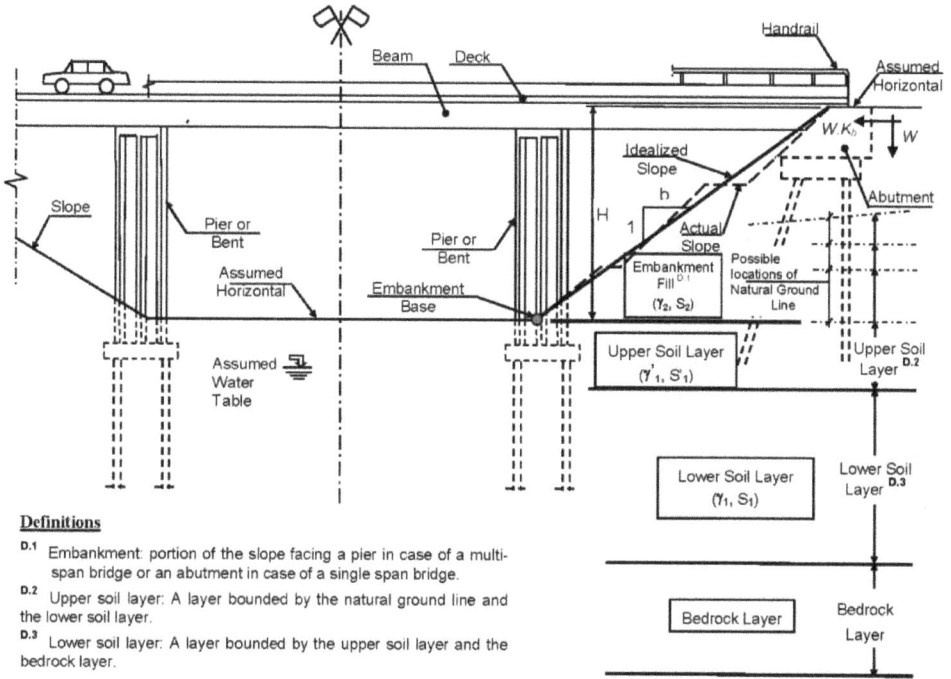

Figure 6. Example of bridge embankment geometry and materials

map is chosen for a given bridge site is based on the level of the *"Natural Ground Line"* (*NGL*) as compared to the respective embankment base (Figures 2*f*, 2*g*, and 3). Whenever the level of the *NGL* is above the level of the embankment base by more than 1.50 m (5 ft), the soil type is solely identified in accordance with the *USGS* maps. Whenever the level of the *NGL* is either above the level of the embankment base by less than 1.50 m (5 ft) or below the level of the embankment base, the soil type is based on both the *USGS* maps, and the *USDA* maps. After specifying the soil type, conservative soil characteristics including shear strength and mass density are estimated. The upper and lower soil layers' types (Figure 5) for embankments in McCracken County are provided in Table 6. Shear strength and mass density for bridge embankments are derived following the guidelines presented earlier. Data regarding the level below which a hard stratum (stiff bedrock layer) exists is not available for the majority of bridge embankment sites along I-24 in western Kentucky. The upper level of the stiff bedrock layer, which falls within the range from the embankment base down to the upper level of the hard stratum, is initially estimated from the *USGS maps*.

District	Status[1]	County[2]	Route	Bridge Bin #[3]	P[4]	Year Built	Main Spans[5]	Approach Spans[6]	Max. Span Length (ft)	Structure Length[7] (ft)	MP[8]	On / Over I-24	Feature Crossed
1	SM	73	131	B00009		1968	2	0	118	294	15.79	Over	I-24
1	SM	73	68	B00060		1968	2	0	92	233	16.16	Over	I-24
1	SM	73	68	B00060	P	1968	2	0	92	233	16.16	Over	I-24
1	SM	73	787	B00064		1966	2	0	94	228	16.88	Over	I-24
1	SM	73	3075	B00065		1966	2	0	92	242	14.09	Over	I-24 @ 14.09
1	SM	73	24	B00100[9]			2	16	730	5634	0.10	On	Ohio River
1	SM	73	24	B00101	P	1968	3	0	54	133	2.21	On	KY 1420
1	SM	73	24	B00101		1968	3	0	54	133	2.21	On	KY 1420
1	SM	73	24	B00102	P	1969	1	0	110	142	2.96	On	KY 305
1	SM	73	24	B00102		1969	1	0	110	142	2.96	On	KY 305
1	SM	73	24	B00103	P	1969	3	0	74	181	3.46	On	P&L Railway
1	SM	73	24	B00103		1969	3	0	74	181	3.46	On	P&L Railway
1	SM	73	24	B00104	P	1968	3	0	69	170	3.69	On	P&L Railway
1	SM	73	24	B00104		1968	3	0	69	170	3.69	On	P&L Railway
1	SM	73	24	B00105		1969	2	0	80	224	4.33	On	US 60
1	SM	73	24	B00105	P	1969	2	0	80	224	4.33	On	US 60
1	SM	73	24	B00107	P	1967	3	0	50	115	4.59	On	Perkins Creek
1	SM	73	24	B00107		1967	3	0	50	115	4.59	On	Perkins Creek
1	SM	73	24	B00111		1971	3	0	51	121	5.60	On	Buchner Lane
1	SM	73	24	B00111	P	1971	3	0	51	121	5.60	On	Buchner Lane
1	SM	73	24	B00112		1971	2	2	85	196	6.87	On	US 45
1	SM	73	24	B00112	P	1971	2	2	85	196	6.87	On	US 45
1	SM	73	24	B00113		1974	4	0	105	337	7.36	Over	I-24 @ Elmdale Rd
1	SM	73	24	B00114		1963	5	0	98	458	9.77	On	P&L Railroad
1	SM	73	24	B00114	P	1963	5	0	98	458	9.77	On	P&L Railroad
1	SM	73	24	B00115	P	1971	3	0	53	143	10.32	On	Island Creek
1	SM	73	24	B00115		1971	3	0	53	143	10.32	On	Island Creek
1	SM	73	24	B00116	P	1975	2	0	96	197	11.04	On	KY 1954
1	SM	73	24	B00116		1975	2	0	96	197	11.04	On	KY 1954
1	SM	73	24	B00117[9]		1972	2	0	15	34	11.44	On	Bee Bridge
1	SM	73	24	B00118	P	1975	3	0	71	191	11.98	On	Old L & N Railroad
1	SM	73	24	B00118		1975	3	0	71	191	11.98	On	Old L & N Railroad
1	SM	73	24	B00119	P	1971	1	2	101	172	12.60	On	KY 450 (Oaks Rd)
1	SM	73	24	B00119		1971	1	2	101	172	12.63	On	KY 450 (Oaks Rd)
1	SM	73	24	B00120	P	1975	3	0	200	486	13.30	On	Clarks River
1	SM	73	24	B00120		1975	3	0	200	486	13.30	On	Clarks River
1	SM	73	62	B00121		1971	2	0	105	260	6.39	Over	I-24
1	SM	73	994	B00122		1971	2	2	108	256	8.61	Over	I-24

[1] Status is defined as SM (State Maintained), RS (Rural Secondary), or County (Locally Maintained). SM bridges are the only bridges to appear in this sample page.

[2] County # 73 stands for McCracken County in western Kentucky.

[3] Bridge Bin # is as appears in the Kentucky Transportation Cabinet bridge inventory.

[4] The letter P, as defined in the Kentucky Transportation Cabinet bridge inventory, stands for a parallel bridge which is located westbound on I-24

[5] Main spans stands for the number of main spans of the designated bridge.

[6] Approach spans stands for the number of approach spans of the designated bridge.

[7] Structure length is the total length of bridge including the approaches.

[8] MP stands for the mile point to which the bridge is logged.

[9] The designated bridge under consideration is out of the scope of the study.

Table 5. Inventory of I-24 bridges in McCracken County, western Kentucky

County	Bridge Number [1,2]	Upper Soil Layer [3,4]		Lower Soil Layer [3,5]	
		Depth [6] (ft)	Soil Type [6]	Depth [7] (ft)	Soil Type [7]
McCracken	73-0024-B00101 & 73-0024-B00101P[2]	0'- 5'	Heavy Silt Loam	5'- 30'	Lacustrine Deposits
	73-0024-B00102 & 73-0024-B00102P[2]	0'- 5.5'	Gravely Sandy Clay Loam	5.5'- 90'	Loess layer followed by Continental Deposits
	73-0024-B00103 & 73-0024-B00103P[2]	0'- 5.5'	Silt Loam	5.5'- 90'	Loess layer followed by Continental Deposits
	73-0024-B00104 & 73-0024-B00104P[2]	0'- 5.5'	Silt Loam	5.5'- 90'	Loess followed by Continental Deposits
	73-0024-B00105 & 73-0024-B00105P[2]	0'- 5.8'	Silty Clay Loam	5.8'- 90'	Loess layer followed by Continental Deposits
	73-0024-B00107 & 73-0024-B00107P[2]	0'- 5'	Silt Loam	5'- 95'	Alluvium
	73-0024-B00111 & 73-0024-B00111P[2]	0'- 5'	Silt Loam	5'- 90'	Loess layer followed by Continental Deposits
	73-0024-B00112 & 73-0024-B00112P[2]	0'- 5.8'	Silt Loam	5.8'- 95'	Alluvium
	73-0024-B00114 & 73-0024-B00114P[2]	0'- 5'	Heavy Silt Loam	5'- 200'	Lacustrine Deposits
	73-0024-B00115 & 73-0024-B00115P[2]	0'- 5.5'	Heavy Silt Loam	5.5'- 200'	Lacustrine Deposits
	73-0024-B00116 & 73-0024-B00116P[2]	0'- 5'	Heavy Silt Loam	5'- 200'	Lacustrine Deposits
	73-0024-B00118 & 73-0024-B00118P[2]	0'- 5.8'	Silty Clay Loam	5.8'- 200'	Lacustrine Deposits
	73-0024-B00119 & 73-0024-B00119P[2]	0'- 5.8'	Silt Loam	5.8'- 200'	Lacustrine Deposits
	73-0024-B00120 & 73-0024-B00120P[2]	0'- 5.5'	Heavy Silt Loam	5.5'- 200'	Lacustrine Deposits
	73-0131-B00009	0'- 5.8'	Silty Clay Loam	5.8'- 90'	Loess layer followed by Continental Deposits
	73-0068-B00060 & 73-0068-B00060P[2]	0'- 6.3'	Silty Clay Loam	6.3'- 90'	Loess layer followed by Continental Deposits
	73-0787-B00064	0'- 5'	Heavy Silt Loam	5'- 90'	Loess layer followed by Continental Deposits
	73-3075-B00065	0'- 5'	Heavy Silt Loam	5'- 93'	Continental Deposits
	73-0024-B00113	0'- 4.5'	Silty Loam	4.5'- 93'	Continental Deposits
	73-0062-B00121	0'- 5.5'	Silt Loam	5.5'- 93'	Continental Deposits
	73-0994-B00122	0'- 5.8'	Silty Clay Loam	5.8'- 93'	Continental Deposits

[1] As defined in the Kentucky Transportation Cabinet bridge inventory.

[2] The letter P, as defined in the Kentucky Transportation Cabinet inventory, stands for a parallel bridge that is located in the westbound lane along the I-24 in western Kentucky.

[3] Upper soil layer and lower soil layer are shown in Fig. 3.

[4] Upper soil layer's data is based on the 'United States Department of Agriculture' (USDA) maps, 'Soil Conservation Service'.

[5] Lower soil layer's data is based on the 'United States Geological Survey' (USGS), 'Geologic Quadrant Maps of the United States'.

[6] Data is obtained from the 'Soil Survey of Ballard and McCracken Counties, Kentucky', the United States Department of Agriculture (USDA) maps, 'Soil Conservation Service', 1976.

[7] Data is obtained from the 'Geology of the Puducah East Quadrangle, Kentucky', Map GQ-531, 'United States Geological Survey' (USGS), 'Geologic Quadrant Maps of the United States'.

Table 6. Types of upper and lower soil layers for embankment sites in McCracken County, western Kentucky

Other upper levels of the bedrock layer within that range are also considered, and the controlling case is the one that results in the lowest seismic slope stability C/D ratio. The input PGA at a designated embankment site is obtained from seismic maps generated by [7] for 50-year, 250-year, and 500-year events. The 50-year, 250-year, and 500-year events are seismic events with a 90% probability of not being exceeded in 50 years, 250 years, and 500 years, respectively. Figure 4 illustrates an example of the anticipated PGA of all counties in the Commonwealth of Kentucky during the 250-year seismic event. The peak ground acceleration for McCracken County during the 250-year event is 0.19 g, where g is the gravitational acceleration. Other anticipated $PGAs$ of all counties in the Commonwealth of Kentucky during the 50-year and 500-year seismic events can be found in the Kentucky Transportation Center report [11]. With the exception of the parallel bridges at the Cumberland River crossing, and at the Tennessee River crossing, each bridge and their embankments along I-24 in western Kentucky is evaluated for the 50-year and 250 year seismic events, for which valuable input data is taken from a study conducted by Street et al. [7]. During the 50-year seismic event, the bridges are expected to behave elastically without any disruption to traffic. During the 250-year seismic event, partial damage to the bridges is permitted, and the bridges are expected to remain accessible to emergency traffic. I-24 parallel bridges at the Cumberland River and at Tennessee River crossings are evaluated for the 250-year seismic event and the maximum credible 500-year seismic event. Detailed evaluation of these bridges and their embankments are presented elsewhere [11].

11. Vulnerability analysis of I-24 bridge embankment in Kentucky

For a bridge on or over I-24 in western Kentucky, the potential of an embankment slope to displace during a designated earthquake event is assessed using the two-dimensional limit equilibrium stability analysis. During the seismic vulnerability evaluation of each embankment, the possibility of occurrence of either circular or wedge–shaped slope failure [11] is investigated and the one that results in the lesser C/D ratio is considered in the ranking process. K_h equals to 2/3 of the PGA. The ranking and prioritization procedure of the embankments is based on three main parameters: (1) seismic slope stability $(C/D)_{min.}$ ratio, (2) embankment displacement, and (3) liquefaction potential at the embankment site. For embankments with $(C/D)_{min.}$ ratio against sliding<1.0, estimation of how far the embankment actually displaces during the ground excitation is necessary. Hence, the displacement of the embankment is calculated. The maximum acceleration $(A_{max.})$ for a specified seismic event is identified for a designated embankment. For slope displacement to occur, the maximum acceleration must exceed the acceleration causing yielding in the embankment slope (A_y). The $(C/D)_{min.}$ ratio is calculated for each embankment, and is used to assign a rank for each embankment relative to the other embankments along I-24 in western Kentucky.

Assuming that the yield displacement is equal to K_{hf}, which corresponds to the $(C/D)_{min.}$ ratios for all the possible failure cases, the resulting 'Yield Factor' (Y) is estimated as the ratio of A_y/A_{max}, where A_y is the acceleration causing yielding in the embankment slope and $A_{max.}$ is equal to the PGA. The displacement of the slope with a specified PGA exceeding the A_y is estimated. At intervals for which the PGA exceeds A_y (Y is less than 1.0), the occurrence of slope displacement is expected. Decreasing A_y results in increasing the magnitude of the embankment displacement, correspondingly. As the seismic slope stability of an embankment decreases, a larger displacement is expected, providing a stronger indication of an at-risk embankment than that obtained from the $(C/D)_{min.}$ ratio analysis. One advantage of this methodology is that the analysis eliminates the misleading condition of how to assess an embankment that has $(C/D)_{min.}$ ratio<1.0, and instead forces a consideration of the possible embankment displacement. The vulnerability rating for a designated soil is based on quantitative assessment of liquefaction susceptibility and the anticipated magnitude of the acceleration coefficient [1]. Bridges subjected to low liquefaction potential shall be assigned a low vulnerability rating.

It is stipulated that it is not necessary to calculate liquefaction potential for the bridge sites, which are required to resist a seismic acceleration of less than 0.09 g [1]. The majority of the area surrounding the fault in the New Madrid Seismic Zone lies on fluvial and alluvial deposits and sandy soils. Defining the liquefaction potential is a matter of considerable concern during the seismic assessment of bridges and their embankments in this region. Western Kentucky encompasses several major bodies of water, including the Ohio River, Mississippi River, Barkley Lake, and Kentucky Lake. These bodies of water cause the saturated soils within the area to be highly susceptible to liquefaction potential. The proximity to these four bodies of water necessitates particular concern when examining the liquefaction potential for bridge sites along I-24 in western Kentucky.

The method to calculate the liquefaction potential is dependent on the availability of the soil boring logs. Whenever the boring logs of an embankment site along I-24 in western Kentucky are not available, the susceptibility of an embankment soil to liquefaction is classified in one of three ways. High susceptibility is associated with saturated loose sands, saturated silty sands, or non-plastic sands. A bridge that crosses a waterway is often constructed on loose saturated cohesionless deposits that are most susceptible to liquefaction. Moderate susceptibility is associated with medium dense soils, such as compacted sand soils. Low susceptibility is associated with dense soils.

Whenever the boring logs of an embankment site along I-24 in western Kentucky are available, the liquefaction potential of the bridge site is accurately determined by the method developed by Seed et al. [9, 10] and reported earlier in this Chapter. This method includes the following four steps: (1) determination of time history of shear stresses induced by the earthquake ground motion; (2) converting the time history to an equivalent number of stress cycles; (3) calculation of the cyclic shear stresses required to cause liquefaction in the same number of stress cycles; and (4) judging the liquefaction potential by comparing the shear stress induced during the earthquake with that required to cause liquefaction.

Liquefaction potential of few embankment sites along I-24 in western Kentucky is estimated using standard penetration tests (SPT) provided by the 'Kentucky Transportation Cabinet, Department of Materials and Geotechnical Testing.' For the rest of the bridge embankments along I-24 in western Kentucky, any judgment of the liquefaction potential is solely based on the surrounding soil type. The soil type is obtained from the USGS and USDA maps. A detailed method to predict the liquefaction potential is shown in Zatar et al. [12, 13].

12. Category identification, ranking, and prioritization of the I-24 bridge embankments in Western Kentucky

In the KESR model, three categories are sought out to specify the failure risk of each embankment during a designated seismic event. A category for each bridge embankment along I-24 in western Kentucky is assigned. The assigned category is based on the three ranking parameters: the $(C/D)_{min}$ ratio, the embankment displacement, and the liquefaction potential. Definition of the three categories (A, B, and C) is provided in Table 3. All 127 bridge embankments along I-24 in western Kentucky were analyzed using the procedures provided in the flowchart of Figure 6. The yield factor, $(C/D)_{min}$ ratio, displacement, and liquefaction potential for each embankment are identified, and a seismic embankment category is assigned. Further prioritization within each category was carried out based on the significance of the three ranking parameters. The embankments are ranked starting from the one with the highest seismic risk. For instance, a bridge embankment in category A with a ranking of A1 is more susceptible to damage than a bridge embankment with a ranking of A2 or A3. The same also applies for categories B and C. The ranking comprises a priority list that will be provided to senior state engineers, who may utilize its information to take appropriate actions. Based on the priority list, accurate soil data for those embankments with the highest risk may be needed in order to accurately identify their risk.

Due to its immense size, the full listing of the 127-embankment ranking and prioritization is not presented. However, a sample ranking and prioritization list for all embankments in McCracken County is presented for the 250-year seismic event (Table 7). Some of the embankments, which are in Category B during the 50-year seismic event, fall in Category A during the 250-year seismic event. For instance, the analysis of Bridge # 73-0024-B00118 in McCracken County to resist the 50-year seismic event results in a displacement of 4.6 centimeters (1.8 inches), and thus falls in category B. The analysis for the same bridge to resist the 250-year seismic event results in a displacement of 27.3 centimeters (10.7 inches) and thus is considered to fall in Category A. None of the embankments in McCracken County fall within category C since the assigned PGA for McCracken County is the highest among all counties along I-24 in western Kentucky, in addition to the associated liquefaction potential. This is not the case for Christian, Lyon, Trigg, and Caldwell counties.

County	Bridge #[1,2]	Peak Ground Acceleration[3] (%g)	Slope Height[4] (m)	(ft)	Yield Factor[5] Y	Horizontal Displacement[6] U (cm)	(in)	Minimum C/D Ratio[7] $(C/D)_{min}$	Liquefaction Potential[8]	Seismic Embankment Category[9]	Seismic Embankment Ranking[10]
McCracken	730024B00104 & 730024B00104P	19	9.1	30	0.084	79.8	31.4	0.75	High	A	A1
	730024B00103 & 730024B00103P	19	8.7	28.5	0.136	39.5	15.6	0.76	High	A	A2
	730024B00120 & 730024B00120P	19	11.6	38	0.166	28.7	11.3	0.67	High	A	A3
	730024B00118 & 730024B00118P	19	8.4	27.5	0.172	27.3	10.7	0.77	Moderate	A	A4
	730068B00060 & 730068B00060P	19	8.4	27.4	0.175	26.3	10.4	0.77	High	A	A5
	730787B00064	19	8.1	26.6	0.177	25.8	10.1	0.78	High	A	A6
	730024B00115 & 730024B00115P	19	7.9	26	0.226	16.8	6.6	0.79	Moderate	A	A7
	730024B00107 & 730024B00107P	19	7.5	24.5	0.236	15.5	6.1	0.76	High	A	A8
	730024B00105 & 730024B00105P	19	7.8	25.5	0.244	14.5	5.7	0.8	High	A	A9
	730024B00112 & 730024B00112P	19	6.9	22.5	0.308	8.9	3.5	0.79	High	A	A10
	730024B00102 & 730024B00102P	19	7	23.1	0.335	7.3	2.9	0.83	High	A	A11
	730131B00009	19	6.9	22.6	0.355	6.4	2.5	0.84	High	A	A12
	730024B00111 & 730024B00111P	19	6.7	22	0.377	5.5	2.2	0.85	High	A	A13
	730024B00119 & 730024B00119P	19	7.3	24	0.3	9.4	3.7	0.82	Moderate	B	B1
	730024B00116 & 730024B00116P	19	7.3	24	0.3	9.4	3.7	0.82	Moderate	B	B1
	730024B00114 & 730024B00114P	19	6.6	21.7	0.388	5.1	2	0.85	Moderate	B	B3
	730024B00101 & 730024B00101P	19	5.8	19	0.495	2.4	0.9	0.9	Moderate	B	B4
	730994B00122	19	7.9	26	1.844	0	0	1.81	Moderate	B	B5
	730062B00121	19	7.3	24	1.997	0	0	1.93	Moderate	B	B6
	733075B00065	19	7.2	23.6	2.029	0	0	1.96	Moderate	B	B7
	730024B00113	19	6.1	20	2.351	0	0	2.24	Moderate	B	B8

[1] As defined in the Kentucky Transportation Cabinet bridge inventory.

[2] The letter P, as defined in the Kentucky Transportation Cabinet inventory, stands for a parallel bridge which is located westbound on I-24.

[3] The peak ground acceleration is the maximum bedrock acceleration at a designated embankment's site ($A_{maximum}$) during a seismic event.

[4] The slope height is the idealized height, at either ends of the embankment that results in the lowest C/D ratio. (Refer to Figure 6).

[5] The yield factor, Y, is the ratio of the minimum yield acceleration (A_y) to the peak ground acceleration ($A_{maximum}$).

A_y is the K_{hf} that corresponds to the obtained $(C/D)_{min}$ ratio (based on Eq. (6) and Eq. (8).

[6] The displacement U of an embankment is calculated by Eq. (9).

[7] $(C/D)_{min}$ ratio is derived from Eq. (1) and Eq. (7).

[8] The liquefaction potential is determined in accordance with the method shown in this chapter.

[9] The category of embankment behavior is defined in accordance with Table 6

[10] In general, a bridge embankment in category A with a ranking of A1 is more susceptible to damage than a bridge embankment with a ranking of A2 or A3. The same applies to categories B and C.

Table 7. Seismic ranking for I-24 bridge embankments in McCracken County for a 250-year event County, western Kentucky

One complete example of the calculation procedures to identify the seismic risk of a bridge embankment in McCracken County is provided in Zatar and Harik [16]. Similar procedures are followed in order to identify the seismic risk of all the 127 bridge embankments in all seven counties along I-24 in western Kentucky. Full details and results of the ranking and prioritization of the bridges along I-24 in western Kentucky are provided in the Kentucky Transportation report [11].

13. Summary and conclusions

This document describes the authors' efforts in addressing the technical component of embankment prioritization, and is well suited to a reliability-based model for seismic risk assessment. A methodology is presented to quickly conduct seismic assessment and ranking of bridge embankments in order to identify and prioritize those embankments that are highly susceptible to failure. The step-by-step methodology is provided in a flowchart that is specifically designed to ensure minimal effort on behalf of the engineer/researcher.

The proposed ranking model is useful for a quick sensitivity assessment of the effect of various site conditions, earthquake magnitudes, and site geometry on possible movement of a designated embankment. The methodology was applied on 127 bridge embankments on a priority route in western Kentucky in order to identify and prioritize the embankments, which are susceptible to failure. Data regarding soil types and depth of bedrock is not available for the majority of the 127 bridge embankments of I-24 in western Kentucky. However, obtaining detailed geo-technical investigations and sophisticated models are typically limited because of the associated cost and effort. The methodology outlines possible approaches to predict the unavailable information regarding a bridge embankment site. The embankment geometry, material, type of underlying soil, elevation of the natural ground line, and upper level of bedrock are the variables of each embankment. Seismic slope stability capacity/demand ratio, displacement, and liquefaction potential of each bridge embankment along I-24 in western Kentucky are estimated. Three categories are presented to identify the failure risk and provide a priority list of the embankments. The seismic vulnerability during projected 50-year, 250-year, and 500-year seismic events are obtained and the associated seismic performance criteria are examined. An example of seismic ranking and prioritization of bridge embankments along I-24 in McCracken County in western Kentucky is presented. The priority list enables decision makers to take appropriate actions.

Author details

Wael A. Zatar[*]
*College of Information Technology and Engineering,
Marshall University, Huntington, West Virginia, USA*

* Corresponding Author

Issam E. Harik

Department of Civil Engineering, University of Kentucky, Lexington, Kentucky, USA

Acknowledgement

The support of the Federal Highway Administration, Transportation Cabinet of the Commonwealth of Kentucky, and Kentucky Transportation Center is gratefully acknowledged.

14. References

[1] Buckle, I. G., and Friedland, I. M. (1995). *Seismic retrofitting manual for highway bridges.* Report No. FHWA-RD-94-052, Federal Highway Administration, May, 309P.

[2] United States Geologic Survey (*USGS*), '*Geologic quadrant maps of the United States*' [map].

[3] United States Department of Agriculture (*USDA*), '*Soil conservation service*' [map].

[4] Sutterer, K., Harik, I., Allen, D., and Street, R., (2000). "Ranking and assessment of seismic stability of highway embankments in Kentucky," Research Report KTC-00-1, Kentucky Transportation Center, University of Kentucky, 98 pages.

[5] Seed, R., and Harder, L. (1990). "SPT-based analysis of cyclic pore pressure generation and undrained residual strength." *Proceedings of the H. Bolton Seed Memorial Symposium,* University of California-Berkeley, Vol. 2, pp. 351-376.

[6] Ambraseys, N. N., and Menu, J. M. (1988). "Earthquake induced ground displacements." *Earthquake Engineering and Structural Dynamics*, volume 16, pp. 985-1006.

[7] Street, R., Wang, Z., Harik, I., Allen, D., and Griffin, J. (1996). *Source zones, recurrence rates, and time histories for earthquakes affecting Kentucky.* Report No. KTC-96-4, Kentucky Transportation Center, University of Kentucky, 194p (Addendum 1998).

[8] Dodds, A. M. (1997). *Seismic deformation analysis for Kentucky highway embankments.* M. Sc. Thesis, University of Kentucky.

[9] Seed, H., Idriss, I., and Arango, I. (1983). "Evaluation of liquefaction potential using field performance data." *ASCE Journal of Geotechnical Engineering,* 109(3), pp. 458-482.

[10] Seed, H. B., Tokimatsu, K., Harder, L. F., and Chung, R. M. (1985). "Influence of SPT procedures in soil liquefaction resistance evaluations." *ASCE Journal of Geotechnical Engineering,* 111(12), 1425-1445.

[11] Zatar, W. A., Yuan, P., and Harik, I. E., "Seismic ranking of bridges on or over I-24 in western Kentucky." Research Report KTC, Kentucky Transportation Center, University of Kentucky, 2007.

[12] Zatar, W. A., Harik, I. E., Sutterer, K. G., Dodds, A., and Givan, G., "Bridge embankments: Part I - Seismic risk assessment and ranking." *ASCE Journal of Performance of Constructed Facilities,* June 2008.

[13] Zatar, W. A., and Harik, I. E., "Bridge embankments: Part II - Seismic risk of I-24 in Kentucky." *ASCE Journal of Performance of Constructed Facilities*, June 2008.

[14] Pflazer, W. J. (1995). "Use of existing geotechnical data to supplement site investigations." *Proceedings of the Ohio River Valley Soils Seminar XXVI*, ASCE Kentucky Geotechnical Engineers Group, Clarksville, Indiana.

Chapter 11

Mechanical Characterization of Laminated Rubber Bearings and Their Modeling Approach

A. R. Bhuiyan and Y.Okui

Additional information is available at the end of the chapter

1. Introduction

Base isolation, also known as seismic base isolation, is one of the most popular means of protecting a structure against earthquake forces. It is a collection of structural elements which should substantially decouple a superstructure from its substructure resting on a shaking ground thus protecting building and bridge structure's integrity. Base isolation is the most powerful tool of earthquake engineering pertaining to the passive structural vibration control technologies. It is meant to enable building and bridge structure to survive a potentially devastating seismic impact through a proper initial design or subsequent modifications. In some cases, application of base isolation can raise both a structure's seismic performance and its seismic sustainability considerably.

An isolation system is believed to be able to support a structure while providing additional horizontal flexibility and energy dissipation. Until the 80th decade of the last century many systems have been put forward involving features such as rollers or rockers bearings, sliding on sand or talc, or complaint first story column, but these have usually not employed in the practice of isolation of engineering structures [1, 2]. The study on the mechanical behavior of the isolation system dates back to 1886, when Professor Milne from Tokyo University, Japan attempted to observe isolation behavior of a structure supported by balls. He conducted an experiment by making an isolated building supported on balls *"cast-iron plates with saucer-like edges on heads of piles. Above the balls and attached to the buildings are cast-iron plates slightly concave but otherwise similar to those below"* [3]. However, another guy J.A. Calantarients in 1909, a medical doctor of the northern English City of Scarborough, was claimed to be the first man who conducted the experiment of isolation behavior of a structure supported by balls [3]. What both guys wanted to get information from their experiments is the global isolation behavior of the structure. The philosophy given by them regarding seismic isolation of a structure is stillin practice. Several mechanisms of investigating the mechanical behavior of isolation systems are developed based on this philosophy which is readily used.

In practice of seismic base isolation of bridge structures, laminated rubber bearings have been popular since the last century. Among many types of laminated rubber bearings, natural rubber bearing (RB) which is formed by alternate layers of unfilled rubbers and steel shims has less flexibility and small damping. It has been used to sustain the thermal movement, the effect of pre-stressing, creep and shrinkage of the superstructures of the bridge or has been used for base isolation practices with additional damping devices [1, 2 and 4]. On the other hand, other two types of bearings possessing high damping were developed and have widely used in the seismic isolation practices [1, 3]. One is the lead rubber bearing (LRB), which additionally inserts lead plugs down the center of RB to enhance the hysteretic damping, and the other is high damping rubber bearing (HDRB), whose rubber material possesses high damping in order to supply more dissipating energy.

Following the same principle as Professor Milne used in his experiments, several authors conducted experimental studies on different bridge structures mounted on laminated rubber bearings. Kelly et al. [5] studied quarter-scale models of straight and skewed bridge decks mounted on plain and lead-filled elastomeric bearings subjected to earthquake ground motion using the shaking table. The deck response was compared to determine the effectiveness of mechanical energy dissipaters in base isolation systems and the mode of failure of base-isolated bridges. Igarashi and Iemura [6] evaluated the effects of implementing the lead-rubber bearing as seismic isolator on a highway bridge structure under seismic loads using the substructure hybrid loading (pseudo-dynamic) test method. The seismic response of the isolated bridge structure was successfully obtained. The effectiveness of isolation is examined based on acceleration and displacement amplifications using earthquake response results. All of their studies were related to observation of the isolation effects on the bridge structures. Very few works were undertaken in the past regarding the mechanical behavior of isolation bearings.

Mori et al. [7, 8] studied the behavior of laminated bearings with and without lead plug under shear and axial loads. They evaluated hysteretic parameters of the bearings: horizontal stiffness; vertical stiffness; and equivalent damping ratio. The similar study was conducted by Burstscher et al. [9]; Fujita et al. [10]; Mazda et al. [11] and Ohtori et al. [12] on lead, natural and high damping rubber bearings. They concluded from the experimental results that the hysteretic parameters have low loading rate-dependence. Furthermore, Robinson[13], a pioneer of developing and introducing the lead rubber bearing (LRB) as an excellent isolation system to be used in seismic design of civil engineering structures, conducted an elaborate experimental tests on LRB in order to describe the hysteretic behavior. From the experiments he concluded that the hysteretic behavior of the LRB can be expressed well by using a bilinear relationship of the force-displacements. In addition, he conducted some tests regarding fatigue and temperature performance of the LRB.

Several authors conducted different loading experiments on laminated rubber bearings (RB, LRB, and HDRB) in order to acquire deep understanding of the mechanical properties. In this case the works of Abe et al. [14]; Aiken et al. [15]; Kikuchi and Aiken [16]; Sano and D Pasquale [17] can be noted. They have applied uni-directional and bi-directional horizontal shear deformations with constant vertical compressive stress. Several types of laminated rubber bearings were used in their experimental scheme. In their investigations they identified some aspects of the bearings such as hardening features and dependence of the restoring forces on the maximum displacement amplitude experienced in the past. Moreover, some of them also identified coupling effects on the restoring forces of the bearings due to deformation in the two horizontal directions. Motivated by the experimental results of the bearings different forms of analytical models of the bearings were proposed by them. However, their studies were mostly related to illustrating the strain-rate independent mechanical behavior of the bearings.

Very few works are reported in literature regarding the strain-rate dependent behavior of the bearings. In this regard, the works of Dall'Asta and Ragni [18]; Hwang et al. [19] and Tsai et al. [20] can be reported. They studied the mechanical behavior of high damping rubber dissipating devices by conducting different experiments such as sinusoidal loading tests at different frequencies, simple cyclic shear tests at different strain-rates along with relaxation tests. From the experiments they have identified the strain-rate dependence of the restoring forces and subsequently developed rate-dependent analytical models of the bearings. Strain-rate dependent Mullin's softening was also identified in the experiments [18]. However, separation of the rate-dependent behavior from other mechanical behavior of the bearings was not elaborately addressed in their studies.

A number of experimental and numerical works on different rubber materials (HDR: high damping rubber and NR: natural rubber) have been performed in the past [21-27]. These works show that the mechanical properties of rubber materials (especially HDR) are dominated by the nonlinear rate-dependence including other inelastic behavior. Moreover, the different viscosity behavior in loading and unloading has been identified [21, 23, 28 and 29].

It is well known that since seismic response of base isolated structures greatly depends on mechanical properties of the bearings, deep understanding of the characteristics of the bearings under the desired conditions is very essential for rational and economic design of the seismic isolation system. The general mechanical behavior of laminated rubber bearings mainly concerns with nonlinear rate-dependent hysteretic property [18, 19] in addition to other inelastic behavior like Mullin's softening effect [30] and relaxation behavior [31]. All these characteristic behaviors generally originate from the molecular structures of the strength elements of rubber materials used in manufacturing of the bearings. Within this context, the chapter is devoted to discuss an experimental scheme used to characterize mechanical behavior and subsequently develop a mathematical model representing the characteristic behavior of the bearings.

2. Experimental observation

The objective of the work is to make qualitative and quantitative studies of strain-rate dependency of the bearings subjected to horizontal shear deformation with a constant vertical compressive load. To this end, an experimental scheme comprised of multi-step relaxation tests, cyclic shear tests, and simple relaxation tests was conducted: the mutli-step relaxation tests were carried out to investigate the strain-rate independent behavior along with viscosity behavior in loading and unloading; the cyclic shear tests, to observe the strain-rate dependency; and the simple relaxation tests, to describe the viscosity behavior of the bearings. To separate the Mullins' effect from other inelastic effects, a preloading sequence was applied on each specimen prior to the actual test. Details of the experiments and the inferences observed therein are described in the following subsections.

2.1. Specimens

The bearings manufactured and used in this study were divided into seven different types of specimens depending on their chemical compositions and damping properties: three types of high damping rubber bearings (HDR1, HDR2, and HDR3); two types of lead rubber bearings (LRB1, LRB2); and two types of natural rubber bearings (RB1, RB2). All the specimens had a square cross-sectional shape with external in-plane dimensions equal to 250 mm x 250 mm. The reinforcing steel plates had similarly a square planar geometry with external dimensions of 240 mm x 240 mm and thickness of 2.3 mm each. The geometry and material properties of these specimens are given in Table 1 and illustrated also in Figure 2. The dimensions of the test specimens were selected following the ISO standard [32]. Due to the space limitation the experimental results of three specimens, one HDRB (HDRB2), one RB (RB2) and one LRB (LRB2) are presented in the subsequent sections. However, the interested readers are requested to refer to the earlier efforts of the author [33-35] for better understanding to the rate-dependent mechanical behavior of the bearings. Figure 3 present typical actual bearing specimens used in the experiment in deformed and unreformed conditions.

2.2. Experimental set-up and loading conditions

A schematic detail of the experimental set-up is presented in Figure 1. The specimens were tested in a computer-controlled servo-hydraulic testing machine at room temperature (23 ^{0}C). Displacement controlled tests, under shear deformation with an average constant vertical compressive stress of 6 MPa, were carried out. This mode of deformation is regarded as the most relevant one for application in seismic isolation [2]. The displacement was applied along the top edge of the specimen and the force response was measured by two load cells. All data were recorded using a personal computer. Throughout this paper, to express the experimental results, the average shear stress and shear strain are calculated using the following two equations

$$\gamma = \frac{u}{h}, \qquad (1)$$

$$\tau = \frac{F_h}{A}, \qquad (2)$$

where u and F_h denote the relative horizontal displacement and applied force, respectively; h stands for the total thickness of rubber layers and A is the area of the cross section.

Specifications	High damping rubber bearing			Natural rubber bearing		Lead rubber bearing	
	HDR1	**HDR2**	HDR3	RB1	**RB2**	LRB1	**LRB2**
Cross-section (mm)	240X240			240X240		240X240	
Number of rubber layers	6			6		6	
Thickness of one rubber layer (mm)	5			5		5	
Thickness of one steel layer (mm)	2.3			2.3		2.3	
Diameter of lead plug (mm)	-			-		34.5	
No of lead plugs	-			-		4	
Nominal shear modulus (MPa)	1.2			1.2		1.2	

Table 1. Geometry and material properties of the bearings

2.2.1. Softening behavior

Virgin rubber typically exhibits a softening phenomenon, known as Mullins' effect in its first loading cycle. Due to presence of this typical behavior, the first cycle of a stress-strain curve differs significantly from the shape of the subsequent cycles [30]. In order to remove the Mullins softening behavior from other inelastic phenomena, all specimens were preloaded before the actual tests. The preloading was done by treating 11 cycles of sinusoidal loading at 1.75 strain and 0.05 Hz until a stable state of the stress-strain response is achieved, i.e., that no further softening occurs. The strain histories as applied in preloading sequence are shown in Figure 4.

Figures 5 (a), (b), and (c) present the typical shear stress-strain responses obtained from pre-loading tests on HDR2, RB2 and LRB2 specimens. The same loading sequence was applied on two types of specimens: virgin specimens and preloading specimens. The virgin specimen was loaded first with the prescribed loading sequence and then the same specimen (known as the preloading specimen) was again loaded with same loading sequence. The time interval between these two loading sequences was 30 min. The softening behavior in the first loading cycle is evident from Figure 5 in each specimen (virgin and preloading) indicating that the Mullins softening effect is not only present in the virgin specimen but also in any preloaded specimen [18]. This implies that Mullins effect can be recovered in quite a very short period. This is certainly due to the 'healing effect' [36, 37]. As can be seen from Figures 5 (a) to (c), the softening behavior is more appreciably well-defined in HDR2 than LRB2 and RB2. All the specimens have shown a repeatable stress-stretch response after passing through 4-5 loading cycles indicating that the Mullin's softening effect of the bearings is removed from the other effects.

2.2.2. Strain-rate dependent behavior

With a view to understanding the mechanical behavior of the bearings regarding the strain rate-dependence, cyclic shear tests (CS tests) at different strain rates were carried out. In the test series, a number of constant strain-rate cases within a range of 0.05/s to 5.5/s were considered as shown in Figure 6. Figures 7 (a), (b), and (c) show the strain-rate dependent shear stress responses of HDR2, RB2, and LRB2 bearings, respectively. For comparison, the equivalent stress responses of the bearings are also represented in each Figure. The equivalent stress responses of the bearings can be identified from MSR test results (Section 2.2.4).

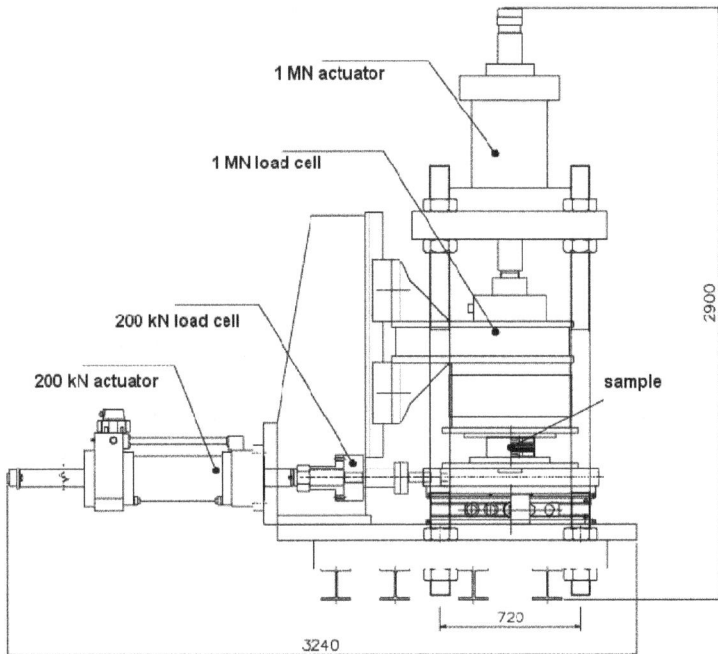

Figure 1. Schematic details of the experimental set-up. All dimensions are in mm

In general, the stress responses in the loading path contain three-characteristic features like a high initial stiffness feature at low strain levels followed by a traceable large flexibility at moderate strain levels as well as a large strain-hardening feature at the end. The untangling and/or the separation of weak bonds between filler particles and long chains are associated with reduced effect of the high initial stiffness [38]. This typical phenomenon is regarded the 'Payne effect' [25]. The final increase of the stiffness is attributed to the limited extensibility of the polymer chains and may be endowed to the 'strain hardening' feature [38]. When compared among the three bearings, the high initial stiffness at a low strain level and the high strain hardening at a high strain level are mostly prominent in HDR2 at a higher strain rates. However, a weaker strain-hardening feature in LRB2 than the other specimen (RB2) at higher strain levels is also noticeable. A comparison of hysteresis loops observed at different strain rates shows that the size of the hysteresis loops increases with increase of strain rates

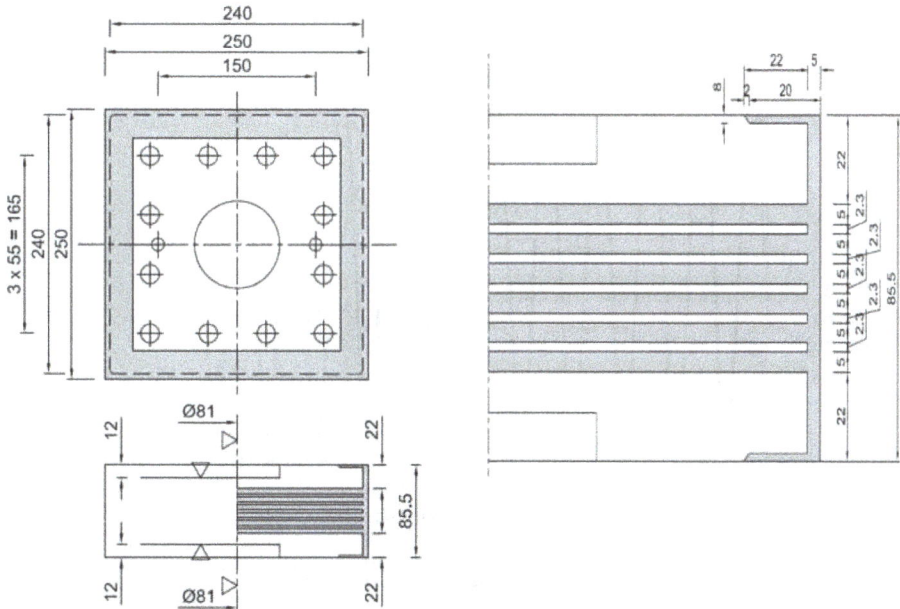

Figure 2. Dimension of a specimens [mm] (a) plan view (b) sectional view

(a)   (b)

Figure 3. Typical bearing specimens used in the experiment in (a) un-deformed and (b) deformed condition

as shown in Figures 7 (a) to (c). While comparing among all the bearings, HDR2 demonstrates a bigger hysteresis loop compared with the other bearings (RB2 and LRB2). This typical behavior can be attributed that the HDR2 inherits relatively high viscosity property than in other bearings. The addition of special chemical compounds in manufacturing process enhances the damping property of the HDRB.

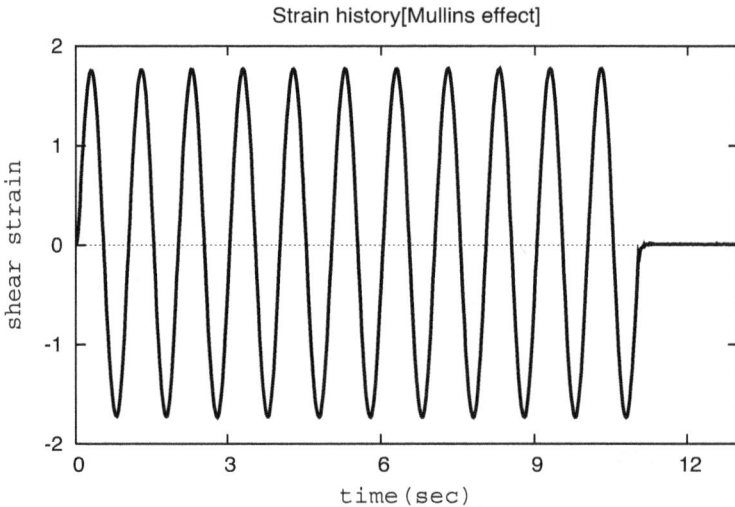

Figure 4. Applied strain histories in preloading test

Another comparison of the shear stress responses at different strain rates of the bearings shown in Figures 6(a) to (c) indicates that the strong strain-rate dependence exists in loading, whereas much weaker strain-rate dependence is observed in unloading. The different viscosity property in loading and unloading is attributed to this typical experimental observation. The basic strength elements of rubber are very long chain molecules, which are cross-linked with each other at some points to form a network [39]. Two types of linkages are occurred in rubber: physical linkages and chemical linkages. Due to the inherent properties of building up the physical and chemical linkages of rubber, the physical linkages are much weaker in stability and strength compared with the chemical linkages [40, 41]. The physical linkages have small energy capacity, which are easily broken; however, the chemical linkages have higher energy capacity, which require external energy to be broken. In loading at a particular strain rate, some of the physical and chemical linkages are broken, however, in unloading at the same strain rate; the breaking up the physical linkages is more prominent than the chemical linkages. These phenomena may be attributed to different viscosity behavior in loading and unloading of the bearings.

A further comparison among the loading-path responses at different strain-rates shows that the stresses increase due to viscosity with the increase of strain-rates. At higher strain rates, however, a diminishing trend in increase of stress responses is observed indicating an approach to the instantaneous state.

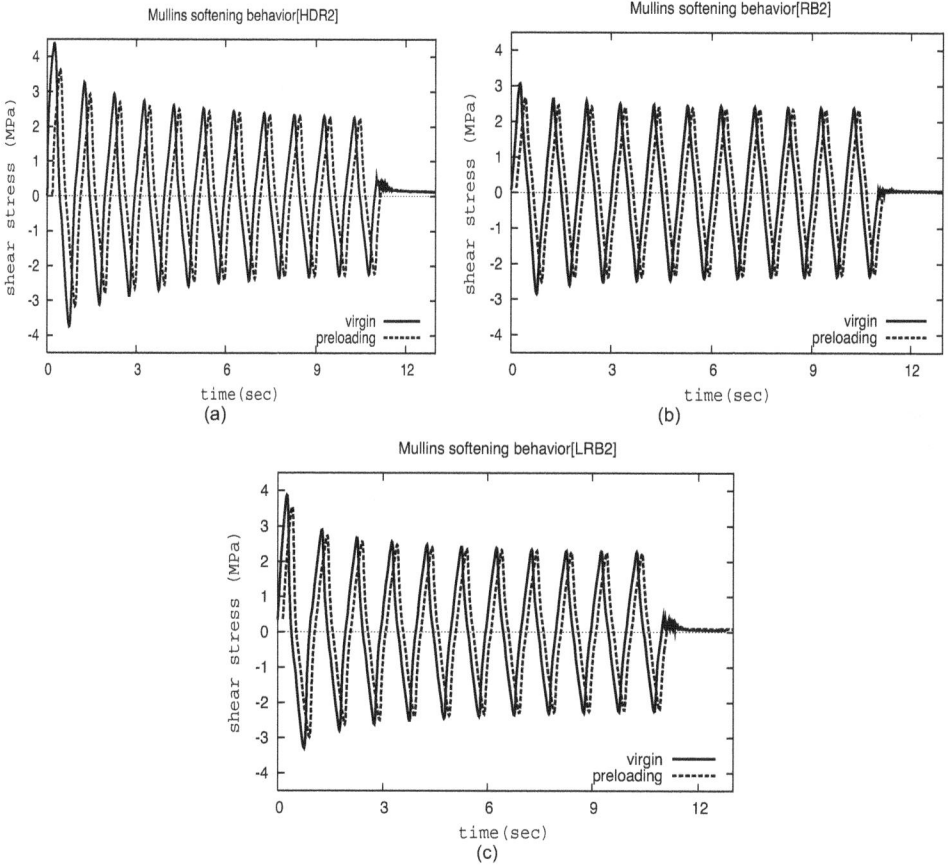

Figure 5. 11-cycle preloading test on the bearings to remove Mullins effect; (a) HDR2, (b) RB2, (c) LRB2; the legend indicates that the solid line in each Figure shows the shear stress histories obtained from the virgin specimens and the dotted line does for the preloading specimens. For clear illustration the shear stress-strain responses are separated by 0.15 sec from each other

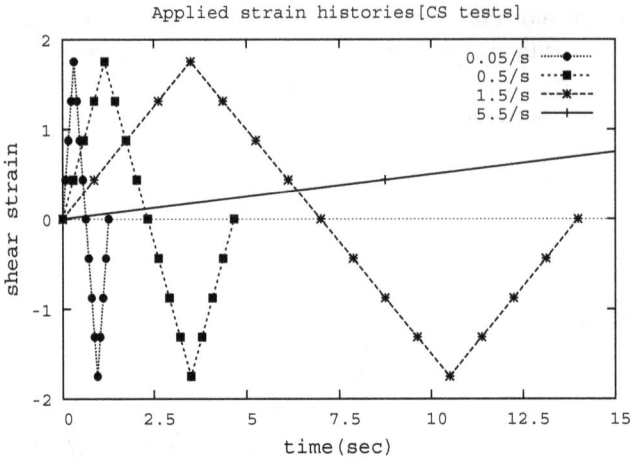

Figure 6. Applied strain histories in CS tests.

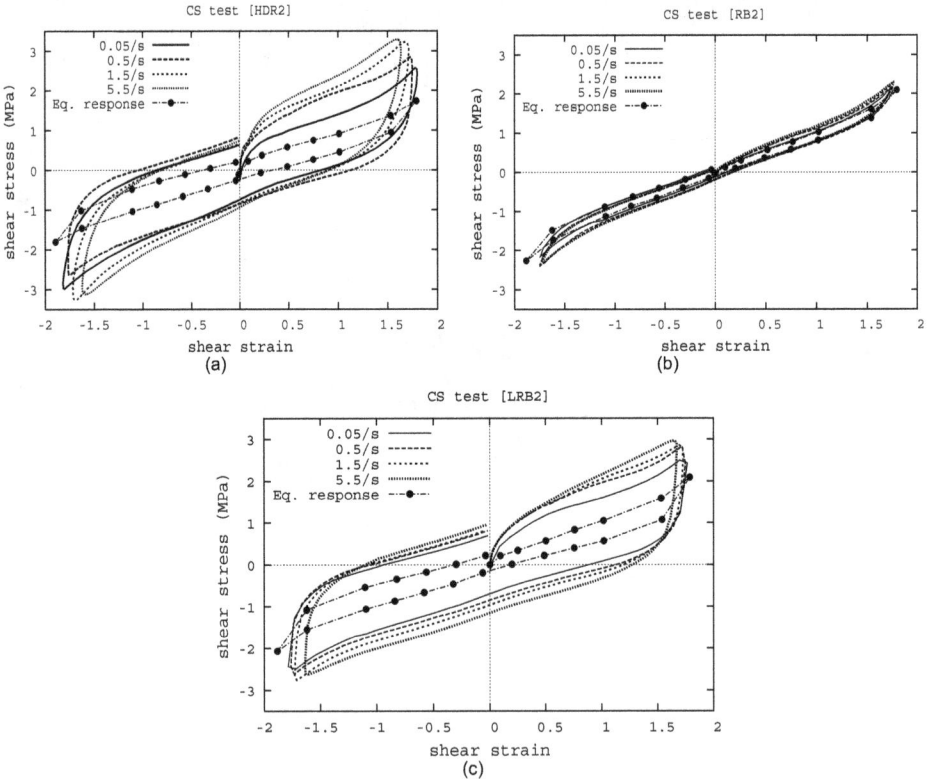

Figure 7. Shear stress-strain relationships obtained from CS tests at different strain rates of the bearings; (a) HDR2, (b) RB2, (c)LRB2; equilibrium response as obtained from MSR tests is also presented for clear comparison.

2.2.3. Viscosity behavior

The cyclic shear tests presented in Section 2.2.2 revealed the existence of viscosity in all specimens. In this regard, simple relaxation (SR) tests were carried out to study the viscosity behavior of the bearings. To this end, a series of SR tests at different strain levels were carried out. Figure 8 shows the strain histories of SR loading tests at three different strain levels of γ = 100, 150, and 175% with a strain rate of 5.5/s in loading and unloading. The relaxation period in loading and unloading was taken 30 min.

The shear stress histories of the bearings as obtained in SR tests are presented in Figures 9 (a) to (c). The stress relaxation histories in each specimen illustrate the time dependent viscosity behavior of the bearings. For all specimens, a rapid stress relaxation was displayed in the first few minutes; after while it approached asymptotically towards a converged state of responses. The stress relaxation was observed in each specimen. The amount of stress relaxation in loading and unloading of HDR2 was found to be much higher than those obtained in other bearings (RB2 and LRB2). As can be seen from Figures 9 (a) to (c), HDR2 shows comparatively high stress relaxation than the other bearings. On the other hand, RB2 shows much lower stress relaxation than that of other bearings. These observations confirm the findings of the cyclic shear loading tests and interpretations as mentioned in the preceding section (Figures 7 (a) to (c)). The stress response obtained at the end of the relaxation can be regarded as the equilibrium stress response in asymptotic sense [25, 42]. The deformation mechanisms associated with relaxation are related to the long chain molecular structure of the rubber. In the relaxation test, the initial sudden strain occurs more rapidly than the accumulation capacity of molecular structure of rubber. However, with the passage of time the molecules again rotate and unwind so that less stress is needed to maintain the same strain level.

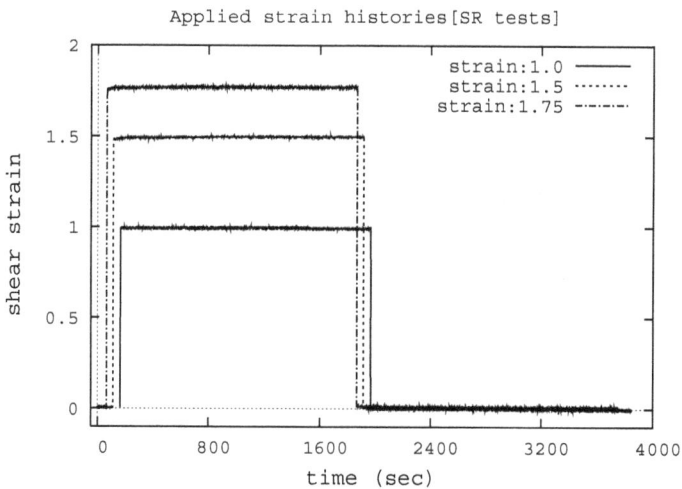

Figure 8. Applied strain histories in SR test.

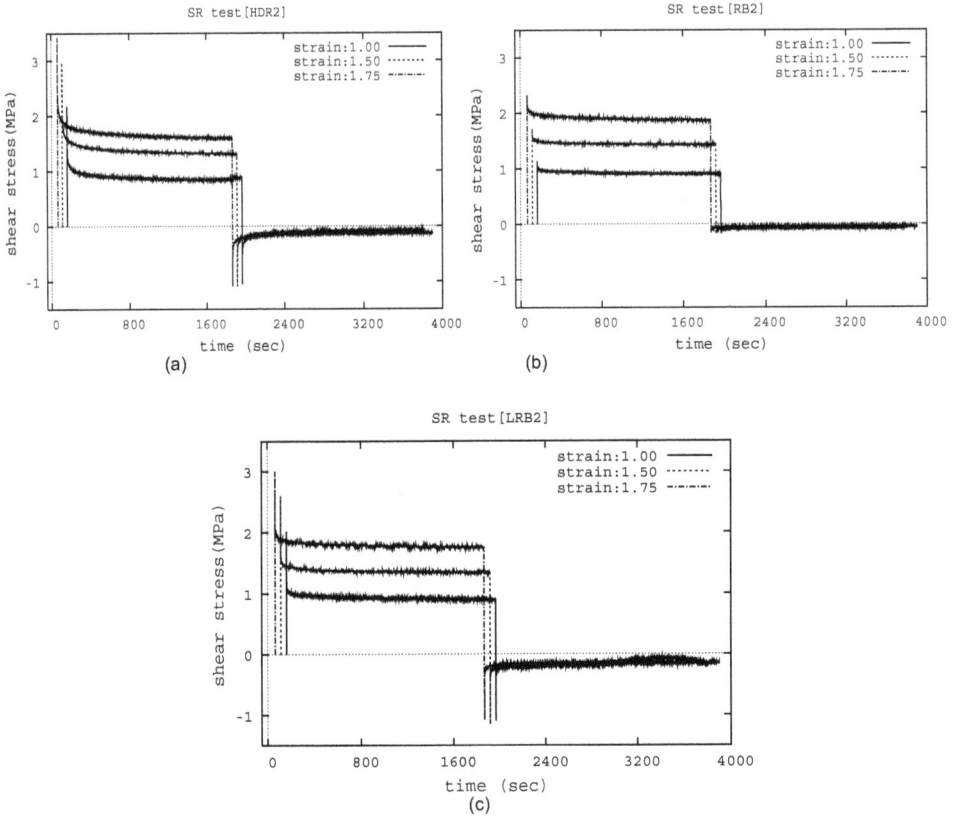

Figure 9. Shear stress histories obtained from SR tests of the bearings at different strain levels (a) HDR2, (b) RB2, (c) LRB2. For clear illustration, the stress histories have been separated by 50 sec to each other.

2.2.4. Static equilibrium hysteresis

The cyclic shear test results presented in Section 2.2.2 illustrated the strain-rate dependent property. The subsequent simple relaxation tests (Section 2.2.3) further explained the property. The tests carried out at different strain levels showed reduction in stress response during the hold time and approached the asymptotically converged states of responses (i.e equilibrium response). In this context, multi-step relaxation (MSR) tests were carried out to observe the relaxation behavior in loading and unloading paths and thereby to obtain the equilibrium hysteresis (e.g. strain-rate independent response) by removing the time-dependent effects.

The shear strain history applied in MSR test at 1.75 maximum strain level is presented in Figure 10, where a number of relaxation periods of 20 min during which the applied strain is held constant are inserted in loading and unloading at a constant strain rate of 5.5/s. Figures 11 to 13 illustrate the shear stress histories and corresponding equilibrium responses obtained in MSR tests of three bearings (HDR2, RB2, and LRB2). It is observed that at the end of each relaxation interval in loading and unloading paths, the stress history converges to an almost constant state in all specimens (Figures 11 to 13). The convergence of the stress responses is identified in an asymptotic sense [25]. The shear stress-strain relationships in the equilibrium state can be obtained by connecting all the asymptotically converged stress values at each strain level as shown in Figures11 (b), 12(b) and 13(b). The difference of the stress values between loading and unloading at a particular shear strain level corresponds to the equilibrium hysteresis, which can be easily visualized in Figures 11 (b), 12(b) and 13(b). This behavior may be attributed due to an irreversible slip process between fillers in the rubber microstructures [30, 43], which is the resulting phenomenon of breaking of rubber-filler bonds [36,37]. Using the stress history data of Figures11 (a), 12(a) and 13(a), the overstress can be estimated by subtracting the equilibrium stress response from the current stress response at a particular strain level.

While comparing the overstress for each specimen as shown in Figures 11 (a), 12(a) and 13(a), the overstress in loading period is seen higher than in unloading at a given strain level. The maximum overstress was observed in HDR2 while in RB2 it was the minimum one. This typical behavior of the bearings is seen comparable with CS test results (Figures 7 (a) to (c)).

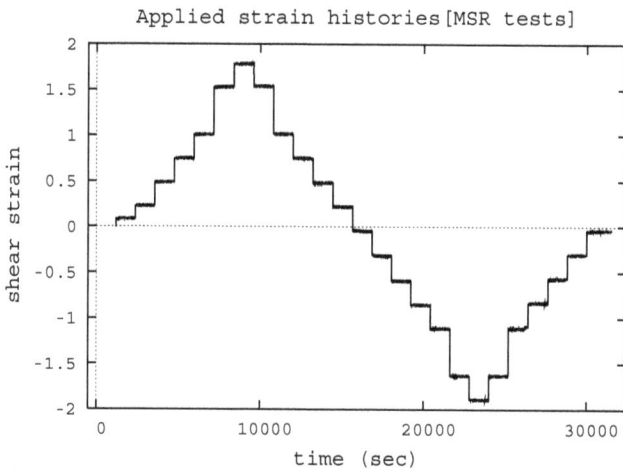

Figure 10. Applied strain histories in MSR test at 1.75 maximum strain level; a shear strain rate of 5.5/s was maintained at each strain step.

Furthermore, with a view to characterizing the strain hardening features along with dependence of the equilibrium hysteresis on loading history of the bearings, another set of multi-step relaxation tests were carried out at different maximum strain level of 2.5. Figures 14(a) and (b) present the results obtained in testing HDR2 due to the strain history of MSR test with maximum strain level of 2.5. Similar to the experiment carried out at the maximum strain level of 1.75, the equilibrium hysteresis effect is also observed in the MSR test; however, the magnitudes were found to increase with increasing strain level with increased supply of energy. Other sets of experiments similar to those in HDR2 were also carried out on other bearings. Figures 15 and 16 present the results on RB2 and LRB2. Although in

Figures 15 and 16 a trend similar to HDR2 in the appearance of equilibrium hysteresis was noticed, the magnitudes were found to differ from bearing to bearing. The comparison of the results indicates strong hardening features to be present at higher strain levels. Moreover, a strong dependence of the equilibrium hysteresis on the past maximum strain level was also appeared in the comparison. In addition, the equilibrium response was also found to be strongly dependent on the current strain values in all bearings.

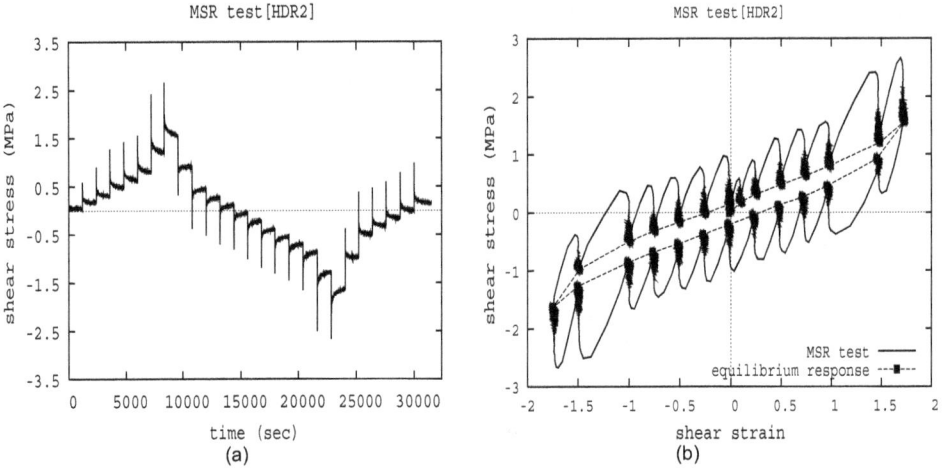

Figure 11. MSR test results of HDR2 (a) stress history (b) equilibrium stress response; equilibrium response at a particular strain level shows the response, which is asymptotically obtained from the shear stress histories of MSR test.

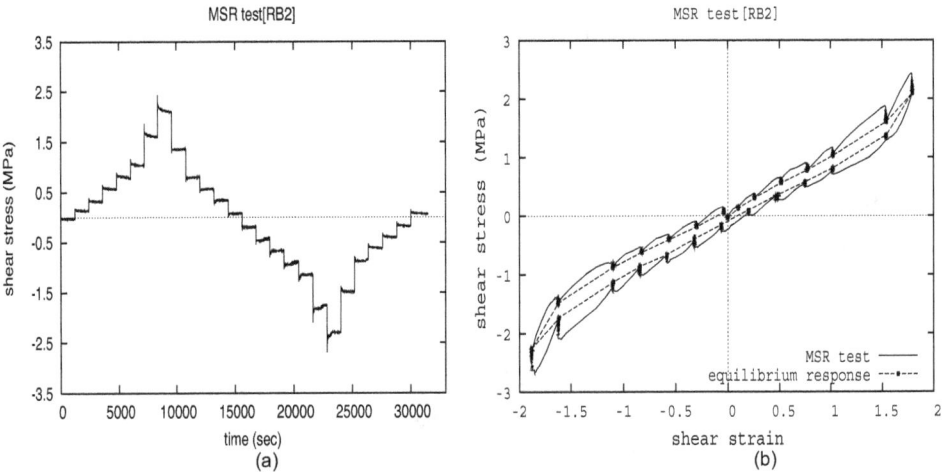

Figure 12. MSR test results of RB2 (a) stress history (b) equilibrium stress response; equilibrium response at a particular strain level shows the response, which is asymptotically obtained from the shear stress histories of MSR test.

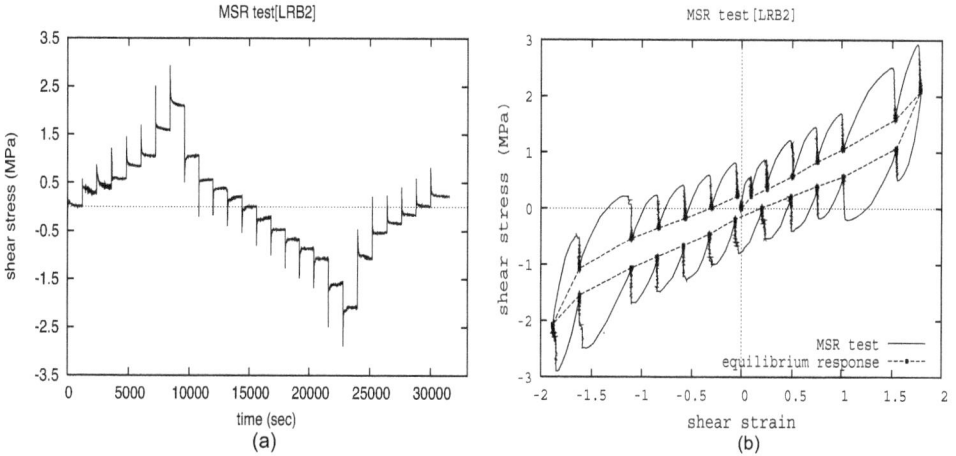

Figure 13. MSR test results of LRB2 (a) stress history (b) equilibrium stress response; equilibrium response at a particular strain level shows the response, which is asymptotically obtained from the shear stress histories of MSR tests.

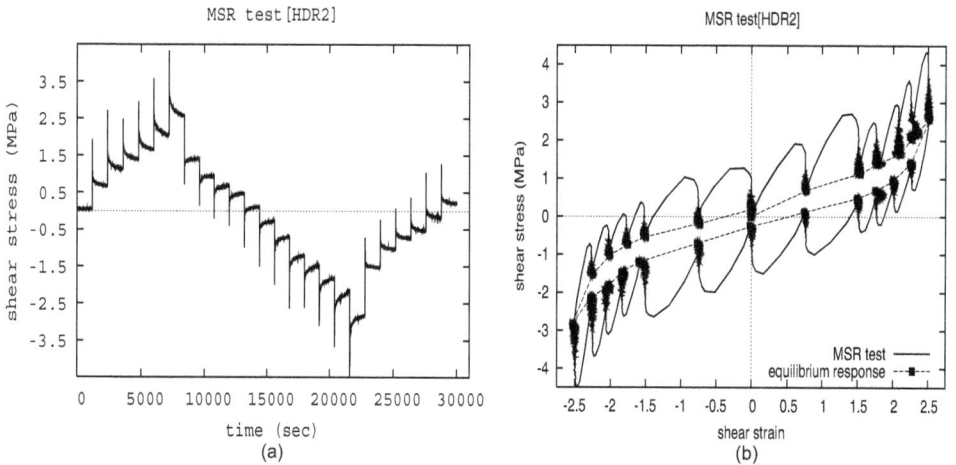

Figure 14. MSR test results of HDR2 at 2.50 maximum strain level (a) stress history (b) equilibrium stress response; equilibrium response at a particular strain level shows the response, which is asymptotically obtained from the shear stress histories of MSR test.

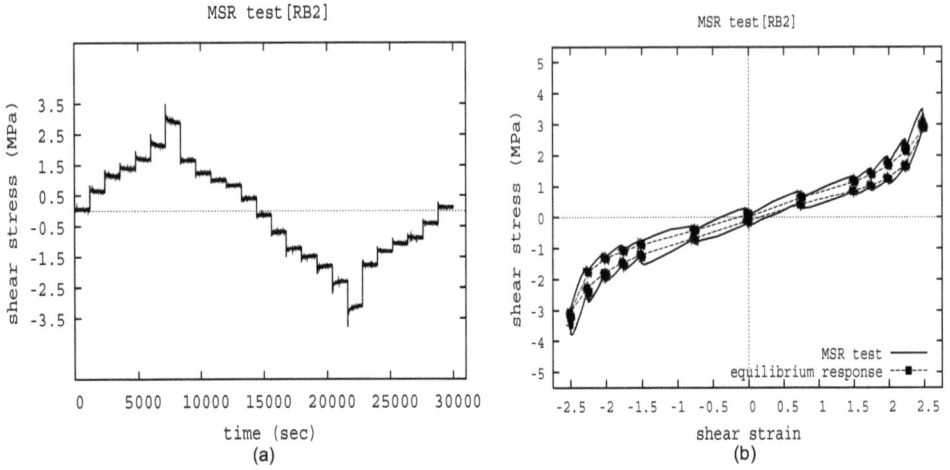

Figure 15. MSR test results of RB2 at 2.50 maximum strain level (a) stress history (b) equilibrium stress response; equilibrium response at a particular strain level shows the response, which is asymptotically obtained from the shear stress histories of MSR test.

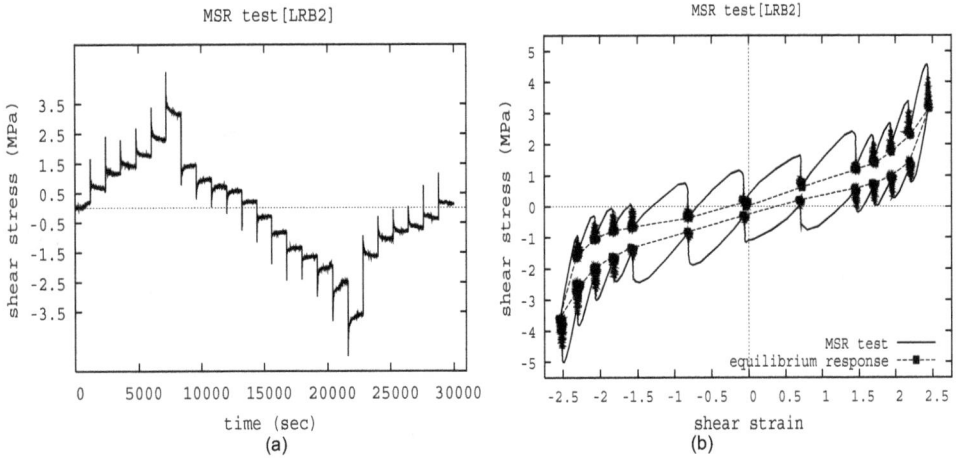

Figure 16. MSR test results of LRB2 at 2.50 maximum strain level (a) stress history (b) equilibrium stress response; equilibrium response at a particular strain level shows the response, which is asymptotically obtained from the shear stress histories of MSR test.

3. Structure of the rheology model

A rheology model for describing the three phenomenological effects of the bearings as mentioned above is constructed in this section. In Section 2, the Mullin's softening effect, strain rate viscosity effect, strain dependent elasto-plastic behavior with hardening effect of the bearings are illustrated. As pointed out earlier that all the experiments were conducted on preloaded specimens and hence the Mullin's effect of the bearing was not considered in deriving the rheology model. The underlying key approach of constructing the model is an additive decomposition of the total stress response into three contributions associated with a nonlinear elastic ground-stress, an elasto-plastic stress, and finally a viscosity induced overstress. This approach has been motivated by the experimental observations (Section 2) of the bearings [33-35]. The decomposition can be visualized for one dimensional analogy of the rheology model as depicted in Figure 17 (a) and (b).The model is the extended version of the Maxwell's model by adding two branches: one branch is the nonlinear elastic spring element and the other one is the elastoplastic spring–slider elements.

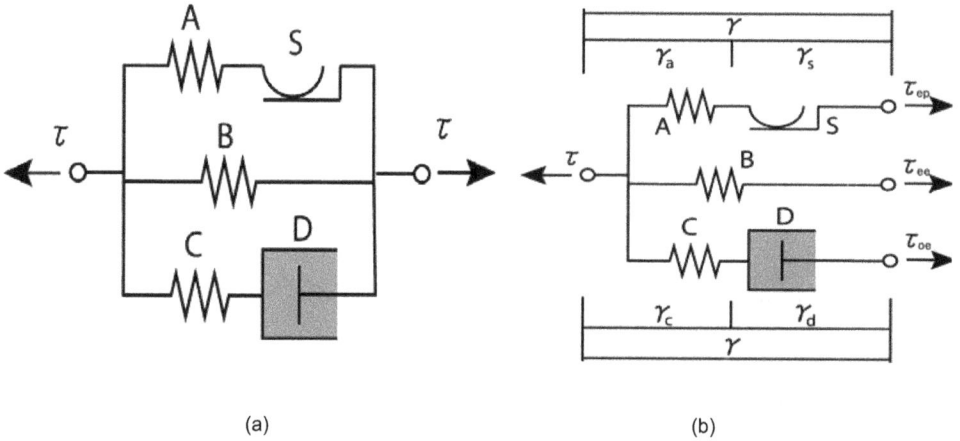

(a)   (b)

Figure 17. Structural configuration of the rheology model.

Motivated by the experimental observations, the mechanical behavior of the bearings can be also described as the sum of the two different behaviors: the rate-independent and the rate-dependent behaviors. The rate-independent behavior comprises the elasto-plastic and the nonlinear elastic response, which are represented in the top two branches of the model (Figures 17(a) and (b)). This phenomenon can be regarded as the equilibrium hysteresis to be identified from the relaxed equilibrium responses of the multi-step relaxations of the bearings. On the other hand, the rate-dependent response becomes very significant in relaxation and cyclic loading tests. The latter showed rate-dependent hysteresis loops where the size of the hysteresis increases with the increase of strain rates (Figures 7 (a) to (c)).

The total stress response of the bearing is motivated to decompose into three branches (Figure 17(b)):

$$\tau = \tau_{ep}\left(\gamma_a\right) + \tau_{ee}\left(\gamma\right) + \tau_{oe}\left(\gamma_c\right),$$

(3)

where τ_{ep} is the stress in the first branch composed of a spring (Element A) and a slider (Element S); τ_{ee} denotes the stress in the second branch with a spring (Element B); τ_{oe} represents the stress in the third branch comprising a spring (Element C) and a dashpot (Element D). The first and second branches represent the rate-independent elasto-plastic behavior, while the third branch introduces the rate-dependent viscosity behavior.

3.1. Modeling of equilibrium hysteresis

From the MSR test data (Figures 11 to 16), an equilibrium hysteresis loop with strain hardening is visible in each bearing. This equilibrium hysteresis loop can be suitably represented by combining the ideal elasto-plastic response (Figure 18a) and the nonlinear elastic response (Figure 18b).

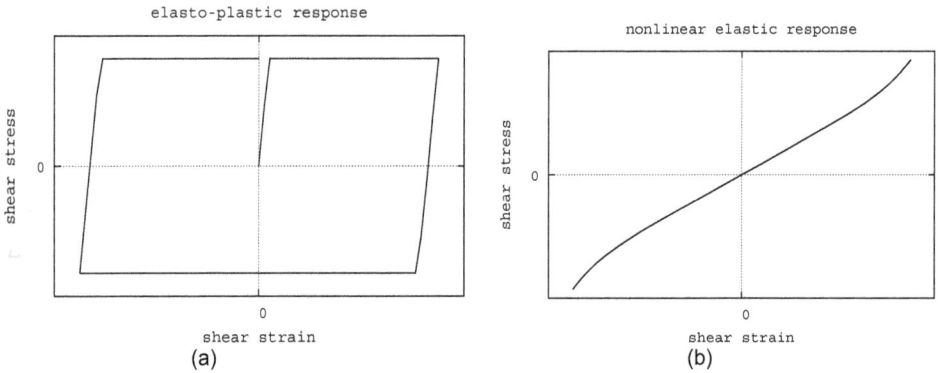

Figure 18. Formation of equilibrium hysteresis (a) elasto-plastic response (b) nonlinear elastic response.

The elasto-plastic response as shown in Figure 18(a) can be idealized by a spring-slider element as illustrated in Figure 19.

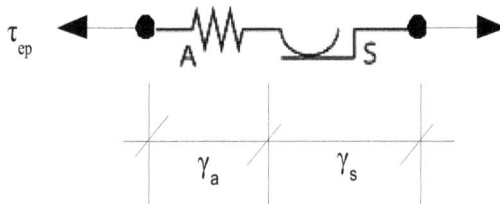

Figure 19. Spring-slider model for illustrating the rate-independent elasto-plasticity.

The total strain can be split into two components:

$$\gamma = \gamma_a + \gamma_s,\tag{4}$$

where γ_a stands for the strain on the spring (Element A), referred to as the *elastic part*, and γ_s is the strain on the slider(Element S), referred to as the *plastic part*.

From equilibrium consideration, the stress on the spring is τ_{ep}, and we have the elastic relationship

$$\tau_{ep} = C_1 \gamma_a ,$$ (5)

where C_1 is a spring constant for Element A.

The mechanical response of the slider (Element S) is characterized by the condition that the friction slider is active only when the stress level in the slider reaches a critical shear stress τ_{cr}, i.e., the stress τ_{ep} in the slider cannot be greater in absolute value than τ_{cr} which can be mathematically expressed as

$$\begin{cases} \dot{\gamma}_s \neq 0 & \text{for} \quad |\tau_{ep}| = \tau_{cr} \\ \dot{\gamma}_s = 0 & \text{for} \quad |\tau_{ep}| < \tau_{cr} \end{cases}$$ (6)

The evolution equation for the elastic strain γ_a can be written using the Eq.(5) as

$$\dot{\tau}_{ep} = \dot{\gamma} \left[U(\dot{\gamma}) U(\tau_{cr} - \tau_a) + U(-\dot{\gamma}) U(\tau_{cr} + \tau_a) \right]$$ (7)

with

$$U(x) = \begin{cases} 1 & : \quad x \geq 0 \\ 0 & : \quad x < 0 \end{cases}$$ (8)

The nonlinear elastic response as shown in Figure 18(b) with strain hardening at higher strain levels can be described by a non-Hookean nonlinear spring (Element B) (Figure 20):

$$\tau_{ee} = C_2 \gamma + C_3 |\gamma|^m \operatorname{sgn}(\gamma)$$ (9)

where C_2, C_3, and m ($m > 1$) are constants with

$$\operatorname{sgn}(x) = \begin{cases} +1 & : \quad x > 0 \\ 0 & : \quad x = 0 \\ -1 & : \quad x < 0 \end{cases}$$ (10)

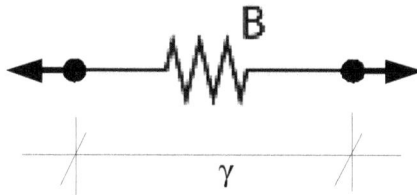

Figure 20. Non-Hookean spring element for illustrating the nonlinear elastic response.

Bearing /Pier	C_1 MPa	C_2 MPa	C_3 MPa	C_4 MPa	τ_{cr} MPa	m
HDR1	2.401	0.535	0.002	2.805	0.205	8.182
HDR2	2.502	0.653	0.006	3.254	0.247	6.621
HDR3	2.101	0.595	0.002	2.653	0.296	7.423
LRB1	4.252	0.710	0.003	2.354	0.190	8.421
LRB2	4.181	0.779	0.010	2.352	0.230	6.684
RB1	1.953	0.798	0.005	0.401	0.130	7.853
RB2	2.051	0.883	0.006	0.402	0.112	7.234

Table 2. Rate-independent response parameters of the bearings

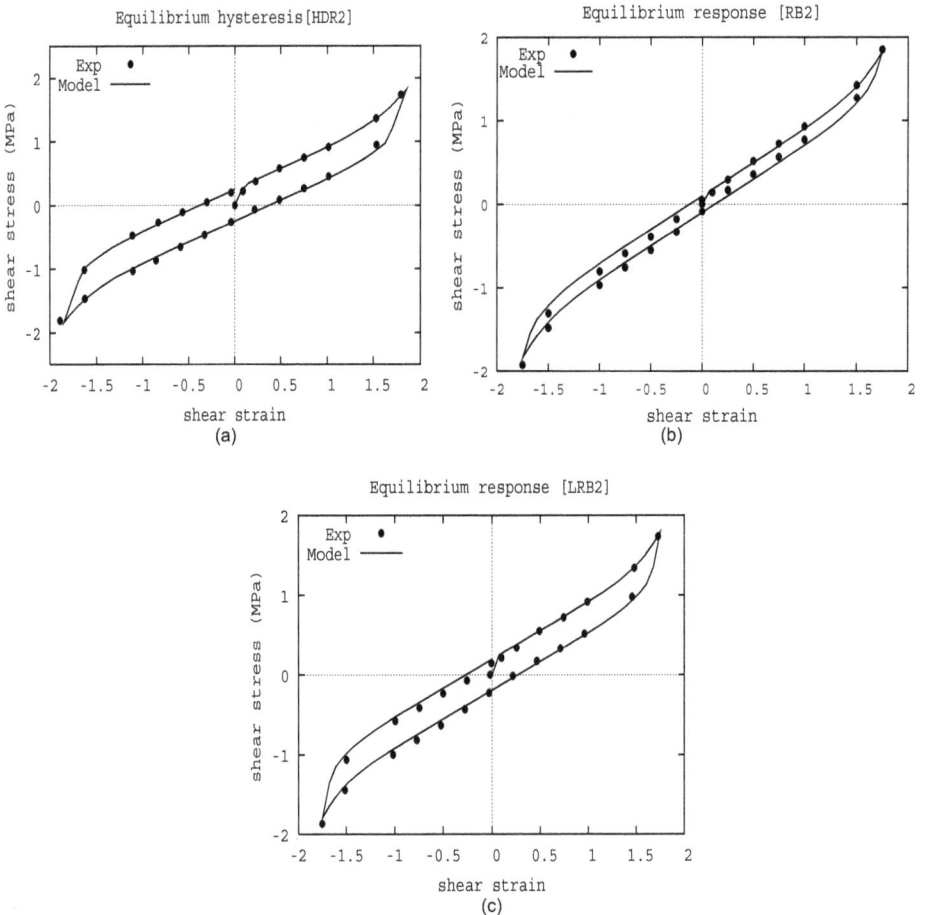

Figure 21. Identification of equilibrium response parameters for (a) HDR2, (b) RB2, and (c) LRB2; the experimental results are obtained from the MSR tests in asymptotic sense and the model results are determined using $\tau = \tau_{ee} + \tau_{ep}$ with parameters given in Table 2.

In order to determine the equilibrium response parameters as presented in Eqs. (5. 6 and 9), the equilibrium hysteresis loops as obtained from the MSR test have been considered. The equilibrium hysteresis loops of all bearings considered in the study are presented in Figures 21 (a) to (c). The experimental data are denoted by solid circular points. The critical shear, τ_{cr} is determined by using the equilibrium hysteresis loop. The difference between loading and unloading stresses in the equilibrium hysteresis loop at each strain level corresponds to $2\tau_{cr}$. Accordingly, τ_{cr} can be determined from the half of the arithmetic average values of the stress differences. The parameter C_1 corresponding to the initial stiffness can then be determined by fitting the initial part as well as the switching parts from loading and unloading in the equilibrium hysteresis loop (see, for example, Figure 18(a)). Finally, the parameters for the nonlinear spring (Element B) are identified. The subtraction of the stress τ_{ep} of Eq.(5) from the equilibrium stress response obtained from the MSR test gives the stress τ_{ee} corresponding to Eq.(9).Parameters C_2, C_3, and m are determined using a standard least square method. The obtained critical stresses τ_{cr} and the equilibrium response parameters C_2, C_3, and m for all specimens are given in Table 2. The equilibrium responses obtained using the proposed model and the identified parameters are presented in Figures 21 (a) to (c). The solid line in each Figure shows the equilibrium responses obtained by the rheology model.

3.2. Modeling of instantaneous response

At the instantaneous state, the structure of the rheology model can be reduced into the same model without the dashpot element (Element D), because the dashpot is fixed ($\dot{\gamma}_d = 0$) owing to infinitely high strain-rate loading. Consequently, the instantaneous response of the rheology model can be obtained by adding τ_{oe} without Element D and the responses obtained from the other two branches as shown in Figure 22.

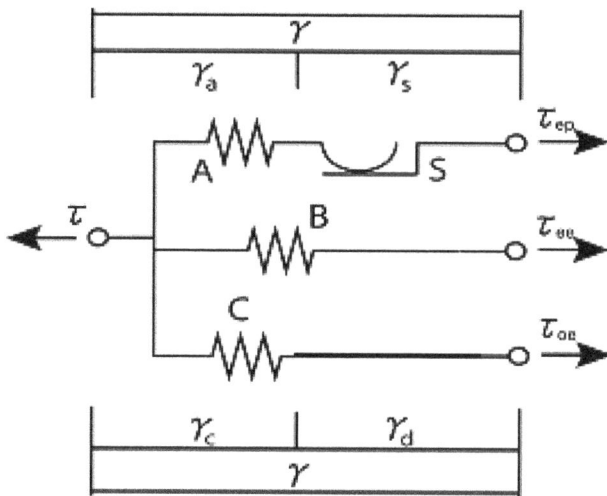

Figure 22. Spring-slider model for illustrating the instantaneous response.

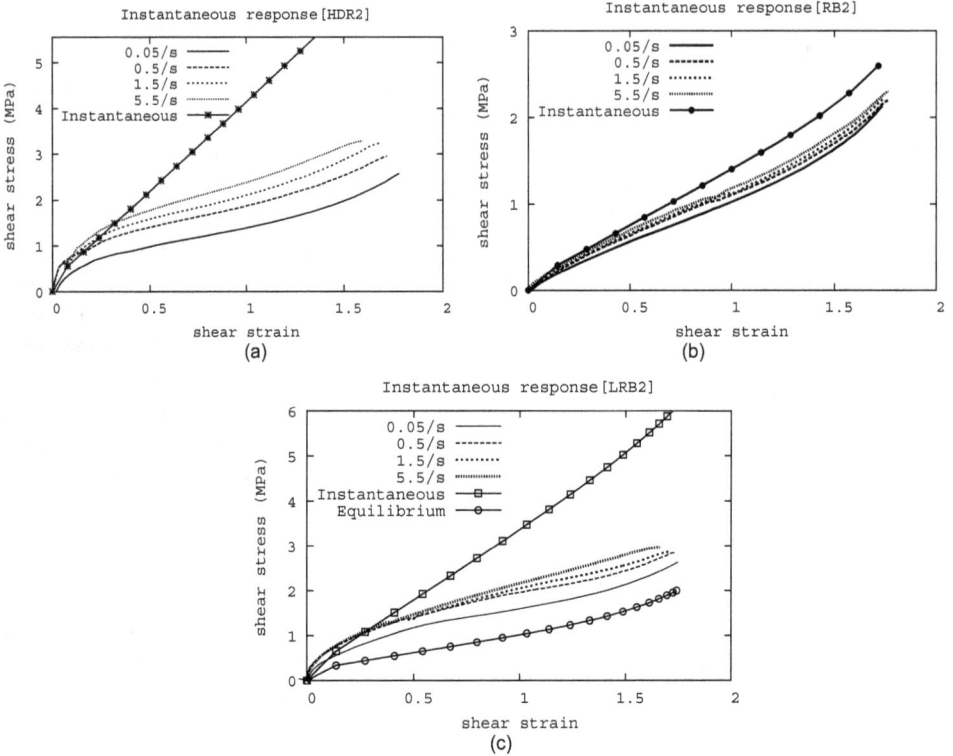

Figure 23. Identification of instantaneous response parameters for (a) HDR2, (b) RB2, and (c) LRB2; the instantaneous response is determined using the model $\tau = \tau_{ee} + \tau_{ep} + \tau_{oe}$(without dashpot element D) and the experimental results represented by different lines are obtained from CS tests at four strain rates of 0.05, 0.5, 1.5, and 5.5 /sec in loading regimes.

From the CS test results, a diminishing trend of the stress responses with increasing strain rates can be observed in all bearings as illustrated in Figures 7 (a) to (c). From these Figures, it has been observed that the instantaneous response lies at the neighborhood of the stress-strain curve at a strain rate of 5.5/s for the HDRB and the LRB; however for the RB, it is around the 1.5 /s strain rate. The instantaneous stress-strain curve, and accordingly the spring C seems to be nonlinear even in loading regime as clearly presented in Figures 23(a) to (c).

For simplicity, a linear spring model is employed for Element C in order to reproduce the instantaneous response of the bearings:

$$\tau_{oe} = C_4 \gamma_c, \tag{11}$$

where C_4 is the spring constant for Element C.

The parameter C_4 is determined so that the instantaneous stress-strain curve calculated from the rheology model ($\tau = \tau_{ee} + \tau_{ep} + \tau_{oe}$ (without the dashpot element)) can envelop the stress-strain curves obtained from the CS test. Figures 23(a) to (c) show comparison between the instantaneous stress-strain curves from the rheology model and those from the CS test at different strain rates up to 5.5/s in loading regime of all bearings. The obtained parameters C_4 for all bearings are listed in Table 2.

3.3. Modeling of nonlinear viscosity

Considering the third branch of the rheology model (Figure 17), the total strain can also be decomposed into two parts (Figure 24):

$$\gamma = \gamma_c + \gamma_d \tag{12}$$

where γ_c and γ_d stand for the strains in the spring (Element C) and the dashpot (Element D), respectively.

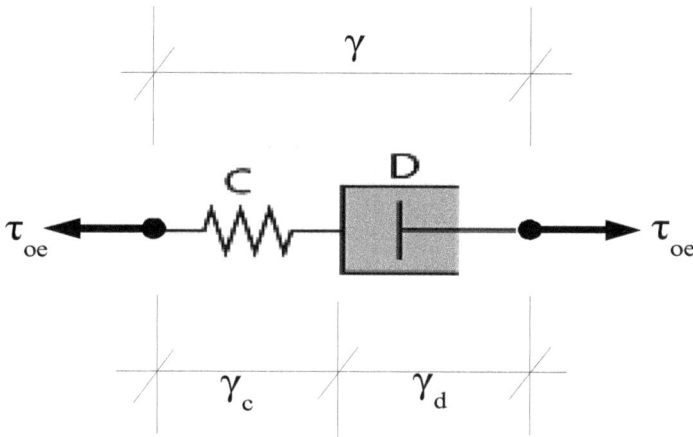

Figure 24. Spring-dashpot model for reproducing the rate-dependent stress response

The equilibrium condition of the stress components in the spring-dashpot elements states that the stress in the Element C must be equal to that in the Element D. The stress component in the Element C is expressed in Eq.(11). The evolution equation for the Element D has been constructed motivating by the experimental results of the bearings to be discussed in the following sub-sections. This section describes the procedure to identify the constitutive relationship of the dashpot (Element D) in the rheology model. To this end, the experimental results obtained from the MSR and the SR tests are analyzed to derive the relationship between the overstress τ_{oe} and the dashpot strain rate $\dot{\gamma}_d$. A schematic diagram to identify $\tau_{oe} - \dot{\gamma}_d$ relationship is presented in Figure 25.

From the stress relaxation results of the MSR and the SR loading tests, the time histories of the total stress τ and the total strain γ are obtained. Assuming that the asymptotic stress response at the end of each relaxation period is the equilibrium stress τ_{eq} at a particular strain level, the over stress history in each relaxation period is obtained by subtracting the equilibrium stress from the total stress. Then, the time history of the elastic strain for Element C is calculated from $\gamma_c = \tau_{oe}/C_4$ in Eq.(11), and consequently the time history of the dashpot strain can be determined as $\gamma_d = \gamma - \gamma_c$ using Eq. (12). In order to calculate the history of the dashpot strain rates, special treatment of the experimental data is required for taking the time derivatives over the experimental data points containing scattering due to noise. In order to reduce the scattering of experimental data, a moving averaging technique was employed in the current scheme before taking time derivatives of the experimental data points. All calculations were done by Mathematica [44].

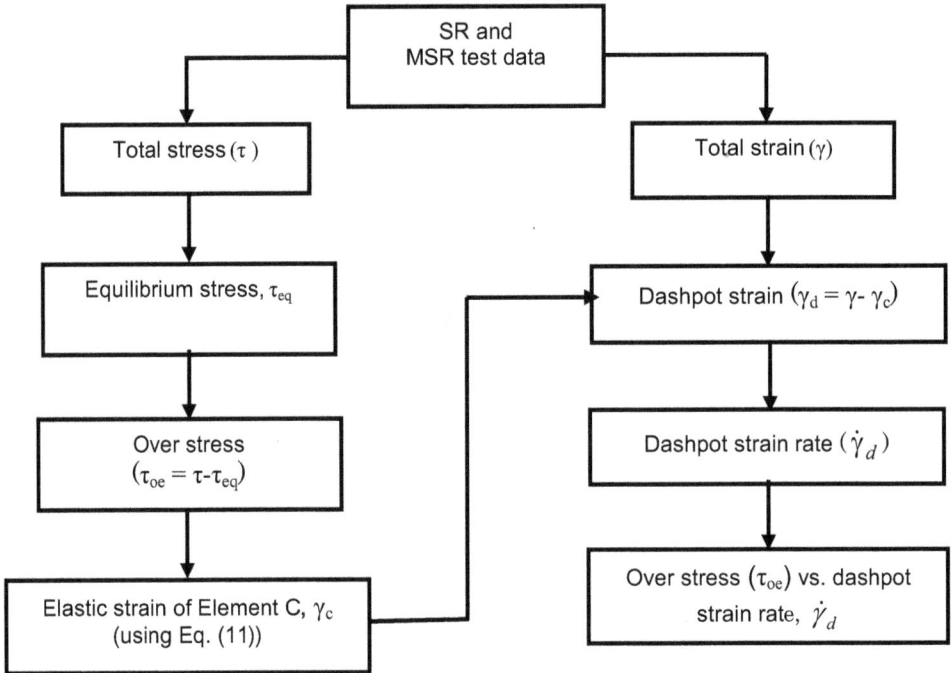

Figure 25. Schematic diagram to determine the analytical relationship between the over stress and the dashpot strain rates.

Figures 26(a) to (c) show the relationships between the overstress and the dashpot strain rates obtained from the MSR test results of the bearings (HDR2, RB2, and LRB2). In these Figures, the positive overstress indicates relaxation after loading, while the negative one does after unloading; (see Figure 10 for the strain histories of the MSR test). Figures 26 (a) to (c) demonstrate nonlinear dependence of the viscosity on the dashpot strain rates for all bearings. Since the gradient of $\tau_{oe} - \dot{\gamma}_d$ curves represents the viscosity, the viscosity decreases with increasing dashpot strain rates. Furthermore, it is found that these relationships depend on the strain levels in the relaxation tests after loading; i.e. the overstress, and accordingly the viscosity increases with increasing the total strains. It should be noted that the dependence of the over stress on the total strain level after unloading is not significant as seen that after loading.

The same tendency of the stress responses have been apparently observed in SR tests which are illustrated in Figures 27(a) to (c). In SR tests, the total strains were assigned from 0 to 100, 150, 175% for loading, and then the strains were reduced to 0 for unloading (see Figure 8 for the strain histories). The values in the legend stand for the total strains in respective relaxation processes, and 100, 150, 175% correspond to relaxation process after loading, and 0% after unloading. While compared among the three bearings regarding the magnitude of the overstress at each strain level, HDR2 shows comparatively high overstress than the other two bearings, which are in agreement with the results of the CS tests (Figures 7 (a) to (c)).

In order to describe the nonlinear viscosity of the dashpot, it is necessary to distinguish loading and unloading with respect to the dashpot. The loading and unloading conditions are defined for the dashpot as follows:

$$\frac{d}{dt}|\gamma_d| > 0 \text{ for loading and } \frac{d}{dt}|\gamma_d| < 0 \text{ for unloading} \tag{13}$$

This loading-unloading condition is identical with

$$\tau_{oe}\gamma_d > 0 \text{ for loading and } \tau_{oe}\gamma_d < 0 \text{ for unloading} \tag{14}$$

Based on the $\tau_{oe} - \dot{\gamma}_d$ relationships obtained form the MSR and the SR test data shown in Figures 26 and 27 for the bearings, the constitutive model for the Element D can be expressed as

$$\tau_{oe} = A_1 \exp(q|\gamma|) \text{sgn}(\dot{\gamma}_d) \left|\frac{\dot{\gamma}_d}{\dot{\gamma}_o}\right|^n \quad \text{for loading,}$$

$$\tau_{oe} = A_u \, \text{sgn}(\dot{\gamma}_d) \left|\frac{\dot{\gamma}_d}{\dot{\gamma}_o}\right|^n \qquad \text{for unloading,} \tag{15}$$

where $\dot{\gamma}_o = 1$ (sec^{-1}) is a reference strain rate of the dashpot; A_l, A_u, q and n are constants for nonlinear viscosity.

In SR and MSR tests, the loading/unloading condition changes abruptly (e.g. Figures 8 to 10). However, under general loading histories, the loading/unloading condition may change gradually. To avoid abrupt change in viscosity due to a shift in the loading and unloading conditions, a smooth function is introduced into the overstress expression, which facilitates the Eq.(15a,b) to be rewritten in a more compact form

$$\tau_{oe} = A \left| \frac{\dot{\gamma}_d}{\dot{\gamma}_o} \right|^n \text{sgn}(\dot{\gamma}_d)$$

with   (16)

$$A = \frac{1}{2}\left(A_1 \exp(q|\gamma|) + A_u\right) + \frac{1}{2}\left(A_1 \exp(q|\gamma|) - A_u\right)\tanh(\xi\tau_{oe}\gamma_d)$$

where ξ is the smoothing parameter to switch viscosity between loading and unloading. Now, in the subsequent paragraphs, the procedure for determining the viscosity constants (A_1, A_u, q and n) will be discussed followed by the smoothing parameter (ξ).

Bearing /Pier	A_1 MPa	A_u MPa	q	n	ξ
HDR1	0.501	0.904	0.532	0.205	1.221
HDR2	0.982	0.952	0.344	0.224	1.252
HDR3	0.754	0.753	0.353	0.213	1.242
LRB1	0.731	0.731	0.0	0.272	0.0
LRB2	0.792	0.792	0.0	0.302	0.0
RB1	0.552	0.552	0.0	0.232	0.0
RB2	0.434	0.434	0.0	0.243	0.0

Table 3. Rate-dependent viscosity parameters of the HDRB

Using the strain histories of the SR loading tests at different strain levels (Figure 8), the overstress-dashpot strain rates relationships are determined (Figures 27 (a) to (c)), which correspond to Eq.16 for both loading and unloading conditions. A standard method of nonlinear regression analysis is employed in Eq. 16 to identify the viscosity constants. As motivated by the relationships of the overstress-dashpot strain rates obtained in the SR/MSR test results, the value of n is kept the same in loading and unloading conditions. The nonlinear viscosity parameters obtained in this way are presented in Table 3. Figures 27(a) to (c) present the overstress-dashpot strain rates relationships obtained using the proposed model and the SR test results; the solid lines show the model results and the points do for the experimental data.

A sinusoidal loading history is utilized to determine the smoothing parameter of the model (Eq.16). The sinusoidal loading history corresponds to a horizontal shear displacement history applied at the top of the bearing at a frequency of 0.5 Hz with amplitude of 1.75. An optimization method based on Gauss-Newton algorithm [45] is employed to determine the smoothing parameter. The optimization problem is mathematically defined as minimizing the error function presented as

$$\text{Minimize}\left\{ E(\xi,t) = \sum_{n=1}^{N}\left(\tau_{exp,n} - \tau_{m,n}\right)^2 \right\},$$ (17)

where N represents the number of data points of interest, $\tau_{exp,n}$ and $\tau_{m,n}$ imply the shear stress responses at time t_n obtained from the experiment and the model, respectively, and ξ stands for the parameter to be identified. Using the Gauss-Newton algorithm, following condition is satisfied for obtaining the minimum error function as

$$\nabla E\left(\xi_i, t\right) + \nabla^2 E\left(\xi_i, t\right)\left(\xi_{i+1} - \xi_i\right) = 0, \tag{18}$$

where ∇ refers the gradient operator. During the iteration process, the updated parameter in each iteration is determined using

$$\xi_{i+1} = \xi_i - \delta \frac{\nabla E\left(\xi_i, t\right)}{\nabla^2 E\left(\xi_i, t\right)}, \tag{19}$$

where δ is the numerical coefficient between 0 and 1 to satisfy the Wolfe conditions at each step of the iteration. The values of ξ determined in this way for the bearings are presented in Table 3.

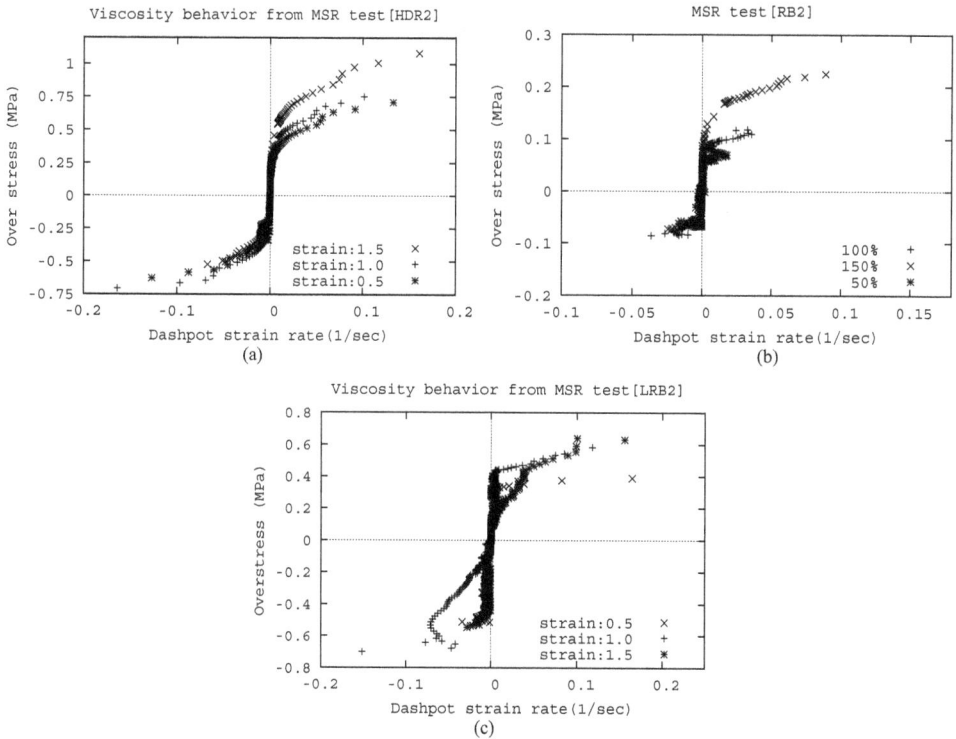

Figure 26. Overstress-dashpot strain rate relation obtained from the MSR tests at different strain levels in loading and unloading regimes of (a) HDR2, (b) RB2, and (c) LRB2; the values in the legend stand for the total strain in respective relaxation processes, and 50,100, 150% correspond to relaxation processes after loading and unloading.

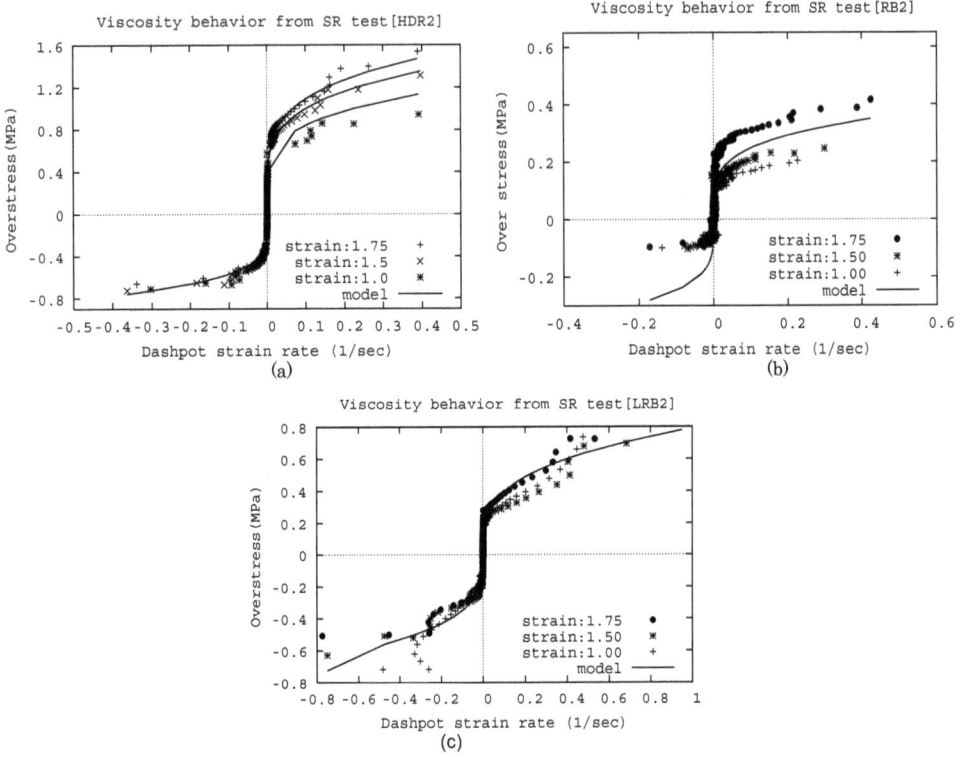

Figure 27. Identification of viscosity parameters of (a) HDR2, (b) RB2, and (c) LRB2; the model results represented by solid lines are obtained by $\tau = \tau_{oe}$ with parameters given in Table 3 and the relations between $\tau_{oe} - \dot\gamma_d$ as calculated from SR test data are shown by points. The values in the legend stand for the total strain in respective relaxation processes, and 100,150, 175% correspond to relaxation processes after loading, and 0% after unloading.

4. Thermodynamic consistency of the rheology model

The Clausisus-Duhem inequality is a way of expressing the second law of thermodynamics used in continuum mechanics. This inequality is particularly useful in determining whether the given constitutive relations of material/solid are thermodynamically compatible [46]. This inequality is a statement concerning the irreversibility of natural resources, especially when energy dissipation is involved. The compatibility with the Clausisus-Duhem inequality is also known as the thermodynamic consistency of solid. This consistency implies that constitutive relations of solids are formulated so that the rate of the specific entropy production is non-negative for arbitrary temperature and deformation processes.

In this context, the Clausius-Duhem inequality reads

$$-\rho\dot{\psi} + \tau\dot{\gamma} \geq 0,$$ (21)

where ρ is the mass density, ψ is the Helmholtz free energy per unit mass, and $\tau\dot{\gamma}$ is the stress power per unit volume. It states that the supplied stress power has to be equal or greater than the time rate of the Helmholtz free energy. For the rheology model, the Helmholtz free energy is the mechanical energy stored in the three springs shown in Figure 17 can be represented as

$$\rho\psi(\gamma,\gamma_a,\gamma_c) = \frac{1}{2}C_1\gamma_a^2 + \frac{1}{2}C_2\gamma^2 + \frac{C_3}{m+1}|\gamma|^{m+1} + \frac{1}{2}C_4\gamma_c^2$$ (22)

Its time-rate reads as follows

$$\rho\dot{\psi}(\gamma,\gamma_a,\gamma_c) = C_1\gamma_a\dot{\gamma}_a + C_2\gamma\dot{\gamma} + C_3|\gamma|^m\dot{\gamma} + C_4\gamma_c\dot{\gamma}_c$$ (23)

The stress power of the model is

$$\tau\dot{\gamma} = \left(\tau_{ee} + \tau_{ep} + \tau_{oe}\right)\dot{\gamma} = \tau_{ee}\dot{\gamma} + \tau_{ep}\left(\dot{\gamma}_a + \dot{\gamma}_s\right) + \tau_{oe}\left(\dot{\gamma}_c + \dot{\gamma}_d\right)$$ (24)

Inserting the stress power (Eq.24) and the time-rate of the Helmholtz free energy (Eq.23) into the 2nd law of thermodynamics (Eq. 21) and rearranging the terms leads to the following expression (Eq.25)

$$\left(\tau_{ep} - C_1\gamma_a\right)\dot{\gamma}_a + \left(\tau_{ee} - C_2\gamma - C_3|\gamma|^m\right)\dot{\gamma} + \left(\tau_{oe} - C_4\gamma_c\right)\dot{\gamma}_c + \tau_{ep}\dot{\gamma}_s + \tau_{oe}\dot{\gamma}_d \geq 0$$ (25)

In order to satisfy this inequality for arbitrary values of the strain-rates of the variables in the free energy, the following equations for the three stress components which correspond to Eqs.(5), (9) ,and (11), respectively, yield

$$\tau_{ep} = C_1\gamma_a, \tau_{ee} = C_2\gamma + C_3|\gamma|^m \text{sgn}(\gamma), \text{ and } \tau_{oe} = C_4\gamma_c$$ (26)

The residual inequality to be satisfied is

$$\tau_{ep}\dot{\gamma}_s + \tau_{oe}\dot{\gamma}_d \geq 0$$ (27)

It states that the inelastic stress-powers belonging to the two dissipative elements (Element S and Element D) have to be non-negative for arbitrary deformation processes. Assuming that the time-derivatives of the inelastic deformations are the same sign as the corresponding stress quantities, each of the product terms of Eq. (27) is ensured to be non-negative with parameters given in Table 3. The non-negative dissipation energy of the bearings is ensured only when all the parameters responsible for expressing the elasto-plastic stress (τ_{ep}) and the viscosity induced overstress (τ_{oe}) are non-negative. The parameters of Table 3 have confirmed this condition.

5. Summary

This chapter discusses an experimental scheme to characterize the mechanical behavior of three types of bearings and subsequently demonstrates the modeling approaches of the stress responses identified from the experiments. The mechanical tests conducted under horizontal shear displacement along with a constant vertical compressive load demonstrated the existence of Mullins' softening effect in all the bearing specimens. However, with the passage of time a recovery of the softening effect was observed. A preloading sequence had been applied before actual tests were carried out to remove the Mullins' effect from other inelastic phenomena. Cyclic shear tests carried out at different strain rates gave an image of the significant strain-rate dependent hysteresis property. The strain-rate dependent property in the loading paths was appeared to be reasonably stronger than in the unloading paths. The simple and multi-step relaxation tests at different strain levels were carried out to investigate the viscosity property in the loading and unloading paths of the bearings. Moreover, in order to identify the equilibrium hysteresis, the multi-step relaxation tests were carried out with different maximum strain levels. The dependence of the equilibrium hysteresis on the experienced maximum strain and the current strain levels was clearly demonstrated in the test results.

The mechanical tests indicated the presence of strain-rate dependent hysteresis with high strain hardening features at high strain levels in the HDRB. In the other bearing specimens, strain-rate dependent phenomenon was seen less prominent; however, the strain hardening features at high strain levels in the RB showed more significant than any other bearings. In this context, an elasto-plastic model was proposed for describing strain hardening features along with equilibrium hysteresis of the RB, LRB and HDRB. The performance of the proposed model in representing the strain rate-independent responses of the bearings was evaluated. In order to model the strain-rate dependent hysteresis observed in the experiments, an evolution equation based on viscosity induced overstress was proposed for the bearings. In doing so, the Maxwell's dashpot-spring model was employed in which a nonlinear viscosity law is incorporated. The nonlinear viscosity law of the bearings was deduced from the experimental results of MSR and SR loading tests. The performance of the proposed evolution equation in representing the rate-dependent responses of the bearings was evaluated using the relaxation loading tests. On the basis of the physical interpretation of the strain-rate dependent hysteresis along with other inelastic properties observed in the bearings, a chronological method comprising of experimentation and computation was proposed to identify the constitutive parameters of the model. The strain-rate independent equilibrium response of the bearing was identified using the multi-step relaxation tests. After identifying this response, the elasto-plastic model (the top two branches of the rheology model) was used to find out the parameters for the elasto-plastic response. A series of cyclic shear tests were utilized to estimate the strain-rate independent instantaneous response of the bearings. A linear elastic spring element along with the elasto-plastic rheology model (the rheology model without the dashpot element) was used to determine the parameters of the instantaneous responses. After determining the elasto-plastic parameters of the bearings, the proposed evolution equation based on the viscosity induced overstress was used to find out the viscosity parameters by comparing the simple relaxation test data. Moreover, a mathematical equation involving smoothing function was proposed to establish loading and unloading conditions of the overstress; and sinusoidal loading data was then used to estimate the smoothing parameter of the overstress. Finally, the thermodynamic compatibility was confirmed by expressing the rheology model by using the Clausius-Duhem inequality equation.

Author details

A. R. Bhuiyan
Department of Civil Engineering, Chittagong University of Engineering and Technology, Chittagong, Bangladesh

Y.Okui
Department of Civil and Environmental Engineering, Saitama University, Saitama, Japan

Acknowledgement

The experimental works were conducted by utilizing the laboratory facilities and bearings-specimens provided by Rubber Bearing Association, Japan. The authors indeed gratefully acknowledge the kind cooperation extended by them. The authors also sincerely acknowledge the funding provided by the Japanese Ministry of Education, Science, Sports and Culture (MEXT) as Grant-in-Aid for scientific research to carry out this research work.

6. References

[1] Skinner R I, Robinson W H, and McVerry G H. An Introduction to Seismic Isolation. DSIR Physical Science, Wellington, New Zealand; 1993.

[2] Kelly J M. Earthquake Resistant Design with Rubber. 2nd edition, Springer-Verlag Berlin Heidelberg, New York;1997.

[3] Naeim F and Kelly J. Design of Seismic Isolated Structures. 1st edition, John Wiley and Sons, New York; 1996.

[4] Priestley M J N, Seible F and Calvi G M. Seismic Design and Retrofit of Bridges. John Wiley and Sons, New York; 1996.

[5] Kelly J M Buckle I G. and Tsai H C. Earthquake Simulator Testing of a Base-Isolated Bridge Deck. UCB/EERC/ 85-09; 1985.

[6] Igarashi A and Iemura H. Experimental and Analytical Evaluation of Seismic Performance of Highway Bridges with Base Isolation Bearings. Proceedings of the 9th World Conference on Earthquake Engineering, Tokyo, Japan, Paper No. 553; 1996.

[7] Mori A, Carr A J, Cooke N and Moss P J. Compression Behavior of Bridge Bearings used for Seismic Isolation. Engineering Structures 1996; 18 351-362.

[8] Mori A, Moss P J, Cooke N, and Carr A J. The Behavior of Bearings used for Seismic Isolation under Shear and Axial Load. Earthquake Spectra 1999; 15 199-224.

[9] Burstscher S, Dorfmann A, and Bergmeister K. Mechanical Aspects of High Damping Rubber. Proceedings of the 2nd International PhD Symposium in Civil Engineering, Budapest, Hungary; 1998.

[10] Fujita T, Suzuki S and Fujita S. Hysteretic Restoring Force Characteristics of High Damping Rubber Bearings for Seismic Isolation. Proceedings of ASME PVP Conference, PVP, 181, 35-41; 1989.

[11] Mazda T, Shiojiri H, Oka Y, Fujita T and Seki M. Test on Large-Scale Seismic Isolation Elements. Transactions of the 10th International Conference on SMiRT-K, 678-685; 1989.

[12] Ohtori Y, Ishida K and Mazda T. Dynamic Characteristics of Lead Rubber Bearings with Dynamic Two Dimensional Test Equipment. ASME Seismic Engineering, PVP Conference, 2,145-153; 1994.

[13] Robinson W H. Lead Rubber Hysteresis Bearings Suitable for Protecting Structures during Earthquakes. Earthquake Engineering and Structural Dynamics 1982; 10 593-604.

[14] Abe M, Yoshida J and Fujino Y. Multiaxial Behaviors of Laminated Rubber Bearings and their Modeling. I: Experimental Study. Journal of Structural Engineering 2004; 130 1119-1132.

[15] Aiken I D, Kelly J M and Clark P W. Experimental Studies of The Mechanical Characteristics of Three Types of Seismic Isolation Bearings. Proceedings of the 10th World Conference of Earthquake Engineering (WCEE), Madrid, Spain, 2280-2286; 1992.

[16] Kikuchi M and Aiken I D. An Analytical Hysteresis Model for Elastomeric Seismic Isolation Bearings. Earthquake Engineering and Structural Dynamics 1997; 26 215-231.

[17] Sano T, Di and Pasquale G. A Constitutive Model for High Damping Rubber Bearings. Journal of Pressure Vessel Technology 1995; 117 53-57.

[18] Dall'Asta A and Ragni L. 2006. Experimental Tests and Analytical Model of High Damping Rubber Dissipating Devices. Engineering Structures 1995; 28 1874-1884.

[19] Hwang J S, Wu J D, Pan T C, and Yang G. A Mathematical Hysteretic Model for Elastomeric Isolation Bearings. Earthquake Engineering and Structural Dynamics 2002; 31 771-789.

[20] Tsai C S, Chiang Tsu-Cheng Chen Bo-Jen band Lin Shih-Bin. An Advanced Analytical Model for High Damping Rubber Bearings. Earthquake Engineering and Structural Dynamics 2003; 32 1373-1387.

[21] Amin A F M S, Alam M S and Okui Y. An Improved Hyperelasticity Relation in Modeling Viscoelasticity Response of Natural and High Damping Rubbers in Compression: Experiments, Parameter Identification and Numerical Verification. Mechanics of Materials 2002; 34 75-95.

[22] Amin A F M S, Wiraguna S I, Bhuiyan, A R and Okui Y. Hyperelasticity Model for Finite Element Analysis of Natural and High Damping Rubbers in Compression and Shear. Journal of Engineering Mechanics 2006; 132 54-64.

[23] Bergstrom J S and Boyce M C. Constitutive Modeling of the Large Strain Time-Dependent Behavior of Elastomers. Journal of the Mechanics and Physics of Solids 1998; 46 931-954.

[24] Haupt P and Sedlan K. Viscoplasticity of Elastomeric Materials: Experimental Facts and Constitutive Modeling, Archive of Applied Mechanics 2001; 71 89-109.

[25] Lion A. A Constitutive Model for Carbon Black Filled Rubber: Experimental Investigations and Mathematical Representation. Continuum Mechanics and Thermodynamics 1996; 8 153-169.

[26] Miehe C and Keck J. Superimposed Finite Elastic-Viscoplastic-Plasto-Elastic Stress Response with Damage in Filled Rubbery Polymers. Experiments, Modeling and Algorithmic Implementation. Journal of Mechanics and Physics of Solids 2000; 48 323-365.

[27] Spathis G. and Kontou E. Modeling of Nonlinear Viscoelasticity at Large Deformations. Journal of Material Science 2008; 43 2046-2052.

[28] Amin A F M S, Lion A, Sekita S and Okui Y. Nonlinear Dependence of Viscosity in Modeling the Rate-Dependent Response of Natural and High Damping Rubbers in Compression and Shear: Experimental Identification and Numerical Verification. International Journal of Plasticity 2006; 22 1610-1657.

[29] Bergstrom J S and Boyce M C. Large Strain Time-Dependent Behavior of Filled Elastomers. Mechanics of Materials 2000; 32 627-644.

[30] Mullins L. Softening of Rubber by Deformation. Rubber Chemistry and Technology 1969; 42 339-362.

[31] Gent A N. Relaxation Processes in Vulcanized Rubber I: Relation among Stress Relaxation, Creep, Recovery and Hysteresis. Journal of Applied Polymer Science 1962; 6 433-441.

[32] International Organization of Standardization (ISO). Elastomeric Seismic-Protection Isolators, Part 1: Test methods, Geneva, Switzerland; 2005.

[33] Bhuiyan A R. Rheology Modeling of Laminated Rubber Bearings for Seismic Analysis. PhD thesis. Saitama University, Saitama, Japan; 2009

[34] Bhuiyan A R, Okui Y, Mitamura H and Imai T. A Rheology Model of High Damping Rubber Bearings for Seismic Analysis: Identification of Nonlinear Viscosity. International Journal of Solids and Structures 2009; 46 1778-1792.

[35] Imai T , Bhuiyan A R, Razzaq M K, Okui Y and Mitamura H. Experimental Studies of Rate-Dependent Mechanical Behavior of Laminated Rubber Bearings. Joint Conference Proceedings of 7th International Conference on Urban Earthquake Engineering (7CUEE) & 5th International Conference on Earthquake Engineering (5ICEE), March 3-5, 2010, Tokyo Institute of Technology, Tokyo, Japan; 2010.

[36] Bueche F. Moelcular Basis for the Mullins Effect. Journal of Applied Polymer Science 1960; 4 107-114.

[37] Bueche F. Mullins Effect and Rubber-Filler Interaction. Journal of Applied Polymer Science 1961; 5 271-281.

[38] Burtscher S and Dorfmann A. Compression and Shear Tests of Anisotropic High Damping Rubber Bearings. Engineering Structures 2004; 26 1979-1991.

[39] Treloar L R G. The Physics of Rubber Elasticity. 3rd edition, Oxford University Press; 1975.

[40] Besdo D and Ihlemann J. Properties of Rubber Like Materials under Large Deformations Explained By Self-Organizing Linkage Patterns. International Journal of Plasticity 2003; 19 1001-1018.

[41] Ihlemann J. Modeling of Inelastic Rubber Behavior under Large Deformations Based on Self-Organizing Linkage Patterns. Constitutive Models for Rubber, Balkema, Rotterdam; 1999.

[42] Lion A. A Physically Based Method to Represent the Thermo-Mechanical Behavior of Elastomers. Acta Mechanica 1997; 123 1-25.

[43] Kilian H G, Strauss M and Hamm W. Universal Properties in Filler-Loaded Rubbers. Rubber Chemistry and Technology 1994; 67 1-16.

[44] Wolfram Research Inc. Mathematica Version 5.2, USA; 2005.

[45] Venkataraman P. Applied Optimization with Matlab Programming. John Wiley and Sons, New York; 2002.

[46] Truesdell C. The Mechanical Foundations of Elasticity and Fluid Dynamics. Journal of Rational Mechanics and Analysis 1952; 1 125–300.

Chapter 12

Finite Element Analysis of Cable-Stayed Bridges with Appropriate Initial Shapes Under Seismic Excitations Focusing on Deck-Stay Interaction

Ming-Yi Liu and Pao-Hsii Wang

Additional information is available at the end of the chapter

1. Introduction

In the last several decades, cable-stayed bridges have become popular due to their aesthetic appeal, structural efficiency, ease of construction and economic advantage. This type of bridge, however, is light and flexible, and has a low level of inherent damping. Consequently, they are susceptible to ambient excitations from seismic, wind and traffic loads. Since the geometric and dynamic properties of the bridges as well as the characteristics of the excitations are complex, it is necessary to fully understand the mechanism of the interaction among the structural components with reasonable bridge shapes, which is used to provide the essential information to accurately calculate the dynamic responses of the bridges under the complicated excitations.

In the previous studies of bridge dynamics, the responses of a cable-stayed bridge can be categorized into global, local and coupled modes [1]. The global modes are primarily dominated by the deformations of the deck-tower system with the quasi-static motions of the stay cables; the local modes predominantly consist of the stay cable motions with negligible deformations of the deck-tower system; the coupled modes have substantial contributions from both the deck-tower system and stay cables. Since the towers are usually designed with a high rigidity to obtain an adequate efficiency of the system, the significant tower deformations do not occur in the lower modes sensitive to the ambient excitations [2]. Consequently, the coupled modes are considered to be dominated by the deck-stay interaction, while the contribution from the towers can be neglected. Numerical approaches based on the finite element method have been widely used to investigate the deck-stay interaction. The finite-element models of a cable-stayed bridge can be classified into two categories [1]: the one-element cable system (OECS), in which each stay cable is represented by a single cable element, and the multi-element cable system (MECS), in which each stay cable is discretized into multiple cable elements.

The deck-stay interaction has attracted much attention, because it not only significantly complicates both the natural frequency and mode shape characteristics of a cable-stayed bridge, but also potentially results in the large-amplitude stay cable vibrations even under the low-level deck oscillations. In the previous literature, the deck-stay interaction is due to the linear coupling (primary resonance) [3-8, 11] or the nonlinear coupling (secondary resonance), which can be further categorized into the subharmonic resonance of order 1/2 (two-to-one resonance) [3-9] and the superharmonic resonance of order 2 (one-to-two resonance) [6, 9, 10]. The primary, two-to-one and one-to-two resonances individually result in the fact that the global modes induce the direct, parametric and angle variation excitations of the local modes. Two types of simplified models: the single cable with moving anchorage [5-7] and the cable-supported cantilever beam [3, 4, 8-11], have been presented to theoretically investigate the deck-stay interaction. To extend the results of the simplified models, the OECS and MECS models of full cable-stayed bridges based on the finite element method have been widely used to explore such coupled phenomena of real structures [1, 11-16]. By focusing on the analytical and numerical study of the linear coupling, the localization factor was introduced to reveal the frequency veering phenomenon and to evaluate the mode hybridization level of a cable-stayed bridge [11]. On the basis of this research, the ambient vibration measurements were conducted to investigate the deck-stay interaction. It was suggested that the nonlinear coupling is not consistent with the measurement data. In contrast, the linear coupling is recognized as the critical excitation source of the coupled modes [16].

In parallel to the previous work [11, 16], the authors of the present paper also studied the deck-stay interaction of cable-stayed bridges based on the analytical and numerical methods as well as the long-term comprehensive full-scale measurements [17]. The measurement data indicated that the deck oscillations of small to moderate amplitudes are coupled with the large-amplitude stay cable vibrations due to the linear coupling between these two components. An analytical model of the single cable with spring-mass oscillator was presented to explain such mechanism attributed to the frequency loci veering and mode localization. Furthermore, the "pure" deck modes, "pure" cable modes and coupled modes are successfully captured by the proposed model. These phenomena are verified by the numerical simulations of the OECS and MECS models of a full cable-stayed bridge. The concepts of the indices for quantitatively assessing the degree of coupling among the structural components were also appeared in this research.

It is important to investigate the deck-stay interaction with the appropriate initial shape of a cable-stayed bridge. This is because such initial shape not only reasonably provides the geometric configuration as well as the prestress distribution of the bridge under the weight of the deck-tower system and the pretension forces in the stay cables, but also definitely ensures the satisfaction of the relations for the equilibrium conditions, boundary conditions and architectural design requirements [18-21]. The computational procedures for the initial shape analyses of the OECS and MECS models were presented for this reason [22, 23]. However, few researchers have studied the deck-stay interaction with the initial shape effect.

The objective of this study is to fully understand the mechanism of the deck-stay interaction with the appropriate initial shapes of cable-stayed bridges. Based on the smooth and convergent bridge shapes obtained by the initial shape analysis [22, 23], the OECS and MECS models of the Kao Ping Hsi Bridge in southern Taiwan are developed to verify the applicability of the analytical model and numerical formulation from the field observations [17]. For this purpose, the modal analyses of the two finite element models are conducted to calculate the natural frequency and normalized mode shape of the individual modes of the bridge. The modal coupling assessment is also performed to obtain the generalized mass ratios among the structural components for each mode of the bridge [24]. To further investigate the deck-stay interaction characteristics of cable-stayed bridges under earthquake excitations, the dynamic displacements and internal forces of the two finite element models are calculated based on the seismic analyses. These results can be used to provide a variety of viewpoints to illustrate the mechanism of the deck-stay interaction with the appropriate initial shapes of cable-stayed bridges.

2. Finite element formulation

On the basis of the finite element concepts, a cable-stayed bridge can be considered as an assembly of a finite number of cable elements for the stay cables and beam-column elements for both the decks and towers. Several assumptions are adopted in this study: the material is homogeneous and isotropic; the stress-strain relationship of the material remains within the linear elastic range during the whole nonlinear response; the external forces are displacement independent; large displacements and large rotations are allowed, but strains are small; each stay cable is fixed to both the deck and tower at their joints of attachment. Based on the system equations with the consideration of geometric nonlinearities, the initial shape analysis, modal analysis, modal coupling assessment and seismic analysis of cable-stayed bridges are conducted in this study.

2.1. Geometric nonlinearities

To reasonably simulate cable-stayed bridges, three types of geometric nonlinearities: the cable sag, beam-column and large displacement effects, are considered in this study.

A stay cable will sag into a catenary shape due to its weight and tensile force. Such cable sag effect has to be taken into consideration when the stay cable is represented by a single straight cable element. A stay cable with tensile stiffness is assumed to be perfectly elastic. The compressive, shear and bending stiffnesses of the stay cable are negligible. The cable sag nonlinearity can be simulated based on the equivalent modulus of elasticity of the stay cable [25]

$$E_{eq} = \cfrac{E_c}{1 + \cfrac{(wl_c)^2 A_c E_c}{12T^3}}, \tag{1}$$

where E_c, A_c and l_c are the effective modulus of elasticity, the cross-sectional area and the horizontal projected length of the stay cable, respectively; w is the weight of the stay cable per unit length; T is the tension in the stay cable. The stiffness matrix of a cable element in Figure 1 can be expressed as

$$KE_{jk} = \begin{cases} \left[\dfrac{E_{eq} A_c}{L_c} \right], & u_1 > 0 \\[4mm] \left[0 \right], & u_1 \le 0 \end{cases} \tag{2}$$

where u_1 is the element coordinate for the relative axial deformation; L_c is the chord length of the stay cable.

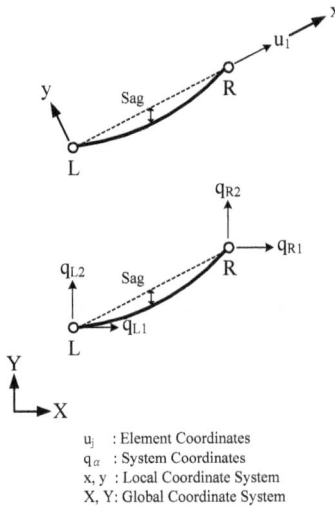

u_j   : Element Coordinates
q_α   : System Coordinates
x, y : Local Coordinate System
X, Y: Global Coordinate System

Figure 1. Cable element for simulating the stiffness of each stay cable.

High pretension forces in the stay cables can result in large compressive forces in the deck-tower system of a cable-stayed bridge. For this reason, the beam-column effect between such compressive forces and bending moments has to be considered when beam-column elements are used to simulate both the decks and towers. For a beam-column element based on the Euler-Bernoulli beam theory in Figure 2, shear strains of the element are neglected. u_1, u_2 and u_3 are the element coordinates for the left end rotation, the right end rotation and the relative axial deformation, respectively. The stiffness matrix of the beam-column element can be written as

$$KE_{jk} = \frac{E_b I_b}{L_b} \begin{bmatrix} C_s & C_t & 0 \\ C_t & C_s & 0 \\ 0 & 0 & R_t A_b / I_b \end{bmatrix}, \tag{3}$$

where E_b, A_b, I_b and L_b are the modulus of elasticity, the cross-sectional area, the moment of inertia and the length of the beam-column element, respectively; C_s, C_t and R_t are the stability functions representing the interaction between the axial and bending stiffnesses of the beam-column element [26].

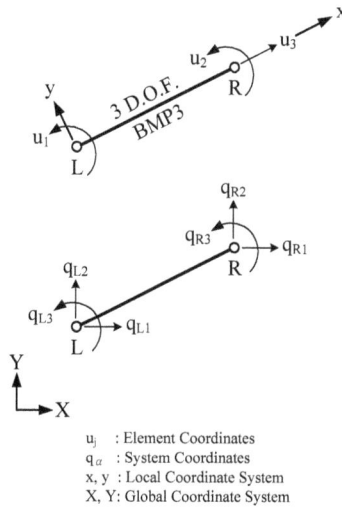

Figure 2. Beam-column element for simulating the stiffness of each deck and tower.

In general, large displacements occur in the deck-tower system due to the large span and less weight of a cable-stayed bridge. Such effect has to be taken into consideration when the equilibrium equations are derived from the deformed position. Under these conditions, the element coordinate u_j can be expressed as a nonlinear function of the system coordinate q_α in both Figure 1 and Figure 2, i.e., $u_j = u_j(q_\alpha)$. By differentiating u_j with respect to q_α, the first-order and second-order coordinate transformation coefficients can be individually written as

$$a_{j\alpha} = \frac{\partial u_j}{\partial q_\alpha}, \tag{4}$$

$$a_{j\alpha,\beta} = \frac{\partial a_{j\alpha}}{\partial q_\beta} = \frac{\partial^2 u_j}{\partial q_\alpha \partial q_\beta}. \tag{5}$$

$a_{j\alpha}$ and $a_{j\alpha,\beta}$ for the stiffness matrices of the cable and beam-column elements can be found in [18], which are provided to develop the tangent system stiffness matrix in Chapter 2.2.

In addition to the element stiffness matrices, the element mass matrices are introduced to fully understand the essential properties of a cable-stayed bridge. Based on the consistent mass model, the mass distribution of each stay cable and that of each deck and tower can be simulated by a cable element and a beam-column element, respectively. The mass matrix of the former with four element coordinates u_j $(j=1-4)$ in Figure 3 and that of the latter with six element coordinates u_j $(j=1-6)$ in Figure 4 can be individually expressed as

$$ME_{jk} = \frac{\rho_c A_c L_c}{6} \begin{bmatrix} 2 & 0 & 1 & 0 \\ 0 & 2 & 0 & 1 \\ 1 & 0 & 2 & 0 \\ 0 & 1 & 0 & 2 \end{bmatrix}, \tag{6}$$

$$ME_{jk} = \frac{\rho_b A_b L_b}{420} \begin{bmatrix} 140 & 0 & 0 & 70 & 0 & 0 \\ 0 & 156 & 22L_b & 0 & 54 & -13L_b \\ 0 & 22L_b & 4L_b^2 & 0 & 13L_b & -3L_b^2 \\ 70 & 0 & 0 & 140 & 0 & 0 \\ 0 & 54 & 13L_b & 0 & 156 & -22L_b \\ 0 & -13L_b & -3L_b^2 & 0 & -22L_b & 4L_b^2 \end{bmatrix}, \tag{7}$$

where ρ_c and ρ_b are the mass densities of the cable and beam-column elements, respectively. The coordinate transformation coefficient $a_{j\alpha}$ connected between u_j and q_α for the mass matrices of the cable and beam-column elements can be found in [20].

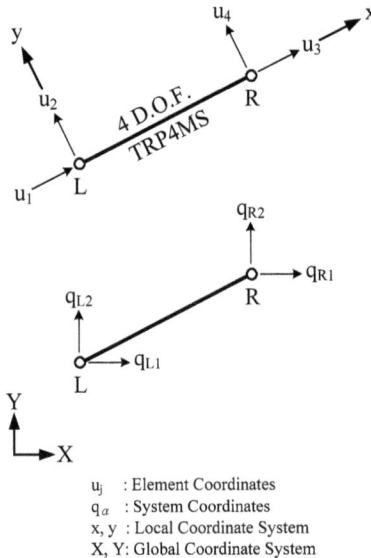

u_j : Element Coordinates
q_α : System Coordinates
x, y : Local Coordinate System
X, Y: Global Coordinate System

Figure 3. Cable element for simulating the mass of each stay cable.

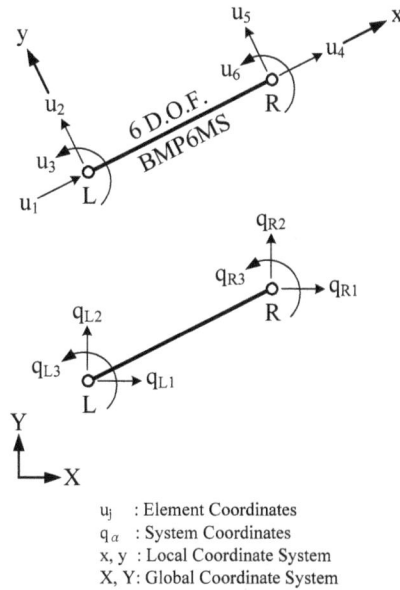

Figure 4. Beam-column element for simulating the mass of each deck and tower.

2.2. System equations

The system equations in generalized coordinates of a nonlinear finite element model of a cable-stayed bridge can be derived from the Lagrange's virtual work principle

$$M_{\alpha\beta}\ddot{q}_{\beta} + D_{\alpha\beta}\dot{q}_{\beta} + \sum_{EL}S_{j}a_{j\alpha} = P_{\alpha}, \quad \alpha = 1,2,3,\cdots,N, \tag{8}$$

$$M_{\alpha\beta} = \sum_{EL}ME_{jk}a'_{j\alpha}a'_{k\beta}, \tag{9}$$

$$S_{j} = KE_{jk}u_{k} + S_{j}^{0}, \tag{10}$$

$$P_{\alpha} = \bar{K}^{j} \cdot \bar{b}^{j}_{\alpha} \tag{11}$$

$$\bar{b}^{j}_{\alpha} = \frac{\partial \bar{W}^{j}}{\partial q_{\alpha}}, \tag{12}$$

$$\dot{q}_{\alpha} = \frac{dq_{\alpha}}{dt} \tag{13}$$

$$\ddot{q}_{\alpha} = \frac{d^{2}q_{\alpha}}{dt^{2}} \tag{14}$$

where $M_{\alpha\beta}$ and $D_{\alpha\beta}$ are the system mass and damping matrices, respectively, which both are assumed to be constant; S_j is the element force vector; P_α is the external force vector; S_j^0 is the initial element force vector; $\bar{K}^j$ is the external nodal force vector; $\bar{b}_\alpha^j$ is the basis vector; $\bar{W}^j$ is the displacement vector corresponding to $\bar{K}^j$; $\dot{q}_\alpha$ and $\ddot{q}_\alpha$ are the system velocity and acceleration vectors, respectively; t is the time; N is the number of degrees of freedom; the subscripts α and β denote the numbers of the system coordinates; the subscripts j and k represent the numbers of the element coordinates; the superscript j denotes the nodal number; $\sum_{EL}$ represents the summation over all elements.

Under consideration of three types of geometric nonlinearities mentioned in Chapter 2.1, KE_{jk} of a cable element and that of a beam-column element can be individually obtained from Eq. (2) and Eq. (3). The former and the latter are due to the cable sag effect and the beam-column effect, respectively. Similarly, ME_{jk} of the cable element and that of the beam-column element can be individually obtained from Eq. (6) and Eq. (7). u_j, $a_{j\alpha}$ and $\bar{b}_\alpha^j$ are nonlinear functions of q_α when the large displacement effect occurs. $\bar{K}^j$ can be written as a function of q_α if they are displacement dependent forces. $M_{\alpha\beta}$ and $D_{\alpha\beta}$ are both assumed to be constant, because only nonlinearities in stiffness are considered in this system.

Eq. (8) is a set of simultaneous second-order nonlinear ordinary differential equations. In order to incrementally solve these equations, the linearized system equations in a small time (or force) interval are derived based on the first-order Taylor series expansion of Eq. (8)

$$M_{\alpha\beta}\Delta\ddot{q}_\beta^n + D_{\alpha\beta}\Delta\dot{q}_\beta^n + {}^2K_{\alpha\beta}^n\Delta q_\beta^n = {}_uP_\alpha^n + \Delta P_\alpha^n, \quad t^n \le t \le t^n + \Delta t^n, \tag{15}$$

$$^2K_{\alpha\beta}^n = \sum_{EL} KE_{jk}^n a_{j\alpha}^n a_{k\beta}^n + \sum_{EL} S_j^n a_{j\alpha,\beta}^n - {}^n\bar{K}^j \cdot {}^n\bar{b}_{\alpha,\beta}^j - {}^n\bar{K}_\beta^j \cdot {}^n\bar{b}_\alpha^j, \tag{16}$$

$$\bar{b}_{\alpha,\beta}^j = \frac{\partial \bar{b}_\alpha^j}{\partial q_\beta}, \tag{17}$$

$$\bar{K}_\alpha^j = \frac{\partial \bar{K}^j}{\partial q_\alpha} \tag{18}$$

$$_uP_\alpha^n = P_\alpha^n - M_{\alpha\beta}\ddot{q}_\beta^n - D_{\alpha\beta}\dot{q}_\beta^n - \sum_{EL} S_j^n a_{j\alpha}^n, \tag{19}$$

$$\Delta P_\alpha^n = P_\alpha^{n+1} - P_\alpha^n, \tag{20}$$

$$\Delta q_\alpha^n = q_\alpha^{n+1} - q_\alpha^n, \tag{21}$$

$$\Delta\dot{q}_\alpha^n = \dot{q}_\alpha^{n+1} - \dot{q}_\alpha^n, \tag{22}$$

$$\Delta\ddot{q}_\alpha^n = \ddot{q}_\alpha^{n+1} - \ddot{q}_\alpha^n, \tag{23}$$

$$\Delta t^n = t^{n+1} - t^n, \tag{24}$$

where $^{2}K_{\alpha\beta}^{n}$ is the tangent system stiffness matrix; $_{u}P_{\alpha}^{n}$ is the unbalanced force vector; ΔP_{α}^{n} is the increment of the external force vector; Δq_{α}^{n}, $\Delta \dot{q}_{\alpha}^{n}$ and $\Delta \ddot{q}_{\alpha}^{n}$ are the increments of the system coordinate, velocity and acceleration vectors, respectively; Δt^{n} is the time increment; the superscript n and $n+1$ denote the numbers of the time (or force) steps; the superscript 2 represents the second-order iteration matrix.

$^{2}K_{\alpha\beta}^{n}$ in Eq. (16) consists of four terms. The first term is the elastic stiffness matrix, while the second and third terms are the geometric stiffness matrices induced by large displacements. Furthermore, the fourth term is the geometric stiffness matrix induced by displacement dependent forces, which is neglected in this study.

Eq. (15) is a set of simultaneous second-order linear ordinary differential equations in a small time interval, which can be solved by the direct integration method [20].

2.3. Initial shape analysis

The initial shape of a cable-stayed bridge provides the geometric configuration as well as the prestress distribution of such bridge under the weight of the deck-tower system and the pretension forces in the stay cables. The relations for the equilibrium conditions, boundary conditions and architectural design requirements should be satisfied. Under consideration of three types of geometric nonlinearities, i.e., the cable sag, beam-column and large displacement effects, the initial shape analyses of an OECS model and a MECS model are presented in this study.

For the initial shape analysis of the OECS model, the weight of the deck-tower system is considered, whereas the weight of the stay cables is neglected. The shape finding computation is performed using a two-loop iteration method: an equilibrium iteration and a shape iteration [18-23]. It can be started with an estimated initial element force (pretension force) in the stay cables. Based on the reference configuration (architectural design form) with no deflection and zero prestress in the deck-tower system, the equilibrium configuration of the whole bridge under the weight of the deck-tower system can be first determined by incrementally solving the linearized system equations

$$^{2}K_{\alpha\beta}^{n}\Delta q_{\beta}^{n} = {}_{u}P_{\alpha}^{n} + \Delta P_{\alpha}^{n}, \quad P_{\alpha}^{n} \leq P_{\alpha} \leq P_{\alpha}^{n+1}, \tag{25}$$

$$_{u}P_{\alpha}^{n} = P_{\alpha}^{n} - \sum_{EL} S_{j}^{n} a_{j\alpha}^{n}, \tag{26}$$

which are individually derived from Eq. (15) and Eq. (19) with negligible inertial and damping effects due to the static case. On the basis of Eq. (25) and Eq. (26), the equilibrium iteration is performed using the Newton-Raphson method [18-23].

After the above equilibrium iteration, the bridge configuration satisfies the equilibrium and boundary conditions, however, the architectural design requirements are, in general, not fulfilled. This is because large displacements and variable bending moments occur in the deck-tower system due to the large bridge span. Under these conditions, the shape iteration is conducted to reduce the displacements and to smooth the bending moments, and the appropriate initial shape can therefore be obtained.

A number of control points are selected for insuring that both the deck and tower displacements satisfy the architectural design requirements in the shape iteration

$$\left|\frac{q_\alpha}{L_r}\right| \le \varepsilon_r, \tag{27}$$

where q_α is the displacement in a certain direction of the control point; L_r is the reference length; ε_r is the convergence tolerance. For checking the deck displacement, each control point is the node intersected by the deck and the stay cable. q_α and L_r individually denote the vertical displacement of the control point and the main span length. Similarly, each node intersected by the tower and the stay cable, or located on the top of the tower is chosen as the control point for checking the tower displacement. q_α and L_r represent the horizontal displacement of the control point and the tower height, respectively.

If Eq. (27) is not achieved, the element axial forces calculated in the previous equilibrium iteration will be taken as the initial element forces in the new equilibrium iteration, and the corresponding equilibrium configuration of the whole bridge under the weight of the deck-tower system will be determined again. The shape iteration will then be repeated until Eq. (27) is reached. Under these conditions, the convergent configuration can be regarded as the initial shape of the OECS model.

The initial shape analysis of the MECS model is also performed to reasonably simulate the bridge configuration. Based on the initial shape of the OECS model obtained previously, the both end coordinates and pretension force in each single stay cable can be used for the shape finding computation of the corresponding stay cable discretized into multiple elements using the catenary function method [22, 23]. Incorporating the interior nodal coordinates and pretension forces in each discrete stay cable into the bridge model, and then conducting the two-loop iteration method again, the convergent configuration can be regarded as the initial shape of the MECS model.

2.4. Modal analysis

Under the assumption that the system vibrates with a small amplitude around a certain nonlinear static state, in which the variation in such state induced by the vibration is negligible, the modal analysis of a cable-stayed bridge can be conducted based on the linearized system equation

$$M_{\alpha\beta}^A \ddot{q}_\beta + {}^2K_{\alpha\beta}^A q_\beta = 0, \tag{28}$$

where $M_{\alpha\beta}^A$ and ${}^2K_{\alpha\beta}^A$ are the system mass and tangent system stiffness matrices with respect to the nonlinear static state q_α^A, respectively. The initial shape obtained in Chapter 2.3 can be regarded as q_α^A. Eq. (28) is derived from Eq. (15) with negligible damping and force effects. On the basis of Eq. (28) representing the free vibration of the undamped system, the natural frequency f_n and the normalized mode shape $\overline{Y}_n$ of the n th mode can be calculated by the subspace iteration method [20].

2.5. Modal coupling assessment

According to the results of both the initial shape analysis (Chapter 2.3) and modal analysis (Chapter 2.4) with the consideration of geometric nonlinearities (Chapter 2.1) in the system equations (Chapter 2.2), three indices for quantitatively assessing the degree of coupling among the stay cables, decks and towers of a cable-stayed bridge in each mode are presented [24] as

$$\overline{M}_n^s = \frac{\left(\overline{Y}_n^s\right)^T M^s \overline{Y}_n^s}{\left(\overline{Y}_n^s\right)^T M^s \overline{Y}_n^s + \left(\overline{Y}_n^d\right)^T M^d \overline{Y}_n^d + \left(\overline{Y}_n^t\right)^T M^t \overline{Y}_n^t}, \tag{29}$$

$$\overline{M}_n^d = \frac{\left(\overline{Y}_n^d\right)^T M^d \overline{Y}_n^d}{\left(\overline{Y}_n^s\right)^T M^s \overline{Y}_n^s + \left(\overline{Y}_n^d\right)^T M^d \overline{Y}_n^d + \left(\overline{Y}_n^t\right)^T M^t \overline{Y}_n^t}, \tag{30}$$

$$\overline{M}_n^t = \frac{\left(\overline{Y}_n^t\right)^T M^t \overline{Y}_n^t}{\left(\overline{Y}_n^s\right)^T M^s \overline{Y}_n^s + \left(\overline{Y}_n^d\right)^T M^d \overline{Y}_n^d + \left(\overline{Y}_n^t\right)^T M^t \overline{Y}_n^t}, \tag{31}$$

where $\overline{M}_n^j$ $(j = s,d,t)$ are the generalized mass ratios of the n th mode; M^j $(j = s,d,t)$ are the submatrices of $M_{\alpha\beta}^A$; $\overline{Y}_n^j$ $(j = s,d,t)$ are the subvectors of $\overline{Y}_n$ in the n th mode; the superscripts s, d and t denote the quantities of the stay cable, the deck and the tower, respectively. The sum of $\overline{M}_n^s$, $\overline{M}_n^d$ and $\overline{M}_n^t$ is 1 for the corresponding n .

2.6. Seismic analysis

According to the assumption that the system is under the uniform earthquake excitation, the seismic analysis of a cable-stayed bridge with respect to the initial shape obtained in Chapter 2.3 can be conducted based on the equivalent difference equations

$$^*Q_\alpha^{n+1} = {}^*K_{\alpha\beta}^n \Delta q_\beta^n, \quad t^n \le t \le t^n + \Delta t^n, \tag{32}$$

$$^*Q_\alpha^{n+1} = P_\alpha^{n+1} - M_{\alpha\beta} {}^*\ddot{q}_\beta^n - D_{\alpha\beta} {}^*\dot{q}_\beta^n - \sum_{EL} S_j^n a_{j\alpha}^n, \tag{33}$$

$$^*K_{\alpha\beta}^n = I_1 M_{\alpha\beta} + I_2 D_{\alpha\beta} + {}^2K_{\alpha\beta}^n, \tag{34}$$

$$P_\alpha^n = -M_{\alpha\beta} I_\beta {}^g \ddot{q}^n, \tag{35}$$

$$^* \dot{q}_\alpha^n = -I_4 \dot{q}_\alpha^n - I_6 \ddot{q}_\alpha^n, \tag{36}$$

$$^* \ddot{q}_\alpha^n = -I_3 \dot{q}_\alpha^n - I_5 \ddot{q}_\alpha^n, \tag{37}$$

$$q_\alpha^{n+1} = q_\alpha^n + \Delta q_\alpha^n, \tag{38}$$

$$\dot{q}_\alpha^{n+1} = {}^* \dot{q}_\alpha^n + I_2 \Delta q_\alpha^n, \tag{39}$$

$$\ddot{q}_\alpha^{n+1} = {}^* \ddot{q}_\alpha^n + I_1 \Delta q_\alpha^n, \tag{40}$$

$$I_1 = \frac{1}{\beta_1 \left(\Delta t^n \right)^2}, \tag{41}$$

$$I_2 = \frac{\gamma_1}{\beta_1 \Delta t^n}, \tag{42}$$

$$I_3 = \frac{1}{\beta_1 \Delta t^n}, \tag{43}$$

$$I_4 = \frac{1}{2\beta_1} - 1, \tag{44}$$

$$I_5 = \frac{\gamma_1}{\beta_1} - 1, \tag{45}$$

$$I_6 = \left(\frac{\gamma_1}{2\beta_1} - 1 \right) \Delta t^n, \tag{46}$$

where these equations are derived from Eq. (15) and Eq. (19) using the Newmark method [27]; $^* Q_\alpha^{n+1}$ is the effective force vector; $^* K_{\alpha\beta}^n$ is the effective system stiffness matrix; $^g \ddot{q}^n$ is the earthquake-induced ground acceleration; I_β is the column vector in which each element is either zero or unity depending on the direction of $^g \ddot{q}^n$; β_1 and γ_1 are the parameters defining the variation of acceleration over a time increment and determining the stability and accuracy characteristics of the Newmark method; $^* \dot{q}_\alpha^n$, $^* \ddot{q}_\alpha^n$ and I_j $(j = 1-6)$ are the coefficients of the seismic analysis.

2. Finite element models

To understand the deck-stay interaction with the appropriate initial shapes of cable-stayed bridges, an OECS model and a MECS model of the full Kao Ping Hsi Bridge are developed, as shown in Figure 5(a) and 5(b), respectively. This bridge is an unsymmetrical single-deck cable-stayed bridge with a main span of 330 m and a side span of 184 m. The deck, which consists of steel box girders in the main span and concrete box girders in the side span, is supported by a total of 28 stay cables (S1-S28), arranged in a central plane originated at the 184 m tall, inverted Y-shaped, concrete tower. A more detailed description of the Kao Ping Hsi Bridge can be found in [28].

Figure 5(a) and 5(b) illustrate the two-dimensional finite element models of the bridge. The OECS and MECS models both contain 48 beam-column elements that simulate the deck and tower. For the MECS model, each stay cable is discretized into 10 cable elements, whereas a single cable element is used to simulate each stay cable in the OECS model. This fact indicates that the OECS and MECS models individually include 28 and 280 cable elements. Figure 5(a) and 5(b) also show that 49 and 301 nodes are involved in the OECS and MECS models, respectively. A hinge, roller and fixed supports are used to model the boundary conditions of the left and right ends of the deck and the tower, respectively, and a rigid joint is employed to simulate the deck-tower connection. On the basis of the OECS and MECS models, the initial shape analysis, modal analysis, modal coupling assessment and seismic analysis of the Kao Ping Hsi Bridge are conducted in this study.

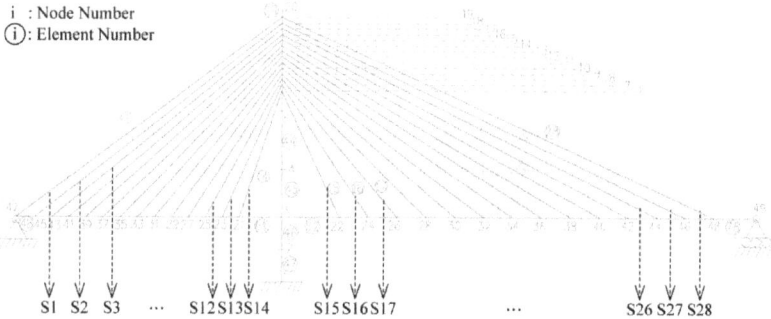

(a) OECS model

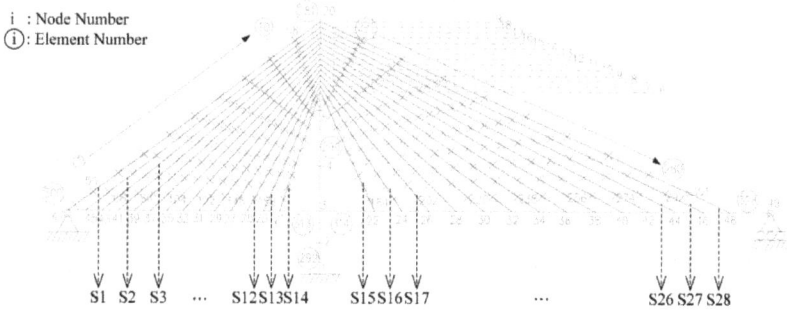

(b) MECS model

Figure 5. Finite element models of the Kao Ping Hsi Bridge.

4. Numerical results

Based on the OECS and MECS models of the Kao Ping Hsi Bridge developed in Chapter 3, the initial shape analysis, modal analysis, modal coupling assessment and seismic analysis are conducted using the finite element formulation presented in Chapter 2. The numerical results can be used to fully understand the mechanism of the deck-stay interaction with the appropriate initial shapes of cable-stayed bridges.

4.1. Initial shape analysis

Based on the finite element procedures presented in Chapter 2.3, the initial shape analyses of the OECS and MECS models are conducted to reasonably provide the geometric configuration of the Kao Ping Hsi Bridge. In both Figure 5(a) and 5(b), nodes 37, 38, 40, 45 and 46 are selected as the control points for checking the deck displacement in the vertical direction, while node 19 is chosen as the control point for checking the tower displacement in the horizontal direction. The convergence tolerance ε_r is set to 10^{-4} in this study.

Figure 6(a) shows the initial shape of the OECS model of the Kao Ping Hsi Bridge (solid line), indicating that the maximum vertical and horizontal displacements measured from the reference configuration (short dashed line) are 0.038 m at node 36 in the main span of the deck and -0.021 m at node 8 in the tower, respectively. The shape of each stay cable represented by a single cable element is straight as expected. Figure 6(a) also illustrates that the overall displacement obtained by the two-loop iteration method, i.e., the equilibrium and shape iterations, is comparatively smaller than that only from the equilibrium iteration (long dashed line). Consequently, the initial shape based on the two-loop iteration method appears to be able to appropriately describe the geometric configurations of cable-stayed bridges.

Figure 6(b) shows the initial shape of the MECS model of the Kao Ping Hsi Bridge (solid line), indicating that the maximum vertical and horizontal displacements measured from the reference configuration (short dashed line) are 0.068 m at node 34 in the main span of the deck and -0.049 m at node 8 in the tower, respectively. The sagged shape occurs in the stay cables due to the fact that each stay cable is simulated by multiple cable elements.

4.2. Modal analysis and modal coupling assessment

According to the results of the initial shape analysis presented in Chapter 4.1, the modal analyses of the OECS and MECS models using the finite element computations developed in Chapter 2.4 are conducted to calculate the natural frequency and normalized mode shape of the individual modes of the Kao Ping Hsi Bridge. The modal coupling assessment based on the proposed formulas in Chapter 2.5 is also performed to obtain the generalized mass ratios among the structural components for each mode of such bridge. These results can be used to provide a variety of viewpoints to illustrate the mechanism of the deck-stay interaction with the appropriate initial shapes of cable-stayed bridges.

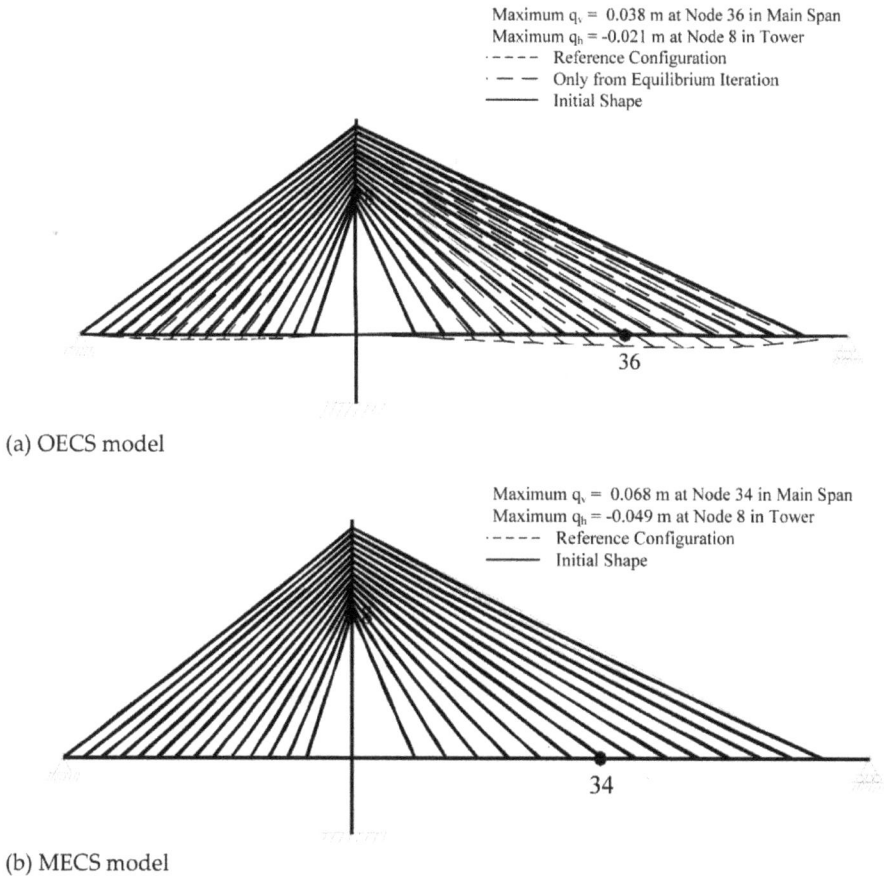

Maximum q_v = 0.038 m at Node 36 in Main Span
Maximum q_h = -0.021 m at Node 8 in Tower
- - - - Reference Configuration
· — —   Only from Equilibrium Iteration
———   Initial Shape

36

(a) OECS model

Maximum q_v = 0.068 m at Node 34 in Main Span
Maximum q_h = -0.049 m at Node 8 in Tower
- - - - Reference Configuration
———   Initial Shape

34

(b) MECS model

Figure 6. Initial shapes of the Kao Ping Hsi Bridge.

Table 1 summarizes the modal properties of the Kao Ping Hsi Bridge based on the OECS model (modes 1 to 3) and the MECS model (modes 1 to 24). In this table, f_n and $\overline{Y}_n$ represent the natural frequency and the normalized mode shape of the n th mode, respectively. As expected, the MECS model reveals the global, local and coupled modes, whereas the OECS model only yields the global modes. The modal properties of modes 1 and 2 in the OECS model are individually similar to those of modes 1 and 12 in the MECS model, because these modes represent the global modes. While mode 3 in the OECS model is identified as the global mode, mode 19 in the MECS model is the coupled mode. The other coupled mode can also be observed in mode 18 in the MECS model. These results suggest that the interaction between the deck-tower system and stay cables can be captured by the MECS model, but not by the OECS model. Also due to the limitations of the OECS model, modes 2 to 11, modes 13 to 17 and modes 20 to 24, which represent the local modes of the stay cables, are successfully captured by the MECS model, but not by the OECS model.

OECS				MECS			
n	f_n (Hz)	$\bar{Y}_n$	Type	n	f_n (Hz)	$\bar{Y}_n$	Type
1	0.2877	1st DT	G	1	0.3053	1st DT	G
				2	0.3382	1st S28	L
				3	0.3852	1st S27	L
				4	0.4274	1st S26	L
				5	0.4554	1st S1	L
				6	0.4653	1st S25	L
				7	0.4899	1st S24	L
				8	0.5067	1st S23	L
				9	0.5269	1st S22	L
				10	0.5378	1st S2	L
				11	0.5471	1st S21	L
2	0.5455	2nd DT	G	12	0.5686	2nd DT	G
				13	0.5944	1st S3	L
				14	0.6040	1st S20	L
				15	0.6333	1st S4	L
				16	0.6346	2nd S28	L
				17	0.6835	1st S5	L
				18	0.6850	3rd DT 1st S19	C
3	0.6854	3rd DT	G	19	0.7171	3rd DT 1st S19	C
				20	0.7269	1st S6	L
				21	0.7500	2nd S27	L
				22	0.7590	1st S7	L
				23	0.8008	1st S8	L
				24	0.8184	1st S18	L

DT: Deck-tower system
S: Stay cable
G: Global mode
L: Local mode
C: Coupled mode

Table 1. Comparisons between corresponding modal properties of the OECS and MECS models of the Kao Ping Hsi Bridge.

Figure 7 shows the relationship between the natural frequency and the mode number for the first 24 modes of the MECS model of the Kao Ping Hsi Bridge. For reference, the fundamental frequency of stay S19 (0.6908 Hz) is also included. This frequency is calculated based on the assumption that stay S19 is clamped at both ends [29].

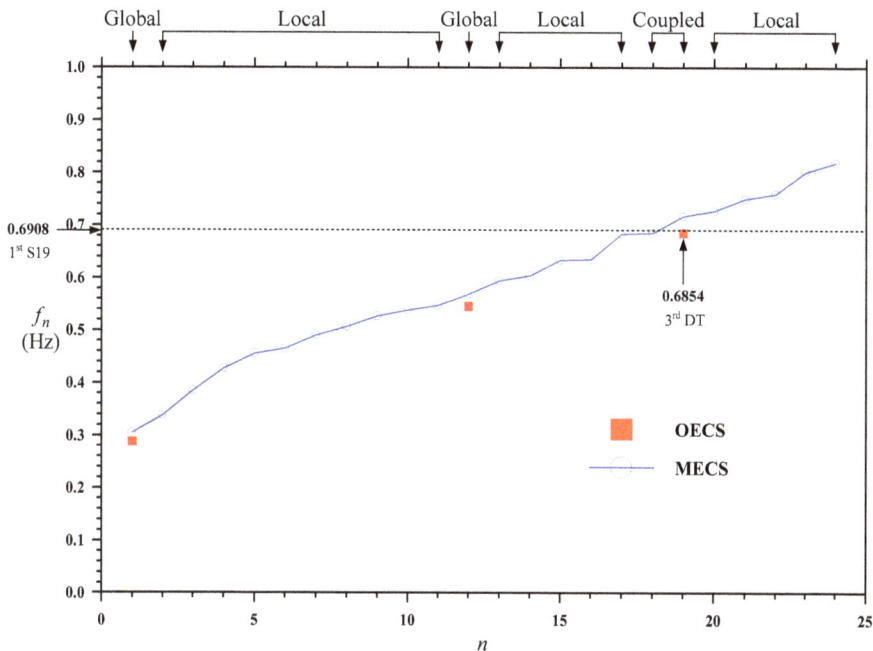

Figure 7. Relationships between natural frequencies and mode numbers of the MECS model of the Kao Ping Hsi Bridge.

Figure 8(a) and 8(b) illustrate the normalized mode shapes of the individual modes of the OECS model (modes 1 to 3) and the MECS model (modes 1 to 24) of the Kao Ping Hsi Bridge, respectively. Each normalized mode shape (solid line) is measured from the initial shape (dashed line) obtained in Chapter 4.1.

To quantitatively assess the degree of coupling for each mode, Figure 9 depicts the variations in the generalized mass ratios with respect to the mode number for the first 24 modes of the MECS model of the Kao Ping Hsi Bridge. In this figure, $\bar{M}_n^s$, $\bar{M}_n^d$ and $\bar{M}_n^t$ represent the generalized mass ratios of the stay cable, the deck and the tower of the n th mode, respectively. The sum of $\bar{M}_n^s$, $\bar{M}_n^d$ and $\bar{M}_n^t$ is 1 for the corresponding n $(n = 1 - 24)$. It is evident that $\bar{M}_n^t$ $(n = 1 - 24)$ approaches 0 for the first 24 modes due to the high rigidity of the concrete tower, resulting in the insignificant tower deformations in the lower modes sensitive to the ambient excitations, as can also be seen in Figure 8(b). These results are in agreement with the literature [2].

Mode 1 $f_1 = 0.2877$ Hz (Global)   Mode 2 $f_2 = 0.5455$ Hz (Global)   Mode 3 $f_3 = 0.6854$ Hz (Global)

(a) OECS model

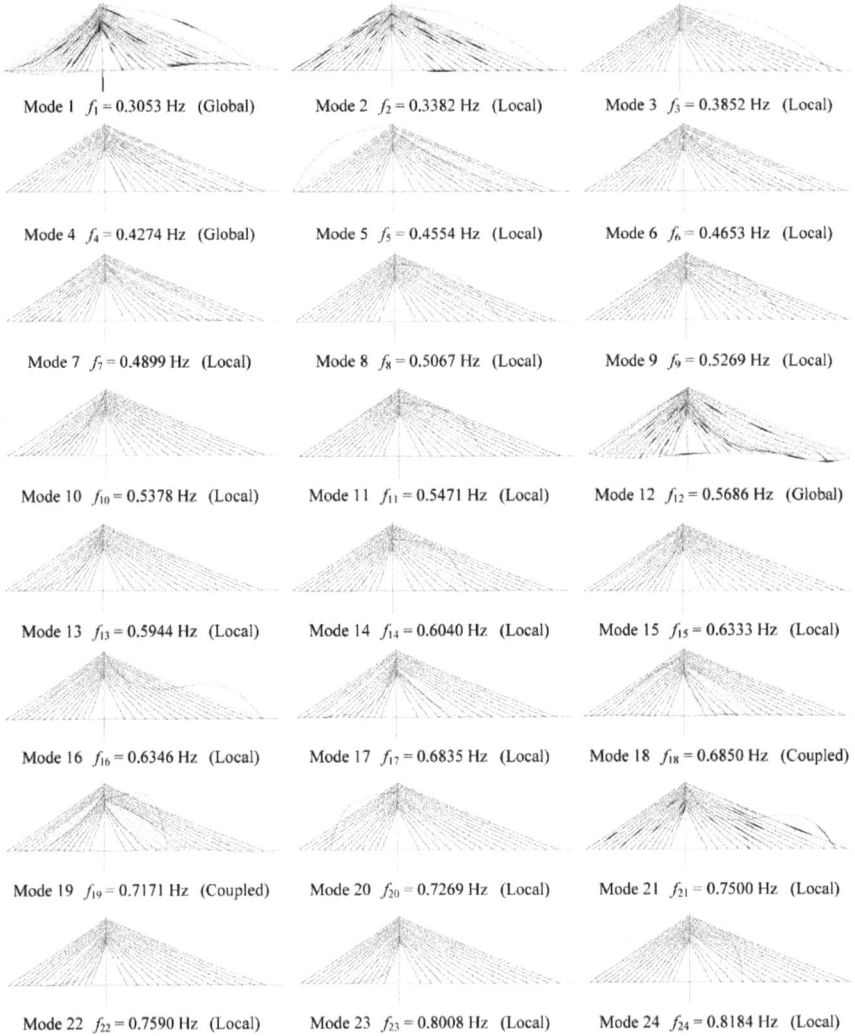

Mode 1 $f_1 = 0.3053$ Hz (Global)   Mode 2 $f_2 = 0.3382$ Hz (Local)   Mode 3 $f_3 = 0.3852$ Hz (Local)

Mode 4 $f_4 = 0.4274$ Hz (Global)   Mode 5 $f_5 = 0.4554$ Hz (Local)   Mode 6 $f_6 = 0.4653$ Hz (Local)

Mode 7 $f_7 = 0.4899$ Hz (Local)   Mode 8 $f_8 = 0.5067$ Hz (Local)   Mode 9 $f_9 = 0.5269$ Hz (Local)

Mode 10 $f_{10} = 0.5378$ Hz (Local)   Mode 11 $f_{11} = 0.5471$ Hz (Local)   Mode 12 $f_{12} = 0.5686$ Hz (Global)

Mode 13 $f_{13} = 0.5944$ Hz (Local)   Mode 14 $f_{14} = 0.6040$ Hz (Local)   Mode 15 $f_{15} = 0.6333$ Hz (Local)

Mode 16 $f_{16} = 0.6346$ Hz (Local)   Mode 17 $f_{17} = 0.6835$ Hz (Local)   Mode 18 $f_{18} = 0.6850$ Hz (Coupled)

Mode 19 $f_{19} = 0.7171$ Hz (Coupled)   Mode 20 $f_{20} = 0.7269$ Hz (Local)   Mode 21 $f_{21} = 0.7500$ Hz (Local)

Mode 22 $f_{22} = 0.7590$ Hz (Local)   Mode 23 $f_{23} = 0.8008$ Hz (Local)   Mode 24 $f_{24} = 0.8184$ Hz (Local)

(b) MECS model

Figure 8. Normalized mode shapes of the Kao Ping Hsi Bridge.

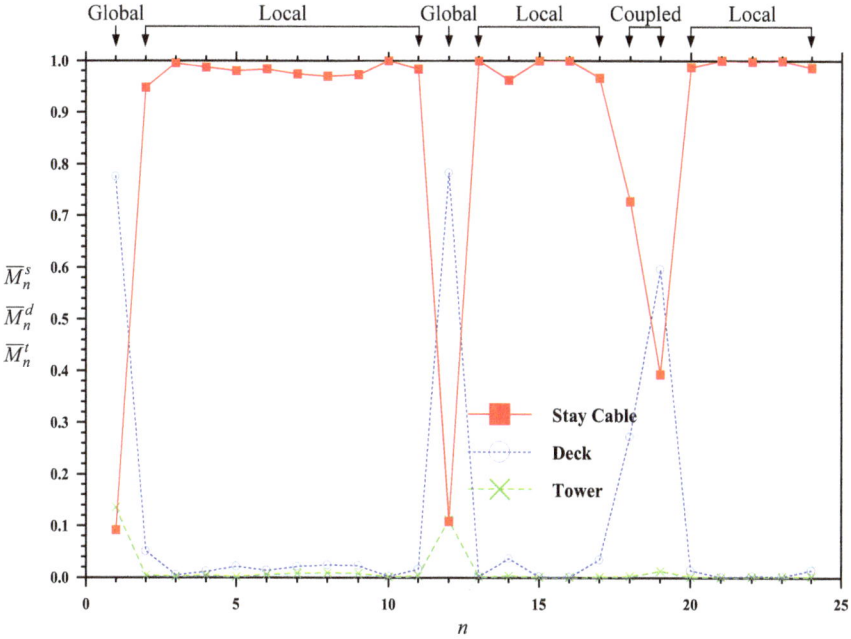

Figure 9. Variations in generalized mass ratios with respect to mode numbers of the MECS model of the Kao Ping Hsi Bridge.

It can be seen in Table 1, Figure 7, Figure 8(a) and 8(b) that for the global modes, f_n and $\overline{Y}_n$ $(n=1,2)$ in the OECS model are individually similar to f_n and $\overline{Y}_n$ $(n=1,12)$ in the MECS model. It is consistent with the results in Figure 9 that for modes 1 and 12 in the MECS model, the sum of $\overline{M}_n^d$ and $\overline{M}_n^t$ $(n=1,12)$ is close to 0.9, whereas $\overline{M}_n^s$ $(n=1,12)$ approaches 0.1. Consequently, these modes are primarily dominated by the deformations of the deck-tower system with the quasi-static motions of the stay cables. This type of response can be identified as the "pure" deck mode in the analytical model [17].

It also can be seen in Figure 9 that for modes 2 to 11, modes 13 to 17 and modes 20 to 24 in the MECS model, $\overline{M}_n^s$ $(n=2-11,13-17,20-24)$ is close to 1, whereas the sum of $\overline{M}_n^d$ and $\overline{M}_n^t$ $(n=2-11,13-17,20-24)$ approaches 0. It is consistent with the results in Table 1, Figure 7 and Figure 8(b) that $\overline{Y}_n$ $(n=2-11,13-17,20-24)$ in the MECS model is the local mode predominantly consisting of the stay cable motions with negligible deformations of the deck-tower system. This type of response can be recognized as the "pure" cable mode in the analytical model [17].

As shown in Table 1, Figure 7, Figure 8(a) and 8(b), the difference between f_{19} in the MECS model (0.7171 Hz) and f_3 in the OECS model (0.6854 Hz) is evident due to the fact that $\overline{Y}_{19}$ in the MECS model is the coupled mode, but $\overline{Y}_3$ in the OECS model is the global mode, i.e., the "pure" deck-tower mode. Similarly, f_{18} in the MECS model (0.6850 Hz) branches from the fundamental frequency of stay S19 clamped at both ends (0.6908 Hz). This is because

$\overline{Y}_{18}$ in the MECS model is the coupled mode, while the fundamental mode shape of stay S19 can be regarded as the "pure" stay cable mode. These observations are attributed to the frequency loci veering when the natural frequency of the "pure" deck-tower mode (0.6854 Hz) approaches that of the "pure" stay cable mode (0.6908 Hz). As illustrated in Figure 9, the sum of $\overline{M}_{19}^{d}$ and $\overline{M}_{19}^{t}$ is relatively higher than $\overline{M}_{19}^{s}$, whereas the sum of $\overline{M}_{18}^{d}$ and $\overline{M}_{18}^{t}$ is comparatively lower than $\overline{M}_{18}^{s}$. Consequently, $\overline{Y}_{18}$ and $\overline{Y}_{19}$ in the MECS model are the pair of coupled modes with the similar configurations, which have substantial contributions from both the deck-tower system and stay cables. These phenomena correspond to the mode localization. This type of response coincides with the coupled mode in the analytical model [17].

In summary, the coupled modes are attributed to the frequency loci veering and mode localization when the "pure" deck-tower frequency and the "pure" stay cable frequency approach one another, implying that the mode shapes of such coupled modes are simply different from those of the deck-tower system or stay cables alone. The distribution of the generalized mass ratios between the deck-tower system and stay cables are useful indices for quantitatively assessing the degree of coupling for each mode. These results are demonstrated to fully understand the mechanism of the deck-stay interaction with the appropriate initial shapes of cable-stayed bridges.

4.3. Seismic analysis

According to the results of the initial shape analysis presented in Chapter 4.1, the seismic analyses of the OECS and MECS models using the finite element computations developed in Chapter 2.6 are conducted to obtain the dynamic responses of the Kao Ping Hsi Bridge. Figure 10 shows the vertical component of the Chi-Chi earthquake accelerogram recorded in Mid-Taiwan on September 21, 1999 [30], which is selected as the earthquake-induced ground acceleration in this study. Under the excitation, the Newmark method $\left(\beta_{1}=1/4, \gamma_{1}=1/2\right)$ is used to calculate the displacement and internal force time histories of the system. The duration of the simulation is set to 30.0 s.

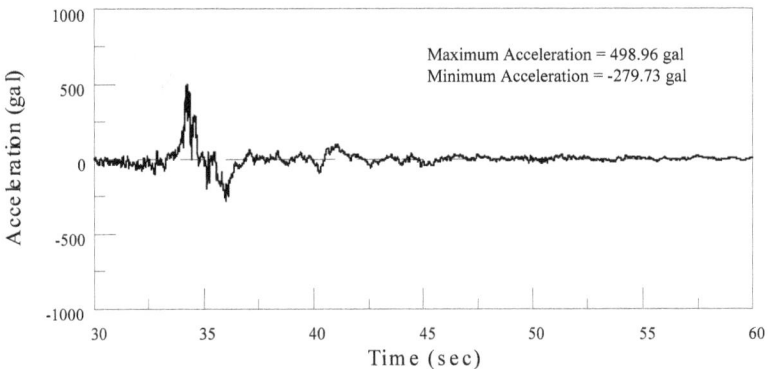

Figure 10. The Chi-Chi earthquake accelerogram.

Figure 11 shows the horizontal and vertical displacement time histories of nodes 295, 297 and 300 in stay S28 for the MECS model. The variations in the dynamic responses among the three nodes for each direction and those between the horizontal and vertical directions for each node are observed in this figure. Consequently, the dynamic displacements of the stay cables are successfully captured by the MECS model, but not by the OECS model. Figure 12 shows the vertical displacement time histories of nodes 35, 36 and 42 in the deck, the horizontal displacement time histories of nodes 8 and 20 in the tower, and the horizontal time history of node 49 in the right end of the deck, for both the OECS and MECS models. The dynamic response of each node in the OECS model coincides with that of the corresponding node in the MECS model. Consequently, the dynamic displacements of the deck-tower system are reasonably simulated by both the OECS and MECS models.

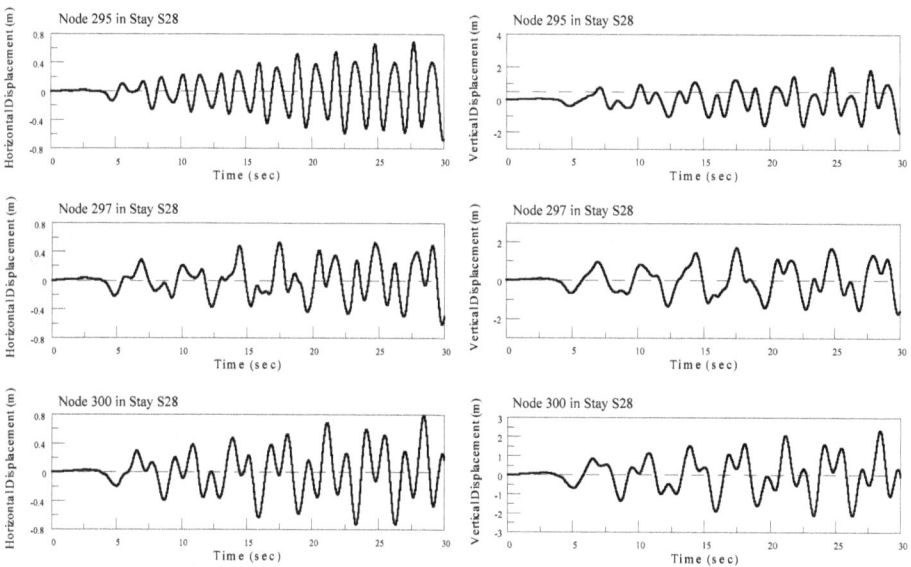

Figure 11. Displacement time histories of the stay cable of the Kao Ping Hsi Bridge.

The axial force, which is in the u_1 coordinate of the cable element in Figure 1, is the unique internal force of the stay cable. Figure 13 shows the internal force time history of element 28 in stay S28 for the OECS model and those of the corresponding elements 271, 275 and 280 in stay S28 for the MECS model. The variations in the dynamic responses among the three elements of the MECS model are negligible. In addition, the dynamic response of each element in the MECS model is in agreement with that of the corresponding element in the OECS model, which can be considered as the "nominal" dynamic axial force of the stay cable. Consequently, the dynamic internal forces of the stay cables are successfully captured by both the OECS and MECS models. The internal forces of the deck-tower system include the left moment, right moment and axial force, which are individually in the u_1, u_2 and u_3 coordinates of the beam-column element in Figure 2. Figure 14 shows the internal force time histories of element 69 (321) in the deck and those of element 40 (292) in the tower for the

OECS (MECS) model. The dynamic responses of each element in the OECS model coincide with those of the corresponding element in the MECS model. Consequently, the dynamic internal forces of the deck-tower system are reasonably simulated by both the OECS and MECS models.

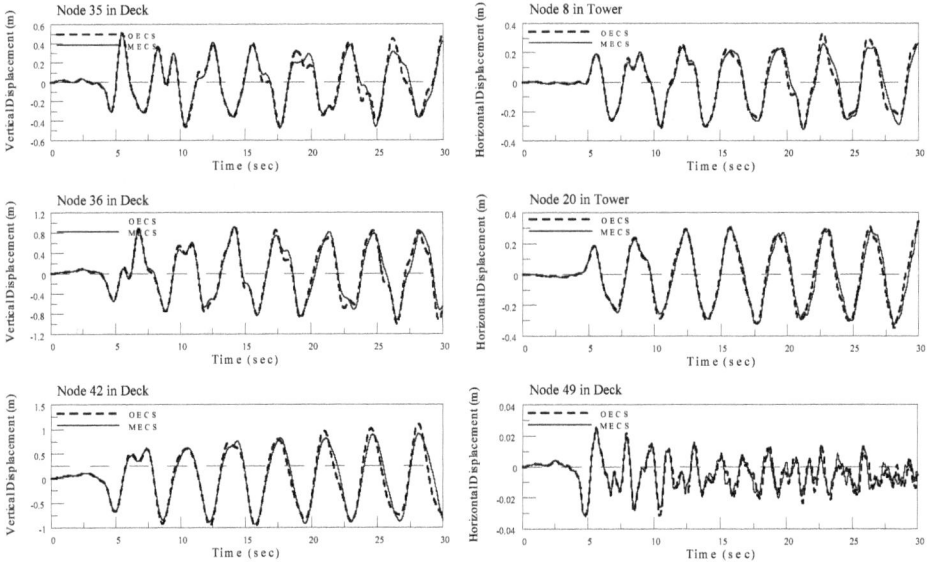

Figure 12. Displacement time histories of the deck-tower system of the Kao Ping Hsi Bridge.

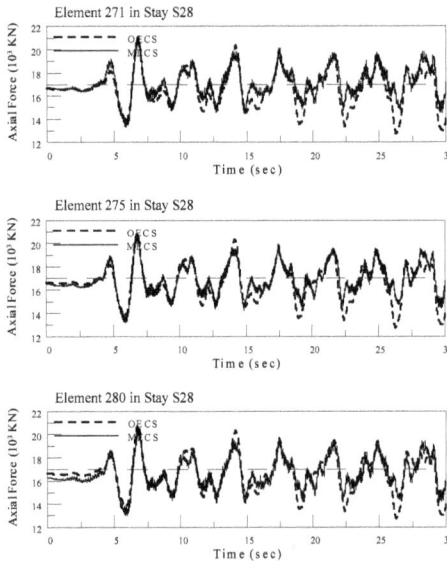

Figure 13. Internal force time histories of the stay cable of the Kao Ping Hsi Bridge.

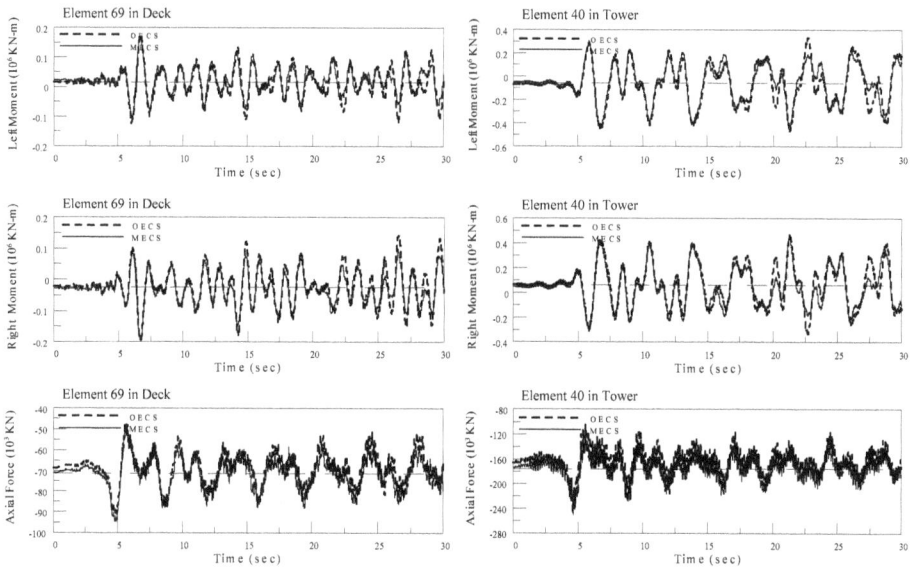

Figure 14. Internal force time histories of the deck-tower system of the Kao Ping Hsi Bridge.

In summary, the dynamic displacements of the stay cables are successfully captured by the MECS model, but not by the OECS model. Furthermore, the dynamic displacements of the deck-tower system as well as the dynamic internal forces of the stay cables and those of the deck-tower system are reasonably simulated by both the OECS and MECS models. These results are demonstrated to fully understand the deck-stay interaction characteristics of cable-stayed bridges under seismic excitations.

5. Conclusions

This study has provided a variety of viewpoints to illustrate the mechanism of the deck-stay interaction with the appropriate initial shapes of cable-stayed bridges. Based on the smooth and convergent bridge shapes obtained by the initial shape analysis, the OECS and MECS models of the Kao Ping Hsi Bridge are developed to verify the applicability of the analytical model and numerical formulation from the field observations in the authors' previous work. For this purpose, the modal analyses of the two finite element models are conducted to calculate the natural frequency and normalized mode shape of the individual modes of the bridge. The modal coupling assessment is also performed to obtain the generalized mass ratios among the structural components for each mode of the bridge. To further investigate the deck-stay interaction characteristics of cable-stayed bridges under earthquake excitations, the dynamic displacements and internal forces of the two finite element models are calculated based on the seismic analyses.

The findings indicate that the coupled modes are attributed to the frequency loci veering and mode localization when the "pure" deck-tower frequency and the "pure" stay cable frequency approach one another, implying that the mode shapes of such coupled modes are simply different from those of the deck-tower system or stay cables alone. The distribution of the generalized mass ratios between the deck-tower system and stay cables are useful indices for quantitatively assessing the degree of coupling for each mode. To extend the two finite element models to be under the seismic excitation, it is evident that the dynamic displacements of the stay cables are successfully captured by the MECS model, but not by the OECS model. In addition, the dynamic displacements of the deck-tower system as well as the dynamic internal forces of the stay cables and those of the deck-tower system are reasonably simulated by both the OECS and MECS models. These results are demonstrated to fully understand the mechanism of the deck-stay interaction with the appropriate initial shapes of cable-stayed bridges.

Author details

Ming-Yi Liu and Pao-Hsii Wang
Department of Civil Engineering, Chung Yuan Christian University, Jhongli City, Taiwan

6. References

[1] Abdel-Ghaffar, A.M., and Khalifa, M.A. (1991). "Importance of cable vibration in dynamics of cable-stayed bridges." *Journal of Engineering Mechanics, ASCE,* 117(11), 2571-2589.

[2] Gimsing, N.J. (1997). "Cable supported bridges: Concept and design." Second Edition, John Wiley & Sons, Ltd, Chichester, UK.

[3] Fujino, Y., Warnitchai, P., and Pacheco, B.M. (1993). "An experimental and analytical study of autoparametric resonance in a 3DOF model of cable-stayed-beam." *Nonlinear Dynamics,* 4(2), 111-138.

[4] Warnitchai, P., Fujino, Y., Pacheco, B.M., and Agret, R. (1993). "An experimental study on active tendon control of cable-stayed bridges." *Earthquake Engineering and Structural Dynamics,* 22(2), 93-111.

[5] Warnitchai, P., Fujino, Y., and Susumpow, T. (1995). "A non-linear dynamic model for cables and its application to a cable-structure system." *Journal of Sound and Vibration,* 187(4), 695-712.

[6] Lilien, J.L., and Pinto da Costa, A. (1994). "Vibration amplitudes caused by parametric excitation of cable stayed structures." *Journal of Sound and Vibration,* 174(1), 69-90.

[7] Pinto da Costa, A., Martins, J.A.C., Branco, F., and Lilien, J.L. (1996). "Oscillations of bridge stay cables induced by periodic motions of deck and/or towers." *Journal of Engineering Mechanics, ASCE,* 122(7), 613-622.

[8] Gattulli, V., Morandini, M., and Paolone, A. (2002). "A parametric analytical model for non-linear dynamics in cable-stayed beam." *Earthquake Engineering and Structural Dynamics,* 31(6), 1281-1300.

[9] Gattulli, V., and Lepidi, M. (2003). "Nonlinear interactions in the planar dynamics of cable-stayed beam." *International Journal of Solids and Structures,* 40(18), 4729-4748.

[10] Gattulli, V., Lepidi, M., Macdonald, J.H.G., and Taylor, C.A. (2005). "One-to-two global-local interaction in a cable-stayed beam observed through analytical, finite element and experimental models." *International Journal of Non-Linear Mechanics*, 40(4), 571-588.

[11] Gattulli, V., and Lepidi, M. (2007). "Localization and veering in the dynamics of cable-stayed bridges." *Computers and Structures*, 85(21-22), 1661-1678.

[12] Tuladhar, R., Dilger, W.H., and Elbadry, M.M. (1995). "Influence of cable vibration on seismic response of cable-stayed bridges." *Canadian Journal of Civil Engineering*, 22(5), 1001-1020.

[13] Caetano, E., Cunha, A., and Taylor, C.A. (2000a). "Investigation of dynamic cable-deck interaction in a physical model of a cable-stayed bridge. Part I: modal analysis." *Earthquake Engineering and Structural Dynamics*, 29(4), 481-498.

[14] Caetano, E., Cunha, A., and Taylor, C.A. (2000b). "Investigation of dynamic cable-deck interaction in a physical model of a cable-stayed bridge. Part II: seismic response." *Earthquake Engineering and Structural Dynamics*, 29(4), 499-521.

[15] Au, F.T.K., Cheng, Y.S., Cheung, Y.K., and Zheng, D.Y. (2001). "On the determination of natural frequencies and mode shapes of cable-stayed bridges." *Applied Mathematical Modelling*, 25(12), 1099-1115.

[16] Caetano, E., Cunha, A., Gattulli, V., and Lepidi, M. (2008). "Cable-deck dynamic interactions at the International Guadiana Bridge: On-site measurements and finite element modelling." *Structural Control and Health Monitoring*, 15(3), 237-264.

[17] Liu, M.Y., Zuo, D., and Jones, N.P. (2005). "Deck-induced stay cable vibrations: Field observations and analytical model." *Proceedings of the Sixth International Symposium on Cable Dynamics*, 175-182, Charleston, South Carolina, USA, September 19-22.

[18] Wang, P.H., Tseng, T.C., and Yang, C.G. (1993). "Initial shape of cable-stayed bridges." *Computers and Structures*, 46(6), 1095-1106.

[19] Wang, P.H., and Yang, C.G. (1996). "Parametric studies on cable-stayed bridges." *Computers and Structures*, 60(2), 243-260.

[20] Wang, P.H., Lin, H.T., and Tang, T.Y. (2002). "Study on nonlinear analysis of a highly redundant cable-stayed bridge." *Computers and Structures*, 80(2), 165-182.

[21] Wang, P.H., Tang, T.Y., and Zheng, H.N. (2004). "Analysis of cable-stayed bridges during construction by cantilever methods." *Computers and Structures*, 82(4-5), 329-346.

[22] Wang, P.H., Liu, M.Y., Huang, Y.T., and Lin, L.C. (2010). "Influence of lateral motion of cable stays on cable-stayed bridges." *Structural Engineering and Mechanics*, 34(6), 719-738.

[23] Liu, M.Y., Lin, L.C., and Wang, P.H. (2011). "Dynamic characteristics of the Kao Ping Hsi Bridge under seismic loading with focus on cable simulation." *International Journal of Structural Stability and Dynamics*, 11(6), 1179-1199.

[24] Liu, M.Y., Zuo, D., and Jones, N.P. "Analytical and numerical study of deck-stay interaction in a cable-stayed bridge in the context of field observations." *Journal of Engineering Mechanics, ASCE.* (under review).

[25] Ernst, H.J. (1965). "Der E-modul von Seilen unter Berücksichtigung des Durchhanges." *Der Bauingenieur*, 40(2), 52-55. (in German).

[26] Fleming, J.F. (1979). "Nonlinear static analysis of cable-stayed bridge structures." *Computers and Structures*, 10(4), 621-635.

[27] Newmark, N.M. (1959). "A method of computation for structural dynamics." *Journal of the Engineering Mechanics Division, ASCE*, 85(EM3), 67-94.

[28] Cheng, W.L. (2001). "Kao Ping Hsi Bridge." Taiwan Area National Expressway Engineering Bureau, Ministry of Transportation and Communications, Taipei, Taiwan.

[29] Irvine, H.M. (1981). "Cable structures." MIT Press, Cambridge, Massachusetts, USA.

[30] Lee, W.H.K., Shin, T.C., Kuo, K.W., Chen, K.C., and Wu, C.F. (2001). "CWB free-field strong-motion data from the 21 September Chi-Chi, Taiwan, earthquake." *Bulletin of the Seismological Society of America*, 91(5), 1370-1376.

Permissions

The contributors of this book come from diverse backgrounds, making this book a truly international effort. This book will bring forth new frontiers with its revolutionizing research information and detailed analysis of the nascent developments around the world.

We would like to thank Halil Sezen, for lending his expertise to make the book truly unique. He has played a crucial role in the development of this book. Without his invaluable contribution this book wouldn't have been possible. He has made vital efforts to compile up to date information on the varied aspects of this subject to make this book a valuable addition to the collection of many professionals and students.

This book was conceptualized with the vision of imparting up-to-date information and advanced data in this field. To ensure the same, a matchless editorial board was set up. Every individual on the board went through rigorous rounds of assessment to prove their worth. After which they invested a large part of their time researching and compiling the most relevant data for our readers. Conferences and sessions were held from time to time between the editorial board and the contributing authors to present the data in the most comprehensible form. The editorial team has worked tirelessly to provide valuable and valid information to help people across the globe.

Every chapter published in this book has been scrutinized by our experts. Their significance has been extensively debated. The topics covered herein carry significant findings which will fuel the growth of the discipline. They may even be implemented as practical applications or may be referred to as a beginning point for another development. Chapters in this book were first published by InTech; hereby published with permission under the Creative Commons Attribution License or equivalent.

The editorial board has been involved in producing this book since its inception. They have spent rigorous hours researching and exploring the diverse topics which have resulted in the successful publishing of this book. They have passed on their knowledge of decades through this book. To expedite this challenging task, the publisher supported the team at every step. A small team of assistant editors was also appointed to further simplify the editing procedure and attain best results for the readers.

Our editorial team has been hand-picked from every corner of the world. Their multi-ethnicity adds dynamic inputs to the discussions which result in innovative outcomes. These outcomes are then further discussed with the researchers and contributors who give their valuable feedback and opinion regarding the same. The feedback is then collaborated with the researches and they are edited in a comprehensive manner to aid the understanding of the subject.

Apart from the editorial board, the designing team has also invested a significant amount of their time in understanding the subject and creating the most relevant covers. They scrutinized every image to scout for the most suitable representation of the subject and create an appropriate cover for the book.

The publishing team has been involved in this book since its early stages. They were actively engaged in every process, be it collecting the data, connecting with the contributors or procuring relevant information. The team has been an ardent support to the editorial, designing and production team. Their endless efforts to recruit the best for this project, has resulted in the accomplishment of this book. They are a veteran in the field of academics and their pool of knowledge is as vast as their experience in printing. Their expertise and guidance has proved useful at every step. Their uncompromising quality standards have made this book an exceptional effort. Their encouragement from time to time has been an inspiration for everyone.

The publisher and the editorial board hope that this book will prove to be a valuable piece of knowledge for researchers, students, practitioners and scholars across the globe.

List of Contributors

Afshin Kalantari
International Institute of Earthquake Engineering and Seismology, Iran

V. B. Zaalishvili
Center of Geophysical Investigations of RAS, Russian Federation

Silvia Garcia
Geotechnical Department, Institute of Engineering, National University of Mexico, Mexico

Haiqiang Lan and Zhongjie Zhang
State Key Laboratory of Lithosphere Evolution, Institute of Geology and Geophysics, Chinese Academy of Sciences, China P.R.

En-Jui Lee and Po Chen
Department of Geology and Geophysics, University of Wyoming, USA

Alexander Tyapin
"Atomenergoproject", Moscow, Russia

Halil Sezen
Department of Civil, Environmental, and Geodetic Engineering, The Ohio State University, Columbus, Ohio, USA

Adem Dogangun
Department of Civil Engineering, Uludag University, Bursa, Turkey

Wael A. Zatar
College of Information Technology and Engineering, Marshall University, Huntington, West Virginia, USA

Issam E. Harik
Department of Civil Engineering, University of Kentucky, Lexington, Kentucky, USA

Ming-Yi Liu and Pao-Hsii Wang
Department of Civil Engineering, Chung Yuan Christian University, Jhongli City, Taiwan

Lan Lin
Department of Building, Civil and Environmental Engineering, Concordia University, Montreal, Canada

Nove Naumoski and Murat Saatcioglu
Department of Civil Engineering, University of Ottawa, Ottawa, Canada

Hakan Yalçiner and Khaled Marar
European University of Lefke, Department of Civil Engineering, Mersin, Turkey

A. R. Bhuiyan
Department of Civil Engineering, Chittagong University of Engineering and Technology, Chittagong, Bangladesh

Y. Okui
Department of Civil and Environmental Engineering, Saitama University, Saitama, Japan